THE EARTH

An Introduction to the Geological and Geophysical Sciences

THE

EARTH

An Introduction to the Geological and Geophysical Sciences

A. LEE McALESTER

Yale University

PRENTICE-HALL, INC., ENGLEWOOD CLIFFS, NEW JERSEY

Library of Congress Cataloging in Publication Data

McAlester, Arcie Lee, 1933–
    The earth.

    Includes bibliographies.
    1.   Earth sciences.   2.   Astronomy.   I.   Title.
QE26.2.M32   551   72-10133
ISBN 0-13-222422-4

This book has been composed on film in News Gothic, with headings in Times Roman. Photographs were obtained by Gabriele Wunderlich. The line illustrations were prepared by B. J. and F. W. Taylor and Graphic Arts International, Inc. The book was designed by Mark A. Binn.

10   9   8   7   6   5   4   3   2   1

Prentice-Hall International, Inc., *London*
Prentice-Hall of Australia, Pty. Ltd., *Sydney*
Prentice-Hall of Canada, Ltd., *Toronto*
Prentice-Hall of India Private Ltd., *New Delhi*
Prentice-Hall of Japan, Inc., *Tokyo*

For Colette, Martine, and Keven

Contents

Preface

This book is an attempt to reinterpret, for the beginning college student, the changing emphases of the geological and geophysical sciences as we approach the last quarter of the twentieth century. In its preparation I have been guided by several strong convictions.

Foremost has been the belief that there is a new unity in earth studies that transcends the traditionally autonomous subjects of physical geology, oceanography, atmospheric science, planetary astronomy, and earth history. At the research level these disciplines now spill into each other's territories continuously: petrologists analyze rocks from the moon which provide clues to early earth history; oceanographers attempt to understand the ocean's interactions with the atmosphere; earth historians worry about climatic change; geomorphologists study the surface of Mars; geochemists pursue the elements from igneous rocks through rivers, the ocean, and the atmosphere into sediments and, ultimately, back into igneous rocks again—such examples can today be multiplied indefinitely. In short, the earth can no longer be understood as a series of independent spheres of rock, water, air, and space, but only as a dynamic whole that mirrors a long sequence of interactions between these units. As a consequence, the text stresses the similarities and interrelationships of the sciences of the earth, not the man-made boundaries that have long separated them.

A second conviction has been that the beginner should be presented with the central triumphs and challenges of the subject, not its endless catalogues of descriptive detail. This conviction has led to an abbreviated treatment of certain traditional bodies of information, among them some aspects of structural geology, geomorphology, stratigraphy, paleontology, meteorology, and climatology. On the other hand, data from some very active subdisciplines, such as solid earth geophysics, aeronomy, sedimentology, and geochemistry, are presented in somewhat greater depth than has been the custom in introductions to the subject.

A third conviction has been that even the most rigorously quantitative sciences of the earth can be understood and appreciated qualitatively. Mathematics is, after all, usually just shorthand for words and ideas, and many beginners lack the background to understand this shorthand. The treatment is therefore nonmathematical. At the same time, every effort has been made to present in words and illustrations, rather than equations, the goals and conclusions of the quantitative aspects of the subject.

Finally, I have been guided by a belief that the contemporary student both desires and deserves to know how science interrelates with day-to-day problems of human existence. To this end a series of brief essays have been included to show how the material treated in each chapter or part is of direct value to man. These are distinguished from the main body of the chapter by page color and format; although closely related to the surrounding material, they can be read as independent introductions to the many applied aspects of the geological and geophysical sciences.

Among the many persons to whom I have become indebted during the years the book has been in preparation are: W. H. Grimshaw, D. R. Esner, and their associates at Prentice-Hall, for enthusiastic support of the project; my colleagues at Yale, for helping me learn more about the earth, especially S. P. Clark, Jr., B. Saltzman, B. J. Skinner, K. K. Turekian, and J. C. G. Walker; Mrs. J. S. Lawless, for solving endless bibliographic, editorial, and illustrative problems; Mrs. M. M. Emons, for typing the manuscript; J. S. Covell of Prentice-Hall, for converting my rough ideas into a finished book; and to the following critics, for reading and improving the manuscript at various stages of its preparation: A. Dolgoff, New York City Community College; W. R. Farrand, University of Michigan; J. W. Harbaugh, Stanford University; J. Jackson, University of Akron; W. M. Jordan, Millersville State College; R. L. Kroll, Newark State College; M. F. Langer, Southwest College, City Colleges of Chicago; J. T. Licari, Cerritos College; W. H. Matthews III, American Geological Institute; L. J. Meyers, Middlesex Community College; P. H. Nichols, Millersville State College; L. V. A. Sendlein, Iowa State University; C. I. Smith, University of Michigan; and D. S. Turner, Eastern Michigan University.

A. L. McA.

THE EARTH

An Introduction to the Geological and Geophysical Sciences

The Study
of the Earth

With the lonely exception of astronomy, all of science is concerned with the earth. Physicists and chemists probe its atoms and molecules, biologists study its organisms, and social scientists try to understand the vagaries of its most complex product: man. How, then, can one group of scientists have the presumption to call their subject *the sciences of the earth*? The answer is that this group seeks to understand the earth as a unified system—a larger whole of which atoms, molecules, and life are but parts. Rocks, rivers, mountains, oceans, atmospheres—even other planets—are their basic units of study and from such study has grown a distinctive body of knowledge about the past and present earth. This knowledge, usually called either *the earth sciences* or *the geological and geophysical sciences,* is the subject of this book.

Modern understanding of the earth is based on thousands of observations on its constituent parts—the varying compositions of its rocks, the motions of its ocean and atmosphere, and the differing properties of its climates and landscapes. Indeed, so large and complex are the major components of the earth that they have traditionally been studied more as independent units, rather than as parts of an interacting whole. One large group of scientists, the *geologists,* has been concerned primarily with its rocks, mountains, landscapes, and the histories they record, and has usually given only passing attention to its ocean and atmosphere. These, in turn, have been the domain of *oceanographers* and *meteorologists,* who seldom considered the underlying solid earth to be relevant to their studies. At the same time, a small group of *astronomers* studied neighboring planets, often with little or no reference to the earth itself. This unnatural separation of earth studies was reinforced by the differing technologies required by each. The working geologist commonly needs

3

little more than strong legs and sharp eyes, whereas oceanographers require ships, meteorologists balloons and rockets, and astronomers telescopes, and all three use elaborate and expensive instrumentation. As a result, research and teaching concerning the ocean, atmosphere, and planets have each been concentrated in a relatively few large institutions where the necessary supporting facilities could be provided, whereas geological research and teaching have prospered on a smaller scale in a much larger number of institutions. Few of these institutions, large or small, have emphasized all aspects of earth study. This physical separation, added to the differences in study techniques, tended to perpetuate geology, oceanography, meteorology, and planetary astronomy as "independent" sciences of the earth.

Fortunately, this long and artificial separation of earth studies has been rapidly disappearing in recent years as scientists in each field have come to appreciate the many similarities and interactions of their studies. Perhaps the greatest contributions to this new unity have come from advances in the exploration of space. We have now become so accustomed to seeing dramatic photographs, such as the one facing page 3, that it is difficult to remember that scarcely a decade ago no one was really certain just what the earth would look like when viewed from distant space! Such views now emphasize, far more graphically than words, that the earth is dominated not by rock, air, or water alone, but by the continuous interaction of all three. On a more fundamental level, increasing knowledge of the earth's nearest neighbors in space—particularly the moon, Mars, and Venus—has provoked comparison with the earth as a planetary whole, and thus led to a healthy reexamination of the interactions of its parts.

Such examination reveals many unifying similarities in our understanding of the earth's rocks, ocean, and atmosphere. For all three there is a large body of information concerning chemical composition. The ocean and atmosphere have relatively uniform compositions because, being turbulent fluids, their materials are continuously stirred and mixed. Over long periods of time the rocks of the solid earth likewise behave as thick, viscous fluids and also tend to become mixed. This process is so slow, however, that rocks of varying composition may persist indefinitely. For this reason, the rocks of the solid earth show far more chemical diversity than either the waters of the ocean or gases of the atmosphere.

In addition to this compositional knowledge, there is a still larger body of information concerning the static structure and dynamic motions of the earth's rock, water, and air. The force of gravity causes the heaviest materials of the earth to be concentrated near its center, and the lightest in the outermost reaches of its atmosphere. Between these extremes, the materials of the solid earth, ocean, and atmosphere are arranged in concentric layers, with heavier materials underlying lighter. Superimposed on this stable, layered structure are continuous smaller-scale disturbances caused by local energy concentrations. These disturbances temporarily upset the gravitational balance and lead to unstable motions within and between the concentric layers. The velocity of these motions is relatively slow in the dense, viscous rocks of the solid earth, is far more rapid in the waters of the ocean, and is still faster in the gases of the atmosphere. In spite of their differing velocities, however, each of the motions has the same underlying cause—the disturbance of the earth's large-scale gravitational layering by smaller-scale concentrations of energy. These similarities among the major components of the earth are emphasized in the first two parts of the book. The composition, structure, and dynamic motions of the solid earth are summarized in the three chapters of Part 1. The three chapters of Part 2 provide a similar survey for the ocean and atmosphere.

The rocks of the solid earth come in contact with the fluids of the ocean and atmosphere only at, or very near, the earth's surface. At this extraordinarily important interface, the solids and fluids interact to produce differing climates, each with a characteristic pattern of landscapes and distinctive accumulations of sedimentary debris. Drawing on the previous discussions, the three chapters of Part 3 consider these fundamental interactions between the solid and fluid earth.

Parts 1, 2, and 3 emphasize our understanding of the earth's rocks, ocean, and atmosphere as they exist today and in the immediate past. In addition, however, there is a large and significant body of information concerning these units and their interactions over several billion years of earth history. This history is summarized in the three chapters of Part 4, which place the earlier chapters in broader perspective by reviewing the long sequence of events that has led to the earth's present configuration.

The final three chapters, making up Part 5, broaden the perspective still farther by contrasting knowledge of the earth with our much more limited understanding of its nearest relatives in space—the sun, planets, satellites, and other bodies of the solar system. The last chapter also moves past the earth's solar neighbors to provide a brief introduction to the countless stars and galaxies of distant space, the study of which lies beyond the outermost limits of the geological and geophysical sciences.

The text gives little attention to defining the rapidly disappearing boundaries between the traditional sciences of the earth, or to cataloging the names of the numerous subdivisions of each. Because the student is likely to encounter many of these names elsewhere, a table has been prepared (opposite) to relate the organization of the chapters and parts to the principal subdisciplines of the geological and geophysical sciences (the table locates only the primary discussions of each subject; additional treatment may occur in other chapters and can be located by use of the Index).

As in all science, much of the knowledge summarized in this book was gathered to help quench man's intellectual curiosity, rather than to satisfy his physical needs. Nevertheless, a great many aspects of the geological and geophysical sciences are of direct and crucial importance to man. The search for mineral deposits, the management of water resources, weather prediction, the control of earthquake damage—these are but a few of the human concerns that lean heavily on the sciences of the earth. Throughout the book, these applied aspects are introduced by a series of illustrated essays, distinguishable by tint of color on their pages, that summarize man's uses of the knowledge covered in each chapter or part. These essays provide an independent survey of the many applications of the geological and geophysical sciences to everyday problems of human existence.

ADDITIONAL READINGS

* Adams, F. D.: *The Birth and Development of the Geological Sciences*, The Williams & Wilkins Co., Baltimore, 1938 (Dover Publications paperback edition, 1954). A standard and detailed history of earth studies from classical times through the early nineteenth century.
* Albritton, C. C., Jr. (ed.): *The Fabric of Geology*, Freeman, Cooper & Co., Stanford, California, 1963. Advanced essays on the history and philosophy of the geological sciences, with extensive bibliographies.

*Cloud, P. E., Jr. (ed.): *Adventures in Earth History*, W. H. Freeman and Co., San Francisco, 1970. An anthology of research on the earth's past, with excellent historical introductions by the editor.

Deacon, G. E. R. (ed.): *Seas, Maps, and Men*, Doubleday & Company, Inc., Garden City, New York, 1962. Popular essays on the history of ocean studies, lavishly illustrated.

Fenton, C. L., and M. A. Fenton: *Giants of Geology*, Doubleday & Company, Inc., Garden City, New York, 1952. Well-written, popular biographies of influential nineteenth-century geologists.

Saltzman, B.: "Meteorology: A Brief History," in *The Encyclopedia of Atmospheric Sciences and Astrogeology*, R. W. Fairbridge (ed.), Reinhold Publishing Corp., New York, 1967. A short, scholarly survey of the history of atmospheric studies.

Vaucouleurs, G. de: *Discovery of the Universe*, The Macmillan Company, New York, 1957. A standard, brief history of astronomy, with good discussions of solar and planetary studies.

* Available in paperback.

Part		Chapter		Principal Subdisciplines Treated	
1	The Solid Earth	1	Materials of the Solid Earth	1	Mineralogy, petrology, petrochemistry
		2	The Earth's Interior	2	Solid-earth geophysics (seismology, geodesy, geomagnetism)
		3	The Earth's Crust	3	Tectonics, structural geology, geological oceanography (in part)
2	The Fluid Earth	4	The Ocean	4	Chemical and physical oceanography
		5	The Atmosphere	5	General atmospheric science, aeronomy
		6	Weather	6	Meteorology
3	Interactions of the Solid and Fluid Earth	7	Climates	7	Climatology, geomorphology (in part)
		8	Landscapes	8	Geomorphology (in part), hydrology
		9	Sediments	9	Sedimentology, low-temperature geochemistry, geological oceanography (in part)
4	The Earth Through Time	10	Earth Chronology	10	Stratigraphy, geochronometry
		11	Early History of the Earth	11	Precambrian geology, paleontology (in part)
		12	Evolving Geography, Life, and Environments	12	Paleogeography, paleoecology, paleontology (in part)
5	The Earth in Space	13	The Sun and Its Satellites	13	Solar physics, solar system astronomy
		14	The Moon	14	Lunar science
		15	The Planets and Beyond	15	Planetary science, stellar and galactic astronomy

PART 1

The Solid Earth

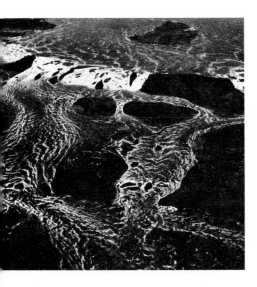 

Part 1 deals with the composition and structure of the solid mass of the earth—the familiar rocks and mountains of the earth's surface and the less familiar but equally important materials that lie beneath it.

Chapter 1 surveys the chemical structure of matter and the organization of matter into minerals and rocks, the basic *materials of the solid earth*.

Chapter 2 considers the materials of, and forces at work in, *the earth's interior*. Most of this vast region lies hidden far beneath the deepest mines and bore holes, and can only be studied by indirect methods. The most useful of these are surface measurements of the earth's deep gravity, magnetism, and shock wave transmission.

Chapter 3 treats the composition and structure of the earth's thin outer skin of rocks that can be studied directly at the surface and in deep mines and bore holes. Called *the earth's crust,* this is a zone of dynamic change where the rocks of the ocean floor are continuously created, moved, and destroyed, while the edges of adjacent continents are deformed into linear mountain chains.

1

Materials
of the
Solid Earth

Rock is the only primary component of the earth that shows wide variation in composition. Air and ocean water, the other principal components, have relatively uniform compositions because, as fluids, they are in almost continuous motion and thus tend to be stirred and mixed together in constant proportions. Rocks, however, are decidedly nonuniform; as stable solids they are composed of thousands of unevenly distributed, complexly patterned materials. The composition of the more important of these many materials will be discussed in this chapter; Chapters 2 and 3 will consider the nature and possible causes of their varied distributions throughout the solid earth.

Earth materials are traditionally and most conveniently considered on two different scales: (1) the units of smaller scale are **minerals,** natural substances of uniform chemical composition possessing a definite crystalline structure. Each mineral has a combination of composition and structure that characterizes it and no other. About 2,000 different minerals have been found, but only about 20 common ones make up the bulk of the earth. Some examples among these common minerals are mica, quartz, and gypsum. (2) The units of larger scale are **rocks,** usually aggregates of several different minerals that have been brought together by one of many rock-forming processes. Examples are granite, basalt, and sandstone. Unlike minerals, rocks are not rigidly defined by their composition and structure since they may be composed of endlessly varying mixtures of minerals. Nevertheless, only about 20 mineral combinations are really common; these 20 rocks make up most of the solid earth.

Rocks and minerals, like all other substances, or **matter,** in the universe, are composed of elemental particles assembled according to certain fundamental chemical principles. Before we look more closely at the common rocks and minerals, it will be useful to turn to this smaller scale and review some of these principles.

MATTER

1.1 Elements and Compounds

The smallest units of matter that normally concern earth scientists are the **elements,** substances that cannot be broken down into other substances by chemical means. There are 88 such elements known to occur naturally; in addition, 15 others have been produced artificially by physicists using atomic reactors and particle accelerators. These 103 elements are the basic building blocks of *all* matter, living and nonliving, on earth and throughout the universe. The interrelationships of the elements are best shown by arranging them into a **periodic table,** which groups together elements with similar physical and chemical characteristics (Figure 1.1). The complete periodic table would be a cumbersome introduction to the structure of earth materials were not most of the elements extremely rare in the earth. Indeed, the rocks and minerals of the solid earth are made up almost entirely of only 10 elements. These are shown in color in Figure 1.1. Five additional elements are of less importance in the solid earth but are abundant in the water and air of the fluid earth. These are shown in gray in Figure 1.1. Our primary concern throughout this book will be with these 15 elements that make up the bulk of the earth.

The properties of the elements are expressed in the periodic table in the following way: the elements become progressively heavier from left to right and from top to bottom. Thus hydrogen (element 1, upper left) is the lightest element and lawrencium (element 103, lower right) the heaviest. Note that the most abundant elements are all relatively light; no element heavier than nickel (element 28) is among the 15. The periodic table also associates in vertical "groups" those elements having similar chemical properties. For example, the six elements in the extreme right-hand column (group VIII) are all very stable, nonreactive gases called the **noble gases.** For the abundant elements of the earth, this columnar grouping shows that sodium and potassium, although differing in weight, have similar chemical properties, as do the pairs magnesium-calcium, carbon-silicon, and oxygen-sulfur. Although it is listed in group I, hydrogen has properties found in no other element.

Pure elements seldom occur in nature; instead, two

Group Period	I	II											III	IV	V	VI	VII	VIII
1	1 H Hydrogen																	2 He Helium
2	3 Li Lithium	4 Be Beryllium											5 B Boron	6 C Carbon	7 N Nitrogen	8 O Oxygen	9 F Fluorine	10 Ne Neon
3	11 Na Sodium	12 Mg Magnesium											13 Al Aluminum	14 Si Silicon	15 P Phosphorus	16 S Sulfur	17 Cl Chlorine	18 Ar Argon
4	19 K Potassium	20 Ca Calcium	21 Sc Scandium	22 Ti Titanium	23 V Vanadium	24 Cr Chromium	25 Mn Manganese	26 Fe Iron	27 Co Cobalt	28 Ni Nickel	29 Cu Copper	30 Zn Zinc	31 Ga Gallium	32 Ge Germanium	33 As Arsenic	34 Se Selenium	35 Br Bromine	36 Kr Krypton
5	37 Rb Rubidium	38 Sr Strontium	39 Y Yttrium	40 Zr Zirconium	41 Nb Niobium	42 Mo Molybdenum	43* Tc Technetium	44 Ru Ruthenium	45 Rh Rhodium	46 Pd Palladium	47 Ag Silver	48 Cd Cadmium	49 In Indium	50 Sn Tin	51 Sb Antimony	52 Te Tellurium	53 I Iodine	54 Xe Xenon
6	55 Cs Cesium	56 Ba Barium	57 La Lanthanum	72 Hf Hafnium	73 Ta Tantalum	74 W Tungsten	75 Re Rhenium	76 Os Osmium	77 Ir Iridium	78 Pt Platinum	79 Au Gold	80 Hg Mercury	81 Tl Thallium	82 Pb Lead	83 Bi Bismuth	84 Po Polonium	85* At Astatine	86 Rn Radon
7	87* Fr Francium	88 Ra Radium	89 Ac Actinium															

58 Ce Cerium	59 Pr Praseodymium	60 Nd Neodymium	61* Pm Promethium	62 Sm Samarium	63 Eu Europium	64 Gd Gadolinium	65 Tb Terbium	66 Dy Dysprosium	67 Ho Holmium	68 Er Erbium	69 Tm Thulium	70 Yb Ytterbium	71 Lu Lutetium
90 Th Thorium	91 Pa Protactinium	92 U Uranium	93* Np Neptunium	94* Pu Plutonium	95* Am Americium	96* Cm Curium	97* Bk Berkelium	98* Cf Californium	99* Es Einsteinium	100* Fm Fermium	101* Md Mendelevium	102* No Nobelium	103* Lw Lawrencium

Figure 1.1 Periodic table of the elements. The 10 abundant solid earth elements are shaded in color; the abundant fluid earth elements are shaded in gray. The steplike line separates metals (below) from non-metals (above). Asterisk denotes a man-made element.

or more elements are usually found chemically combined in substances called **compounds.** Within the solid earth, only carbon, sulfur, copper, and iron are found in any abundance as free elements, and even these four are much more common in compounds such as *calcium carbonate,* a compound of calcium, carbon, and oxygen that makes up ordinary limestone, or *iron sulfide,* the familiar "fool's gold" that is a compound of iron and sulfur. With few exceptions, then, the bulk of the earth is composed of chemical compounds.

1.2 Atoms and Molecules

Our discussions of earth materials will deal primarily with elements and compounds, but in order to understand better the behavior of these materials, we shall now briefly consider the components of the elements themselves and the ways they combine to make compounds. The smallest particles of an element that still possess its characteristic properties are called **atoms.** Reactions between atoms form compounds. Just as elements are made up of individual atoms, so are compounds composed of **molecules,** which are the smallest particles of the compound that still possess *its* characteristic properties. In common salt (sodium chloride), for example, the molecule is composed of one sodium atom combined with one chlorine atom. If the

13

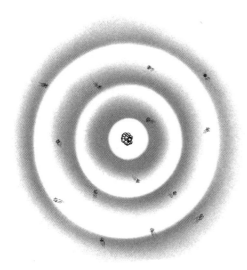

Figure 1.2 A schematic representation of the concentric electron shells of the silicon atom. Silicon, with atomic number 14, has 14 electrons distributed in three shells—two in the inner, eight in the middle, and four in the outer shell.

atoms were separated to yield uncombined sodium and chlorine, then the properties of the compound would be lost, for chlorine alone is a poisonous, greenish gas and sodium is a gray, explosively reactive metal.

Until the beginning of this century, scientists believed that the atoms of the chemical elements were indivisible units of matter. Then, in the early 1900s, it was discovered that atoms are themselves composed of still smaller units: a central **nucleus** containing most of the mass of the atom, surrounded by one or more extremely tiny, rapidly moving, and electrically charged particles called **electrons.** Researchers also found that the manner in which the elements combined to make compounds is determined by the number of electrons surrounding the nucleus. Each element has a characteristic number of electrons, ranging from only one in hydrogen to 103 in lawrencium. The normal number of electrons corresponds to the **atomic number** of the element (atomic numbers are shown above the element symbols in Figure 1.1).

The positions of the electrons around the nucleus are not random; instead, the rapidly moving electrons occur in concentric, cloudlike clusters called **shells** (Figure 1.2). The innermost shell always contains only two electrons (or one in hydrogen) and electrons are added in complex cycles of eight to form additional shells. The electron arrangement is important because the manner in which elements combine to form compounds is largely determined by the number of electrons in the *outermost* shell. The number of these outer-shell electrons for the 15 abundant elements are shown arranged around their element symbols in Figure 1.3.

The most stable elements are those with completely filled outer shells containing eight electrons. Examples are the previously mentioned noble gases (helium, neon, argon, krypton, xenon, and radon) occupying group VIII in the

Figure 1.3 Electron-dot formulas of the 15 abundant elements. The dots show the number of electrons in the outermost electron shell.

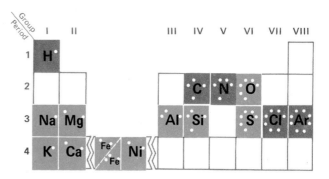

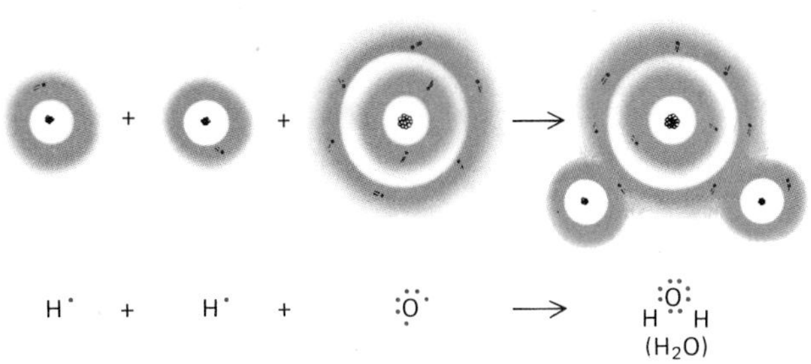

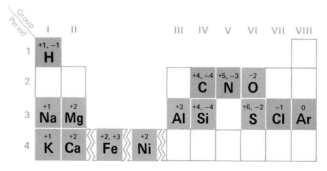

Figure 1.4 *A schematic representation of the combination of two hydrogen atoms with one oxygen atom to form a molecule of water sharing eight outer-shell electrons.*

H· + H· + :Ö· → H:Ö:H (H₂O)

periodic table. These gases almost never react with other elements to form compounds. In contrast, the most reactive elements are those that are only one or two electrons short of having a completely filled outer shell, that is, those having one, two, six, or seven electrons in that shell. These elements occur in groups I, II, VI, and VII of the periodic table. In general, elements with one or two electrons in the outer shell tend to react with those having six or seven electrons to form stable compounds having complete outer shells containing eight electrons. Thus two atoms of hydrogen, with one outer electron each, combine with one atom of oxygen, with six outer electrons, to form a molecule of water (Figure 1.4).

The number of electrons gained, lost, or shared by an element when it forms compounds determines the *oxidation number* of the element. By convention, elements that tend to *receive* electrons in forming compounds are given negative oxidation numbers while those that tend to *donate* electrons are given positive numbers. All compounds contain elements having both positive and negative oxidation numbers; furthermore, the sum of all positive oxidation numbers in the compound must always equal the sum of all the negative oxidation numbers. The normal oxidation numbers of the 15 abundant elements are shown in Figure 1.5. Note that some elements may

Figure 1.5 *The normal oxidation numbers of the 15 abundant elements.*

Period \ Group	I	II			III	IV	V	VI	VII	VIII
1	+1, −1 **H**									
2						+4, −4 **C**	+5, −3 **N**	−2 **O**		
3	+1 **Na**	+2 **Mg**			+3 **Al**	+4, −4 **Si**		+6, −2 **S**	−1 **Cl**	0 **Ar**
4	+1 **K**	+2 **Ca**	+2, +3 **Fe**	+2 **Ni**						

15

have more than one oxidation number, depending on how the electrons are gained, lost, or shared in forming a particular compound.

Although the pattern of electrons determines most of an element's chemical properties, many physical properties, including the element's weight, are determined not by the electrons but by the massive nucleus around which the almost weightless electrons move. Atomic nuclei are themselves composed of two principal kinds of particles: **protons** and **neutrons.** Protons carry a positive electrical charge; an element always has the same number of protons as electrons, for the positively charged protons in the nucleus serve to balance the negative electrical charges of the surrounding electrons. Thus the atomic number of an element corresponds not only to the number of electrons in that element, but also to the number of nuclear protons.

Neutrons, in contrast, have no electrical charge and may be thought of as being composed of both a proton and an electron that balance each other in charge, but are grossly unequal in weight, the much larger proton contributing almost all the mass of the neutron. The important point about neutrons is that in a random sample of atoms of a single element, the individual atoms may have *differing numbers of neutrons* in their nuclei, whereas each atom will have the same *fixed number of nuclear protons and surrounding electrons* (see Figure 1.6). Because an element may have atoms with differing numbers of heavy neutrons, each element occurs in differing weight configurations called **nuclides.** Nuclides of a single element differ in number of neutrons and in weight, but not in ordinary chemical properties.

In principle, any element might have any number of neutrons in its nucleus, thus leading to an infinite number of nuclides. In fact, however, most nuclides are *unstable;* that is, the nuclear forces of attraction are insufficient to hold the neutrons together. In these circumstances the neutrons "decay" by breaking into separate electrons, protons, or other particles; this decay process is called **radioactivity.** Many thousands of nuclides are known but only about 270 are stable, an average of less than three stable nuclides per element. All other nuclides are radioactive and tend to decay at rates that are constant for a particular nuclide but vary greatly between nuclides. For most of these nuclides the decay rate is so rapid that they do not occur naturally on earth (any that were present early in earth history having long since de-

Figure 1.6   Schematic representations of four nuclides: hydrogen-1, carbon-12, carbon-14, and silicon-28.

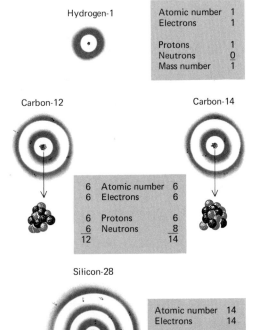

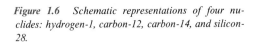

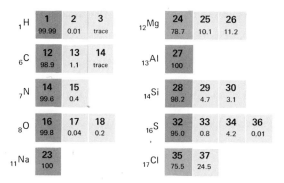

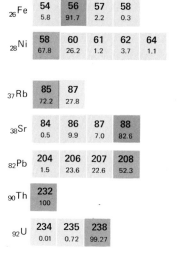

Figure 1.7 Natural nuclides of the 15 common elements and of 5 additional elements (*lower right*) important in earth chronology (*see* Chapter 10). Radioactive nuclides are shown in color. The lower numbers of each box show the percentage abundance of each nuclide in natural mixtures of the element; darker shading indicates dominant nuclides. (*Data from Wedepohl,* Handbook of Geochemistry, *vol. 1, 1969*)

cayed), but they can be created artificially in nuclear reactors and particle accelerators.

Because almost all the weight of an atom comes from its heavy neutrons and protons, nuclides are distinguished by the total number of *both* types of particles in the nucleus. This total is known as the **mass number** of the nuclide. All of the important naturally occurring nuclides and their mass numbers are shown in Figure 1.7. Most elements occur in nature not as pure nuclides, but as relatively uniform mixtures containing all the natural nuclides of that element. The proportions of each nuclide in these mixtures are also shown in Figure 1.7.

1.3 Gases, Liquids, and Solids

Let us now turn from the structure of atoms and molecules to a more general consideration of the three *states* or *phases*—gaseous, liquid, and solid—in which all matter can occur. Most elements and compounds can exist in each of the three phases, depending on the physical conditions, particularly temperatures and pressures, to which they are subjected. At room temperature and normal atmospheric pressure, 90 of the 103 elements are solids,

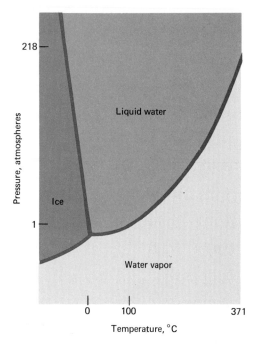

Figure 1.8

Figure 1.8 Phase diagram for water, showing the relation of the gaseous, liquid, and solid phases to pressure and temperature (not to scale).

11 are gases, and only 2 (bromine and mercury) are liquids. With changing temperatures and pressures, however, the elements, and compounds made up of them, may pass from one state to another. The most familiar example is water: at normal atmospheric pressure it is a solid below 0°C (32°F), a liquid between 0 and 100°C (32 and 212°F), and a gas (water vapor) above 100°C (212°F). We are all aware of these temperature effects on water, but less familiar are the effects of changes in pressure. At high pressures water remains liquid at temperatures as great as 371°C (700°F) and solid ice persists to temperatures as high as 204°C (400°F). The combined effects of both temperature and pressure on water are shown in Figure 1.8, an example of a **phase diagram.**

At the atomic level, the principal differences among the three states of matter relate to the *degree of ordering* of the constituent atoms. In typical solids, the atoms have a very regular arrangement; most commonly they are joined in rigid geometrical frameworks (Figure 1.9). Such complete frameworks seldom exist in liquids, although the atoms are still held in close association. In gases the atoms (or molecules, in the case of gaseous compounds) are completely separated from each other and move about freely. Many processes important to an understanding of the earth are concerned with transformations among these three states. Liquid-gas transitions are particularly crucial in the interaction of the ocean and atmosphere whereas liquid-solid transitions, and the resulting solid state, are of primary importance for the solid earth.

1.4   The Solid State

Solids having a regular, orderly arrangement of their internal atoms are said to have *crystalline structure* and are

Figure 1.9 Solid, liquid, and gaseous states: a schematic representation of the degree of atomic ordering in the three states of matter.

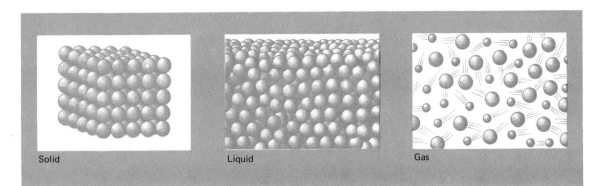

known as **crystals.** Most solid substances, including rocks and minerals, are made up of aggregates of many small crystals. Crystals are characteristically bounded by flat surfaces, called *crystal faces*, which are large-scale reflections of the internal arrangement of the atoms in the crystal. Long before the principles of atomic structure were understood, study of the arrangement of faces in natural crystals showed that there were six basic groups, each with a characteristic symmetry of the faces (Figure 1.10). These six groups of natural crystals, each of which could be divided into additional subgroups, provided the principal clues to the structure of solid matter until 1912. Then it was discovered that the orderly layers of atoms in crystals altered X-ray beams passing through the crystal in such

Figure 1.10 *The six basic systems of crystal symmetry. Drawings at left show the simplest crystal form of each system, with symmetry axes in color. At right are typical crystal forms of the common minerals in each system.*

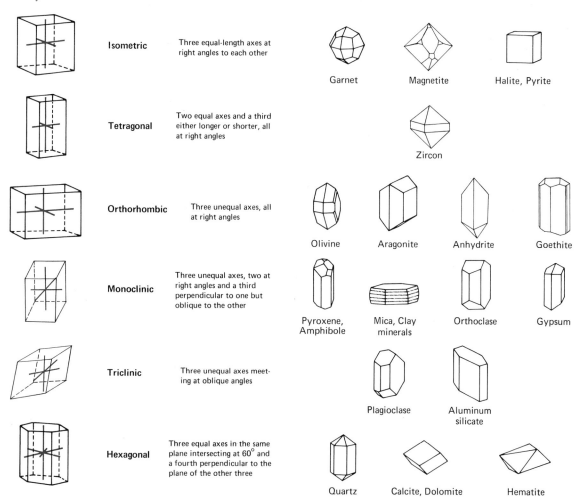

Isometric	Three equal-length axes at right angles to each other	Garnet — Magnetite — Halite, Pyrite
Tetragonal	Two equal axes and a third either longer or shorter, all at right angles	Zircon
Orthorhombic	Three unequal axes, all at right angles	Olivine — Aragonite — Anhydrite — Goethite
Monoclinic	Three unequal axes, two at right angles and a third perpendicular to one but oblique to the other	Pyroxene, Amphibole — Mica, Clay minerals — Orthoclase — Gypsum
Triclinic	Three unequal axes meeting at oblique angles	Plagioclase — Aluminum silicate
Hexagonal	Three equal axes in the same plane intersecting at 60° and a fourth perpendicular to the plane of the other three	Quartz — Calcite, Dolomite — Hematite

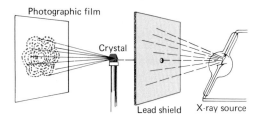

Figure 1.11 X-ray determination of crystal structure. The X-ray pattern shown indicates a hexagonal arrangement of the internal atoms.

Figure 1.12 Five types of atomic coordination. As the positive and negative (color and gray) atoms become more similar in size, additional negative (gray) atoms can be packed around a single positive (color) atom.

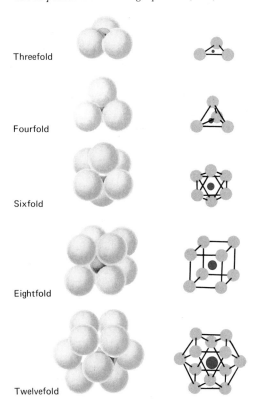

Threefold

Fourfold

Sixfold

Eightfold

Twelvefold

a way that it became possible to determine precisely the internal arrangement of atoms (Figure 1.11). This X-ray technique quickly supplanted older methods of measuring crystal faces and became the principal tool for studying the structure of crystalline solids. X-ray methods now also serve as a quick and reliable means of identifying even the smallest grains of a crystalline substance, for each element or compound has a distinctive pattern of X-ray transmission.

X-ray studies of atomic structure, combined with other lines of evidence, have shown that the arrangement of atoms in crystals is largely determined by two properties: the *size* and *oxidation number* of the constituent atoms. In order to visualize how these properties affect crystal structure, individual atoms can be thought of as small spheres which become packed together to form crystals. These atomic spheres tend to arrange themselves so that positive and negative oxidation numbers are balanced and so that there is a minimum distance between the spheres. When the positive and negative atomic spheres differ greatly in size, a triangular packing arrangement is favored in which each atom shares its electrons with three adjacent atoms of opposite charge, giving rise to what is known as *threefold* **coordination** (Figure 1.12). As the difference in size between positive and negative atoms decreases, *fourfold, sixfold,* and *eightfold* coordinations are favored; finally, when the size of the positive and negative atomic spheres are equal, a *twelvefold* coordination with a maximum density of closely packed spheres occurs. Figure 1.13 summarizes the normal oxidation numbers and comparative atomic sizes of the atoms that are common in rock-forming minerals.

MINERALS

1.5 Chemistry of the Solid Earth

As noted earlier, the relative abundances of the 103 chemical elements are such that only 10 elements make up most of the earth's minerals and rocks. This composition has been established by chemical analyses of many thousands of surface rocks, combined with less direct evidence about the materials of the earth's vast interior. The darker bars of Figure 1.14 present a modern estimate of the relative abundances of the elements most common in the entire solid earth. Note that only 8 elements constitute

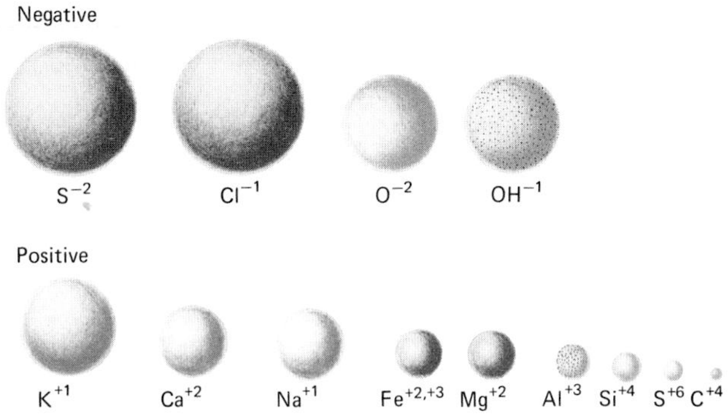

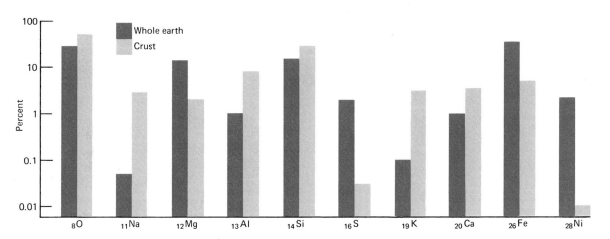

Negative

S^{-2} Cl^{-1} O^{-2} OH^{-1}

Positive

K^{+1} Ca^{+2} Na^{+1} Fe$^{+2,+3}$ Mg^{+2} Al^{+3} Si^{+4} S^{+6} C^{+4}

Figure 1.13 *Oxidation numbers and relative sizes of atoms found in common rock-forming minerals. Carbon, chlorine, and the oxygen-hydrogen atom group, while less abundant in the solid earth than the other atoms, are included because they occur in specific minerals discussed later in this chapter.*

over 99 percent of the total earth; all the other 80 naturally occurring elements combine to make up less than 1 percent. Of these 8 elements, only 4—iron, oxygen, silicon, and magnesium—constitute over 90 percent of the solid earth.

In Chapter 2 we shall see that there is good evidence that the rocks exposed on the earth's surface are part of a thin "skin," called the **crust,** that formed at some earlier stage of the earth's development by a separation of lighter and more volatile elements from heavier materials of the interior. Because of this separation, the distribution of elements in the crust differs from that in the earth's deeper interior. (The crust makes up less than 1 percent of the earth's total mass, and thus the abundances of elements in the entire solid earth primarily reflect the composition of the interior where heavy elements, particularly iron and nickel, dominate.) The most abundant ele-

Figure 1.14 *Abundances of elements (weight percent) in the whole earth and in the surficial crust. (Data from Mason,* Principles of Geochemistry, *1966)*

ments in the earth's crust are shown in the lighter bars of Figure 1.14. Once again, 8 elements constitute the bulk of the whole (in this case 98 percent of the crust). Note that oxygen, sodium, aluminum, silicon, potassium, and calcium are enriched in the crust, whereas magnesium and sulfur, along with iron and nickel, are more abundant in the interior.

It is now possible to apply some of the chemical principles discussed earlier to predict the *combinations* of elements present in the minerals of the solid earth. Refer again to Figure 1.13, which shows the normal oxidation numbers and atomic sizes of the principal elements. The only really abundant element with a *negative* oxidation number is oxygen. This means that in most compounds making up the solid earth oxygen supplies the negative charge, and the other elements supply the balancing positive charges. Note also the very large size of the oxygen atom relative to all the common positive atoms (except potassium). This leads us to predict that minerals are, *by volume,* mostly composed of large spheres of atomic oxygen interspersed with smaller, positively charged atoms of other elements. The character of this oxygen-positive atom interaction provides a useful means of classifying many of the 2,000 minerals that have so far been discovered in the rocks of the solid earth (Figure 1.15).

In the simplest case, oxygen combines directly with various positive atoms to give **oxide** minerals such as *hematite* (iron oxide, Fe_2O_3). More commonly, the positive atoms combine not with oxygen alone, but with combinations of oxygen and additional positively charged atoms that function together as a unit in carrying the negative charge of the mineral. These combinations, called **atom groups,** have an overbalance of negative charges and thus form compounds just as if they were single, negatively charged elements. Three such atom groups are of great importance in minerals: the **silicate group,** in which oxygen combines in various proportions with silicon; the **sulfate group,** in which four oxygen atoms combine with one sulfur atom; and the **carbonate group,** in which three oxygen atoms combine with one carbon atom.[1]

Sulfur is one of the few common elements that can have either a positive or negative oxidation number (see

---

[1]Carbon is not one of the more abundant elements of the solid earth although it makes up a small but significant fraction of the atmosphere. Even though it comprises much less than 1 percent of the crust, carbon is locally concentrated into accumulations of carbonate minerals that are of great importance in earth chemistry.

MINERAL GROUP		MINERAL	GENERALIZED CHEMICAL COMPOSITION						
			Positive Atoms						Negative Atom Group or Atoms
			Sodium	Magnesium	Aluminum	Potassium	Calcium	Iron	
Silicates	Isolated tetrahedra	Olivine		Mg				Fe	(SiO$_4$)
		Aluminum silicates			Al				
		Garnets		Mg	Al		Ca	Fe	
	Chain	Pyroxenes	Na	Mg	Al		Ca	Fe	(SiO$_3$)
		Amphiboles	Na	Mg	Al		Ca	Fe	(Si$_4$O$_{11}$), (OH)
	Sheet	Micas		Mg	Al	K		Fe	(Si$_2$O$_5$), (OH)
		Clay minerals			Al	K			
	Framework	Quartz		[Si only]					(SiO$_2$)
		Orthoclase feldspar			Al	K			
		Plagioclase feldspar	Na		Al		Ca		
Carbonates		Calcite					Ca		(CO$_3$)
		Aragonite					Ca		
		Dolomite		Mg			Ca		
Sulfates		Gypsum					Ca		(SO$_4$) + H$_2$O
		Anhydrite					Ca		(SO$_4$)
Oxides		Hematite						Fe	O, (OH)
		Goethite						Fe	
		Magnetite						Fe	
		Aluminum oxides			Al				
Halide		Halite	Na						Cl
Sulfide		Pyrite						Fe	S

Figure 1.15 Classification and generalized chemical composition of the principal rock-forming minerals.

Figures 1.5, 1.13). When it combines with oxygen in the sulfate atom group its oxidation number is positive. With a negative oxidation number it provides the sole negative charge in **sulfide** minerals.

The relatively rare element chlorine occurs concentrated as the sole negative atom in the mineral *halite* (sodium chloride, NaCl). Such minerals in which chlorine, or other closely related elements, provide the only negative charge are called **halides.**

These six principal kinds of negative chemical charges define six principal groups of rock-forming minerals: silicates, carbonates, sulfates, oxides, halides, and

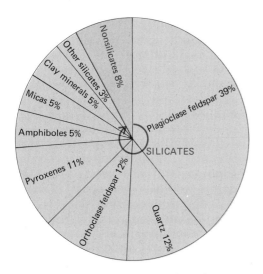

Figure 1.16 Percentage (*by volume*) *of common minerals in the earth's crust.* (*Data from Ronov and Yoroshevsky,* American Geophysical Union Monograph *13, 1969*)

sulfides (Figure 1.15). A seventh mineral group can be added for those few elements, primarily carbon, iron, nickel, and sulfur, that may occur not only as compounds, but also as free, uncombined elements. Such relatively rare minerals are called **native elements.** These mineral groups contain most of the 2,000 known minerals, of which only about 20 are really common in the solid earth (see Figure 1.15); the others are very rare and restricted in occurrence. These common rock-forming minerals are those that we shall consider here and be referring to in later chapters. Discussions of many of the less common minerals, including the economically important *gem* and *ore minerals,* may be found in several of the books listed in the "Additional Readings" at the end of this chapter.

1.6 Silicate Minerals

Of the principal mineral groups, the silicates are by far the most important because they make up about 92 percent of the earth's crust (Figure 1.16). In addition to being an extremely abundant element, silicon also has unique chemical properties that permit it to link with oxygen to form large structural networks of atoms. Silicon has an oxidation number of +4 and a very small atomic size (Figure 1.13); a silicon atom is always surrounded in crystal structures by four much larger oxygen atoms in a fourfold coordination that resembles a geometric tetrahedron (Figure 1.12).

In the simplest case, these silicon-oxygen tetrahedra occur as isolated units, bound together only by positive metallic atoms attached to some of the oxygen atoms. More commonly, silicon atoms share a linkage to one or more oxygen atoms so that more complex structures—chains, sheets, or completely interlocking frameworks—are formed; these are illustrated in Figure 1.17. These structural differences define four principal groups of silicate minerals: *silicates with isolated tetrahedra; chain silicates; sheet silicates;* and *framework silicates.*[2] The minerals within each of these groups may be further classified by the positive atoms that combine with the negative silicon-oxygen tetrahedral groups. These are principally atoms of sodium, magnesium, aluminum, potassium, calcium, and iron, the dominant positive elements, other than silicon, found in the earth's crust.

---

[2] Several minor groups of silicates, whose minerals are relatively uncommon, have been omitted.

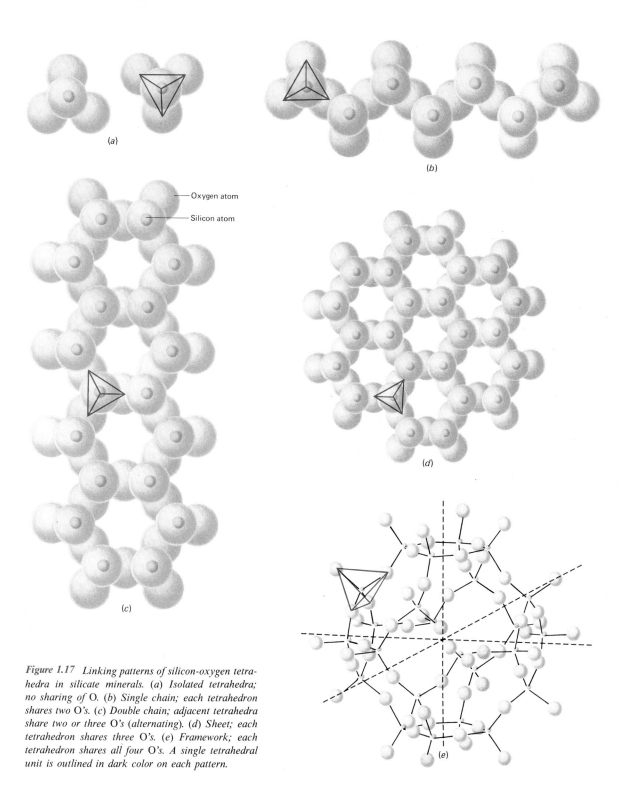

Oxygen atom

Silicon atom

(a)

(b)

(c)

(d)

(e)

Figure 1.17 Linking patterns of silicon-oxygen tetrahedra in silicate minerals. (a) Isolated tetrahedra;
no sharing of O. (b) Single chain; each tetrahedron
shares two O's. (c) Double chain; adjacent tetrahedra
share two or three O's (alternating). (d) Sheet; each
tetrahedron shares three O's. (e) Framework; each
tetrahedron shares all four O's. A single tetrahedral
unit is outlined in dark color on each pattern.

Figure 1.18 *Isolated tetrahedra silicate minerals.*
(a) Olivine; (b) aluminum silicate; (c) garnet; (d)
atomic structure of olivine. (a–c: Yale Peabody
Museum)

Silicates with Isolated Tetrahedra In this group the ratio of silicon to oxygen atoms is 1:4 (SiO_4) because none of the oxygen atoms is shared among adjacent silicon atoms (Figure 1.17). The basic oxidation number of the silicon-oxygen group is -4 since the oxidation number of the silicon atom is $+4$ and that of each of the four oxygen atoms is -2. This large excess negative charge serves to bind a high proportion of other positive atoms into the structure so that the charges are balanced. Three important mineral subgroups have this structure: the **olivines,** in which magnesium and iron provide the principal additional positive atoms; the **aluminum silicates,** in which aluminum is the principal positive atom; and the **garnets,** which have various complex combinations of positive atoms, principally magnesium, aluminum, calcium, and iron (Figure 1.18).

(a)

(b)

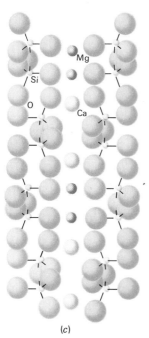

(c)

Figure 1.19 Chain silicate minerals. (a) Pyroxene; (b) amphibole; (c) atomic structure of a calcium- and magnesium-bearing pyroxene. (a, b: Yale Peabody Museum)

Chain Silicates In this structure oxygen atoms are shared between silicon atoms in such a way that long silicon-oxygen chains of two kinds are formed. In one, each silicon atom shares two of its four oxygen atoms with adjacent silicon atoms, thus giving a ratio of one silicon atom for each three oxygen atoms (SiO_3); this is the *single chain* structure. The principal minerals with this structure are the **pyroxenes** (Figure 1.19), important rock-forming minerals that may contain all the common positive atoms except potassium, which is prevented from entering the chain structure by its large atomic size. The other group of chain silicates has a *double chain* structure in which adjacent silicon atoms alternately share two and then three of their oxygen atoms to give a ratio of four silicon atoms to 11 oxygen atoms (Si_4O_{11}). The principal double chain silicates are the **amphiboles** (Figure 1.19). Like the closely related pyroxenes, these may contain various combinations of all the common positive atoms except potassium. Because of their chain structure, amphiboles and pyroxenes tend to crystallize as elongate minerals whose external shape reflects their internal atomic arrangement.

Sheet Silicates In this structure each silicon atom shares three of its oxygen atoms to form a sheetlike, two-dimensional network having a ratio of two silicon atoms for every five oxygen atoms (Si_2O_5). Because of the atomic arrangement, minerals with this structure tend to split into thin parallel sheets. The **micas,** one of two important groups of sheet silicates, are familiar examples (Figure 1.20). Micas all contain aluminum atoms, some of which replace

(a)

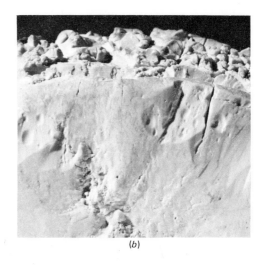

(b)

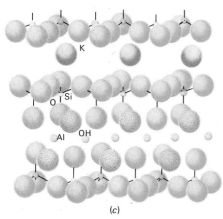

(c)

Figure 1.20 Sheet silicate minerals. (a) Mica; (b) clay mineral [earthy mass about 2 cm (³⁄₄ in.) across; the extremely tiny individual crystals are not visible]; (c) atomic structure of a mica. The large potassium atoms fit between the structural sheets of silicon, oxygen, and aluminum. (a, b: Yale Peabody Museum)

some silicon atoms in the structural tetrahedron. They are also characterized by potassium atoms, whose large size permits them to fit *between* the sheetlike layers of silicon-aluminum-oxygen tetrahedra (Figure 1.20c). In addition, some micas contain iron and magnesium. The other important group of sheet silicates, the **clay minerals,** are similar, but differ in structural detail. Clay minerals never occur as large crystals but are always found in earthy masses of extremely tiny crystalline particles, particles that are usually so small that they can be seen only with an electron microscope. Both the clay minerals and micas have water-derived oxygen-hydrogen (OH) atomic groups in their structures in addition to the principal oxygen-silicon groups.

Framework Silicates In this last group every silicon atom shares each of its four oxygen atoms with adjacent silicon atoms to give a three-dimensional network with a ratio of one silicon atom for every two oxygen atoms (SiO_2). They are thus the most silicon-rich of the silicate minerals. Note that the basic charge of the SiO_2 atomic group is neutral since silicon's oxidation number of $+4$ is balanced by the two oxygen atoms, each of which has an oxidation number of -2. Because of this neutral charge, framework silicates have the most stable structural arrangement and are by far the most abundant of all the minerals in the earth's crust. The two principal types are **quartz,** which is composed *only* of silicon and oxygen, and the **feldspars,** a group of minerals in which sodium, aluminum, potassium, and calcium occur in the framework in varying combina-

(a)

(b)

(c)

K

O   Al

Si

(d)

Figure 1.21 *Framework silicate minerals. (a) Quartz;*
(b) orthoclase feldspar; (c) plagioclase feldspar; (d)
atomic structure of orthoclase feldspar. The unbal-
anced charges caused by the substitution of aluminum
for some silicon atoms serve to hold potassium in the
structure (see text). (a–c: Yale Peabody Museum)

tions and proportions (Figure 1.21). The bulk of the earth's crust is composed of feldspar (51 percent) and quartz (12 percent) (Figure 1.16).

Because the basic charge of the SiO_2 atomic group is neutral, we need to explain how additional positive atoms, such as sodium, aluminum, potassium, and calcium, can occur in the structure of the feldspars. The reason for this is related to the presence of aluminum, which is unique among all the common elements in being able to substitute for silicon in some of the silicon-oxygen tetrahedra. This is possible because the small atomic size of the aluminum atom is close to that of silicon (see Figure 1.13). Aluminum, however, has an oxidation number of $+3$ where silicon's is $+4$; thus when an aluminum atom

(a)

(b)

(c)

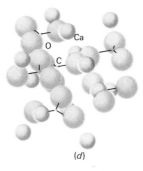
(d)

Figure 1.22 Carbonate minerals. (a) Calcite; (b) aragonite; (c) dolomite; (d) atomic structure of aragonite. Each carbon atom is linked to three oxygen atoms. (a–c: Yale Peabody Museum)

substitutes for a silicon atom, it leaves an unbalanced negative charge which serves to bind other positive atoms into the structure (Figure 1.21d). Two groups of feldspars are defined by these extra atoms: in **orthoclase** feldspars potassium is the dominant additional atom; in **plagioclase** feldspars varying proportions of sodium and calcium dominate.

A summary of the compositions of all the major silicate minerals is given in Figure 1.15. Note again the predominance of only a few elements in the atomic structure of these common minerals; these elements are the principal constituents of the solid earth.

1.7 Nonsilicate Minerals

Although only 8 percent of the earth's crust is composed of minerals that lack the characteristic silicon-oxygen tetrahedra of the silicates, several of these *nonsilicate minerals* are common enough in certain geologic settings to require consideration.

(a)

(b)

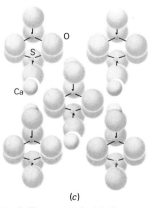

(c)

Figure 1.23 Sulfate minerals. (a) Gypsum; (b) anhy-drite; (c) atomic structure of anhydrite. Each sulfur atom is linked to four oxygen atoms. (a, b: Yale Peabody Museum)

Carbonates and Sulfates In carbonate and sulfate minerals the negative charges are provided by atomic groups of oxygen combined with either carbon or sulfur, rather than with silicon as in the silicates. Carbon, like silicon, has a normal oxidation number of +4; unlike silicon, however, it combines with oxygen in only one arrangement in forming carbonate minerals. Because of the extremely small size of the carbon atom, it has only three attached oxygen atoms arranged in a triangular, threefold coordination (Figures 1.12, 1.22).

There are three common carbonate minerals. Two of these, **calcite** and **aragonite,** have the same chemical composition in which the only positive atom is calcium ($CaCO_3$). The two minerals differ only in crystal structure (minerals which differ only in structure but not in chemical composition are termed *polymorphs* of each other). In calcite the atoms are arranged with hexagonal symmetry; in aragonite the atoms have orthorhombic symmetry (Figures 1.10, 1.22). The other important carbonate mineral, **dolomite** [$CaMg(CO_3)_2$], is similar to calcite except that it's crystal structure contains varying amounts of magnesium in addition to calcium.

In sulfate minerals four oxygen atoms combine with one positive sulfur atom to form tetrahedral atomic groups with an overall oxidation number of −2 (Figure 1.23). There are only two common sulfate minerals, and both have calcium as the only additional positive atom: **gypsum** ($CaSO_4 \cdot 2H_2O$) has water bound into the crystal structure, while **anhydrite** ($CaSO_4$) lacks water and has a different structural arrangement.

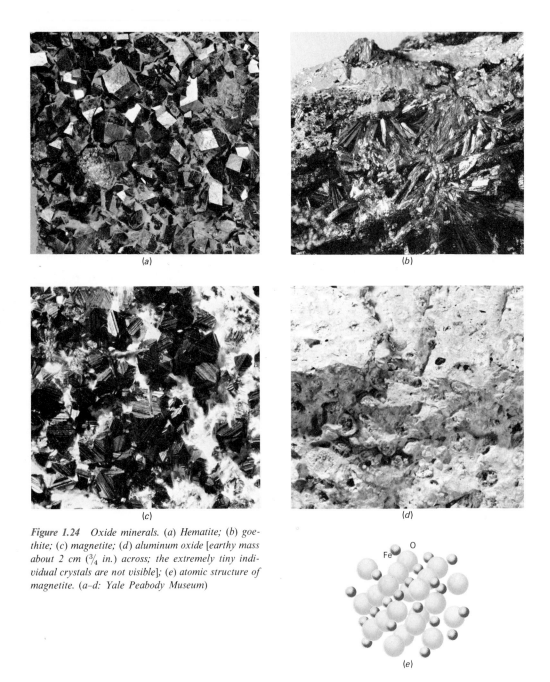

Figure 1.24   Oxide minerals. (a) Hematite; (b) goethite; (c) magnetite; (d) aluminum oxide [earthy mass about 2 cm (³⁄₄ in.) across; the extremely tiny individual crystals are not visible]; (e) atomic structure of magnetite. (a–d: Yale Peabody Museum)

(a)

(b)

(c)

(d)

O

Fe

(e)

Oxides   A few minerals have oxygen *alone* as the negative atom, rather than oxygen combined into atom groups with silicon, carbon, or sulfur. The most important such minerals are the *iron oxides*. Iron is one of several elements that may exist with either of two different oxidation num-

(a)

(b)

(c)

Figure 1.25 Halide and sulfide minerals. (a) Halite; (b) pyrite; (c) atomic structure of halite. (a, b: Yale Peabody Museum)

bers, depending on how the electrons in the outer shell combine with other elements. For iron these oxidation numbers are +2 and +3; because of them, iron combines directly with oxygen in either of two arrangements. In one the iron has an oxidation number of +3 and forms the mineral **hematite** (Fe_2O_3); in the other the iron has an oxidation number of +2 and forms the mineral **goethite** [$FeO(OH)$], which also contains an oxygen-hydrogen atomic group (from water) in its crystal structure (Figure 1.24). In addition, iron of both oxidation numbers mixes and combines with oxygen in the ratio of Fe_3O_4 to make the important iron oxide mineral **magnetite.**

A second important group of oxide minerals contains aluminum as the only positive atom. Like the silicon-bearing clay minerals, these **aluminum oxides** occur mostly as earthy masses in which the tiny individual crystals can be seen only under an electron microscope (Figure 1.24).

Halides and Sulfides Two other common minerals that need to be considered are the only ones made up of compounds which lack oxygen in their structure. In **halite** (NaCl), the negative charge is carried by the relatively rare element chlorine, with sodium as the positive atom (Figure 1.25). The other important oxygen-free mineral is **pyrite** (FeS_2), in which sulfur supplies the negative charge and iron the positive (Figure 1.25). Although pyrite (iron sulfide) is the only *common* mineral in the group, many *rare* sulfide minerals are of great economic importance as ore minerals for lead, copper, and zinc.

(a)

(b)

(c)

(d)

Figure 1.26 *Native element minerals. (a) Gold; (b) copper; (c) sulfur; (d) diamond (carbon). (a–c: American Museum of Natural History; d: DeBeers Consolidated Mines)*

Native Elements The final group of nonsilicate minerals comprises the elements that are *not* combined into compounds but occur instead as uncombined "native" elements (Figure 1.26). Such elements are rare in the earth's crust although some, particularly gold, silver, copper, sulfur, and carbon (as either *graphite* or the high-pressure polymorph *diamond*), are found in local concentrations that are of great economic importance. Uncombined iron, although very rare in the crust, is believed to be present in large quantities deep within the interior of the earth, where it occurs mixed with nickel and other uncombined metallic elements.

Man
and Earth
Materials

Natural resources used by man can be generally divided into two classes: those that are renewable and those that are not. A large proportion of the *nonrenewable* natural resources used by industrialized nations is derived from the rocks and minerals of the solid earth. These have accumulated in the crust over many millions of years of earth history and, once consumed, cannot be replaced. In contrast, *renewable* resources—principally the products of living animals and plants—can be replaced indefinitely if they are managed properly.

The hundreds of different minerals and rocks that comprise nonrenewable resources range in abundance and value from common gravel, clay, and other sedimentary building materials, costing a few dollars per ton, to rare igneous metals and gemstones, such as gold, platinum, diamonds, and emeralds, whose value may be hundreds or even thousands of dollars per ounce. Even though many different rocks and minerals are of value to man, only a few are consumed in large quantities by industrialized societies. The principal rocks and minerals from which the major nonrenewable resources are extracted are summarized in the accompanying table.

Although much has been written about the need to conserve the earth's finite supply of nonrenewable mineral resources, many modern planners feel that concern for future shortages of most of these resources has been premature. We have seen that elements such as iron, aluminum, magnesium, and potassium make up a large fraction of the earth's crust; with improved methods of extraction (and perhaps recycling) they will probably last indefinitely. There are also large potential sources of manganese, phosphorous, coal, building materials, halite, and sulfur. Among the principal nonrenewable resources, only copper, zinc, lead, and petroleum appear to have relatively limited sources, yet it will probably be several generations before even these are exhausted. In addition, the history of technology suggests that satisfactory alternative resources are usually developed as the supply of one resource drops and, thus, even these seemingly essential materials may become obsolete in the future. Aluminum, for example, is already replacing copper in many applications.

Perhaps more serious than the ultimate shortage of essential nonrenewable resources is the immediate problem of extracting and using them in ways that do not threaten other aspects of human existence. For example, unreclaimed quarries, mines, oil fields, and gravel pits blight large areas of the land surface, while even larger areas are scarred by the abandoned debris resulting from processing and consuming mineral resources which, unlike most animal and plant materials, do not quickly disappear through organic decay. In growing recognition of these problems, many conservationists now feel that the emphasis over the next decades should be shifted from merely searching for *new* mineral resources, to perfecting techniques that minimize the damage of extracting and consuming them, a prospect that we shall be exploring, in different contexts, in later chapters.

A salt mine in the Dominican Republic. (*United Press*)

A copper mine in Arizona. (*Tad Nichols*)

Major Nonrenewable Resources		Principal Rock or Mineral Sources
Industrial metals	Iron	Sed. concentrations of hematite, magnetite, goethite; less common ig. or meta. concentrations of magnetite or hematite
	Manganese	Sed. concentrations of uncommon manganese oxide minerals
	Magnesium	Extracted directly from sea water; also from sed. dolomites
	Aluminum	Sed. aluminum oxide ("bauxite") deposits
	Copper	Rare ig. concentrations of native copper and copper-bearing sulfide minerals
	Lead	Rare ig. concentrations of lead-bearing sulfide minerals
	Zinc	Rare ig. concentrations of zinc-bearing sulfide minerals
Fertilizers	Potassium	Sed. evaporite deposits
	Phosphorus	Phosphorus-rich sed. rocks associated with limestones, dolomites
Fuels	Coal	Sed. coal deposits interlayered with sandstones, shales, limestones
	Petroleum	Extracted from interstitial spaces in sed. sandstones, limestones
	Uranium	Rare sed. concentrations of uranium-bearing minerals
Building materials	Sand and gravel	Sed. deposits of quartz sand or pebbles
	Brick clay	Sed. deposits of mud, shale
	Limestone	Sed. limestones
	Gypsum	Sed. evaporite deposits
	Building stone	Sed. limestones, sandstones; ig. granites; meta. schists, marbles
Industrial chemicals	Halite (common salt)	Sed. evaporite deposits
	Sulfur	Native sulfur formed by alteration of sulfate minerals in sed. evaporite deposits

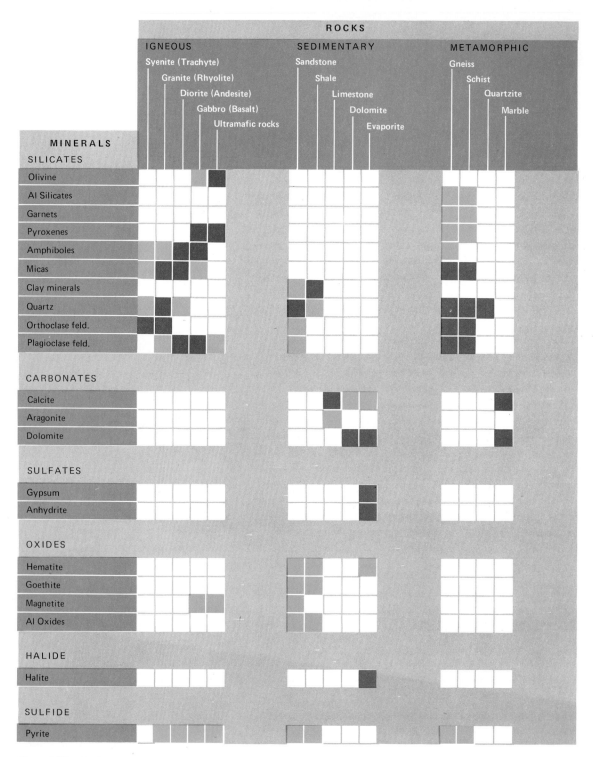

Figure I.27 The distribution of the common minerals in the principal rocks. Major occurrences are noted in dark color, and secondary occurrences in light color.

Just as the 10 most abundant elements of the earth's crust are combined to make only about 20 common minerals, so these minerals in turn are associated into a relatively few kinds of common rocks, the next larger structural units of the solid earth. In general, rocks are aggregates of several minerals; that is, rock-forming processes do not ordinarily completely segregate different minerals but, instead, lead to characteristic associations of minerals (Figure 1.27). Granite, for example, is a rock made up primarily of three minerals: mica, quartz, and feldspar. Under special circumstances *monomineralic rocks* (composed of only one mineral) may be formed. Ordinary limestone is such a rock; it is composed mostly of the mineral calcite.

There are three universally recognized classes of rocks, each of which has a different origin: **igneous rocks,** which form by crystallization from a hot molten mass called a **magma; sedimentary rocks,** which form on the earth's surface when fragmentary particles or dissolved substances from preexisting rocks are transported and deposited by water, ice, or wind; and **metamorphic rocks,** which are made up of either igneous or sedimentary rocks that have been profoundly modified, but not completely melted, by high temperatures and pressures.

1.8   Igneous Rocks

Igneous rocks make up about two-thirds of the earth's crust because during our planet's long history most parts of the crust have at some time been heated to form a molten liquid and subsequently cooled to crystallize as igneous rocks (Figure 1.28). As might be predicted from our earlier discussions of the abundant elements and minerals, the rocks produced by these meltings and coolings are composed almost entirely of a few common silicate minerals,—feldspars, quartz, micas, amphiboles, pyroxenes, and olivine (Figure 1.27). Not all these minerals occur together in a single igneous rock, for when magmas crystallize only certain combinations of minerals are formed. These combinations are the basis for the classification of igneous rocks shown in Figure 1.29. Note that the varying compositions of igneous rocks form a gradational series: from silicon-rich **syenites** and **granites,** composed mostly of quartz and feldspars, at one end of the

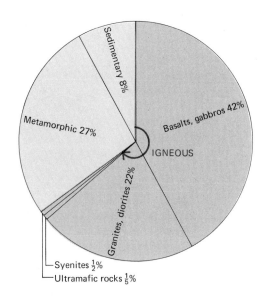

Figure 1.28 *Abundances of the principal rock types in the earth's crust. Igneous basalts and granites are the dominant crustal rocks. (Data from Ronov and Yoroshevsky,* American Geophysical Union Monograph *13, 1969)*

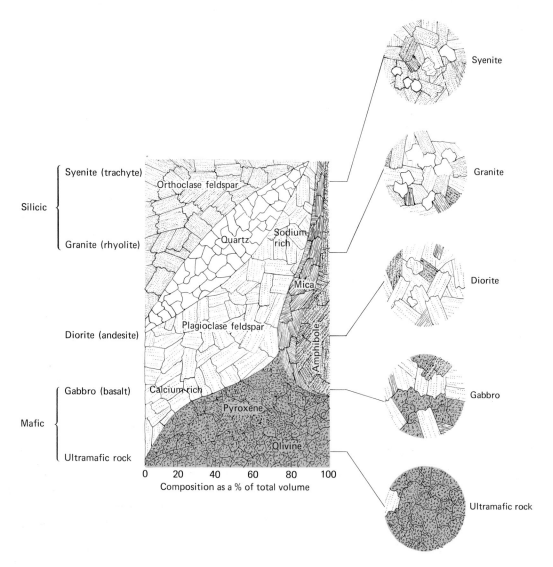

Figure 1.29 *Classification of the common igneous rocks by mineral composition. The graph at left shows the proportions of common minerals occurring in each rock type (names in parentheses are fine-grained equivalents; see text). At the right are enlarged drawings and photographs of the coarser-grained (intrusive or plutonic) rock types. The enlargements are of an area about 3 mm (⅛ in.) in diameter; the photographs show an area of rock at actual size. Dark-colored iron- and magnesium-bearing minerals are shown in color in the graph and enlargements. (Photographs: B. M. Shaub)*

scale to magnesium- and iron-rich **gabbros** and **ultramafic rocks,** composed largely of pyroxenes and olivines, at the other end. (Recall that pyroxenes and olivines both have iron and magnesium as their dominant positive atoms, whereas iron and magnesium do not occur in quartz or feldspars.) The silicon-rich syenites, granites, and diorites are commonly referred to as **silicic** igneous rocks, whereas the iron- and magnesium-rich gabbros and ultramafic rocks are called **mafic** igneous rocks.

Silicic and mafic igneous rocks differ not only in mineral composition, but also in certain gross physical

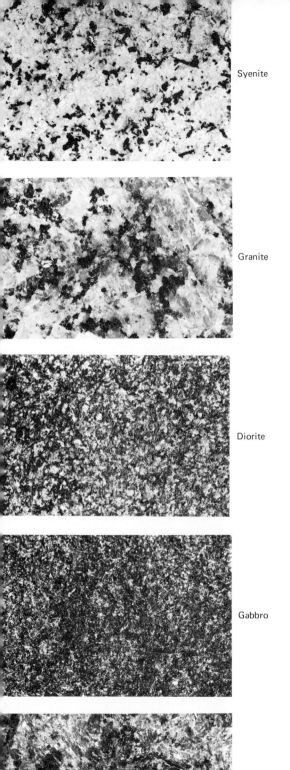

Syenite

Granite

Diorite

Gabbro

Ultramafic rock

characteristics, such as color and density. Silicic rocks are typically light in color—white, pink, or light gray as in the familiar granites used as building stone (Figure 1.30a). Mafic rocks, on the other hand, are usually very dark in color—dark gray, dark green, or even black—because of the predominance of dark-colored pyroxenes and olivines (Figure 1.30b). Mafic rocks are also heavier and denser than silicic rocks. This density difference has important implications for the distribution and origin of igneous rocks because the light, silicic rocks, particularly granite, make up most of the earth's continents, whereas the ocean basins, and probably much of the earth's interior as well, are composed of heavier mafic rocks.

Figure 1.29 also indicates which silicate structures are dominant in silicic and mafic rocks. In silicic rocks, the stable framework structures of quartz and feldspars prevail, with only minor amounts of minerals with sheet or chain structures (some micas and amphiboles). In contrast, the silicon-poor mafic rocks are dominated by chain structures (pyroxenes) and isolated tetrahedral structures (olivine). The ratio of silicon to other elements is much lower in these structures than in the framework structures, and this accounts for the relative depletion of silicon in mafic rocks.

The arrangement of igneous rocks in Figure 1.29 emphasizes another important characteristic that depends on *where* in the earth's crust the igneous magma crystallized. Magmas that crystallize deep within the crust tend to cool slowly, and this slow cooling gives rise to relatively large crystals of the minerals making up the rock. Such rocks are called **intrusive** or **plutonic** igneous rocks, and their names are given first at the left of the graph in Figure 1.29. At the opposite extreme, magmas that pour out on the earth's surface as lava, or that crystallize at very shallow depths, tend to cool much more rapidly, leading to very fine-grained textures of the enclosed minerals. Such rocks are called **extrusive** or **volcanic** igneous rocks, and their names are noted in parentheses at the left of the graph in Figure 1.29. Both plutonic and volcanic rocks may have any of the compositions shown in Figure 1.29, but because of the difference in size of the mineral grains, the two groups are usually given different names. Of the intrusive rocks named in Figure 1.29 (syenite, granite, etc.), granite is by far the most common in the earth's crust (Figure 1.28). The parenthetical names in Figure 1.29 apply to the four fine-grained, extrusive equivalents of the five major intrusive types (extrusive ultramafic rocks are

41

(a)

(b)

Figure 1.30 Exposures of igneous rocks. (a) Granite exposed in a massive dome in the mountains of central California. (b) Basalt exposed along the Snake River in southeastern Washington. (a: Josef Muench; b: U.S. Geological Survey)

extremely rare). Basalt is the most important of these because most volcanic rocks have this composition.

1.9 Origin of Igneous Rocks

The classification of igneous rocks shown in Figure 1.29 was worked out long ago from the observed associations and abundance of minerals in common rocks. Earth scientists, however, want to do more than just classify these rocks—they want to explore and seek answers to more fundamental questions of rock origin such as: Why are some rocks dominated by silicon-rich minerals and others by magnesium- and iron-rich minerals? Why can certain combinations of minerals coexist in the same rock while other combinations never occur? Under what conditions of temperature and pressure do the common igneous rocks originate? Many clues to such questions have been provided by careful field and laboratory studies of natural igneous rocks, but perhaps the most important clues have come from laboratory studies of *silicate melts*—artificial, man-made "magmas" heated and cooled under varying conditions in an attempt to simulate the natural settings in which igneous rocks form. Such studies were begun over 50 years ago by pioneering workers, particularly N. L. Bowen, at the Geophysical Laboratory of the Carnegie Institution in Washington, D.C. This laboratory, still one of the leaders in this work, is today augmented by many university and government laboratories that are also engaged in experimental studies of igneous rock origins.

Laboratory experiments can seldom *precisely* duplicate natural conditions because most natural magmas have from six to 10 or more chemical constituents, which is too many for accurate, controlled laboratory analysis. Instead, most laboratory experiments are confined to simpler systems containing only two to four elements or compounds. These studies have shown that solid minerals do not "freeze" all at once from the liquid as a magma cools, but instead go through complex series of interactions with the liquid magma; that is, certain minerals crystallize at high temperatures, but become unstable as the temperature drops and react with the magmatic liquid to give still other minerals.

Among the first minerals to be investigated experimentally were the feldspars and pyroxenes, two of the most common constituents of igneous rocks. Figure 1.31 shows the interrelationships of a pyroxene and calcium-rich plagioclase feldspar at different temperatures; these two minerals are the dominant constituents of gabbro and its fine-grained extrusive equivalent, basalt. The relations are expressed in a *phase diagram* which shows the composition of the various liquid and solid components in the artificial magma at differing temperatures. In Section 1.3 we discussed a phase diagram for water (Figure 1.8) which showed its gaseous, liquid, and solid phases at various temperatures and pressures. That phase diagram presented the behavior of a rather uncomplicated one-component system. Studies of the more complex phases of multicomponent silicate melts are often conducted at atmospheric pressure in unsealed containers in which the gaseous phases are lost. Phase diagrams of such systems show only the changing proportions of the solid and liquid

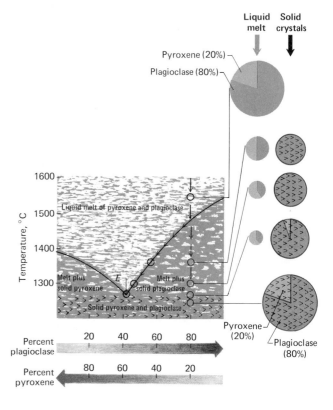

Figure 1.31 Pyroxene–plagioclase feldspar phase diagram, showing relations between cooling liquid melt and solid crystals for melts containing various proportions of pyroxene and plagioclase (bottom bars). The circles on the right and arrows on the diagram illustrate the cooling sequence in a melt containing 80 percent plagioclase (gray) and 20 percent pyroxene (color). As the melt cools, plagioclase crystals form while the melt changes composition along the curved line. On reaching the composition and temperature at point E, pyroxene and the small amount of remaining plagioclase crystallize simultaneously to give the same mineral proportions as in the original melt.

phases at differing temperatures. The two-component system in Figure 1.31 is of this sort. It was worked out over 40 years ago by Bowen and his associates and is still among the most fundamental because it deals with abundant rock-forming minerals. Recent experimental studies have concentrated on more complex three- and four-component systems, and on simpler systems at high pressures such as those to be expected deep within the earth. In addition, the great influence of *volatile constituents,* such as water and dissolved gases, in the course of magmatic crystallization has come to be recognized, and most experiments now examine the gaseous as well as the liquid and solid phases.

From his experimental studies, combined with observations on natural igneous rocks, Bowen in 1922 proposed the following theory to account for the varying compositions of igneous rocks. If a mafic magma (rich in magnesium and iron) crystallizes directly, it will form gabbroic or ultrabasic rocks made up of calcium feldspar, pyroxene, and olivine. But if, as crystallization begins, the first-formed crystals of olivine, pyroxene, and calcium feldspar are somehow separated from the remaining liquid (for example, by sinking to the bottom of a chamber filled with the liquid magma), then the remaining magma will become progressively more silicon-rich. If the separation of iron and magnesium minerals continues long enough, the remaining silicon-rich liquid magma will crystallize as a granite or syenite made up largely of quartz and potassium feldspar. Bowen called the sequences of different minerals formed in such a magma of changing composition a **reaction series;** he showed that two such series, one for feldspars, the other for olivines, pyroxenes, amphiboles, and micas, coexist in magmas of changing composition (Figure 1.32). These two series correspond, in a general way, to the changing compositions of igneous rocks shown in Figure 1.29. These findings led Bowen to postulate that all igneous rocks form from "primary" magmas of mafic composition. Silicic rocks originate when these mafic magmas change composition, or "evolve," as early-formed, silicon-poor crystals are separated from the main magmatic mass. Field studies of igneous rocks have shown that such magmatic evolution does indeed occur, and there are other lines of evidence suggesting that many silicic igneous rocks did originate from mafic materials, but in a more complex manner than visualized by Bowen. This subject will be discussed again in Chapter 3 when we look at the overall distribution of different igneous rock types in the earth's crust.

1.10 Sedimentary Rocks

The silicate minerals of igneous rocks normally form at high temperatures and pressures and are generally unstable when exposed to the lower temperatures and pressures of the earth's surface. There the action of air and water tends to break down the common igneous minerals and convert them into other minerals that are stable under surface conditions. This process, known as **weathering,** provides the raw materials for sedimentary rocks, which are merely the products of rock weathering transported and deposited by the action of water, wind, or ice. Sedimentary rocks thus originate from the interaction of the minerals of the solid earth with the waters and atmosphere of the fluid earth, and for this reason a complete discussion of their origin will be postponed until Chapter 9. Sedimentary rocks constitute only about 8 percent of the earth's crust (Figure 1.28), yet they are of great importance as the primary documents for understanding the

Figure 1.32 Bowen's reaction series. If crystals of olivine, pyroxene, and calcium-rich plagioclase are removed (small arrows) from a cooling mafic magma, the remaining melt becomes more silicic in composition. Crystallization of amphiboles, micas, and ultimately orthoclase feldspars and quartz results from this silicon enrichment. These changes correspond to the changing compositions of the principal igneous rocks (Figure 1.29).

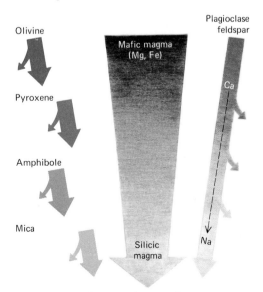

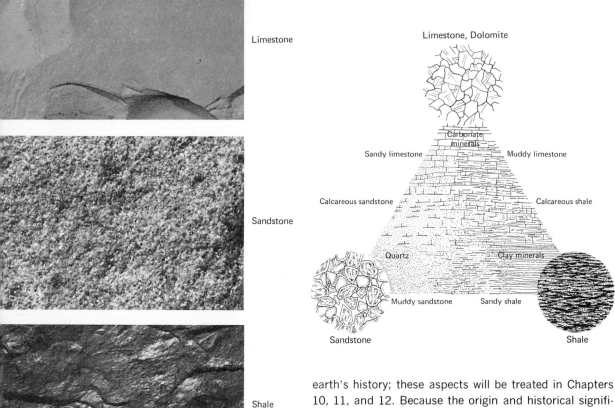

Limestone

Sandstone

Shale

Limestone, Dolomite

Carbonate minerals

Sandy limestone · Muddy limestone

Calcareous sandstone · Calcareous shale

Quartz · Clay minerals

Muddy sandstone · Sandy shale

Sandstone · Shale

Figure 1.33 Classification of the common sedimentary rocks by mineral composition. All gradations exist between pure sandstones, shales, and limestones-dolomites, which are shown at the points of the triangular diagram and in the photographs and enlarged drawings. The photographs show an area of rock at actual size; the enlargements are of an area about 3 mm (⅛ in.) in diameter. (Limestone: B. M. Shaub; sandstone, shale: Yale Peabody Museum)

earth's history; these aspects will be treated in Chapters 10, 11, and 12. Because the origin and historical significance of sedimentary rocks will be discussed later, we need mention here only the principal kinds that we shall be referring to in Chapters 2 and 3.

About 99 percent of all sedimentary rocks fall into one of four intergrading groups: sandstones, shales, carbonate rocks, or evaporites (Figures 1.33, 1.34, 1.35). **Sandstones,** as the name implies, are rocks make up of particles of sand compressed and cemented to form rock. Usually the sand particles are composed of quartz because, of all the common igneous minerals, only quartz is stable and resists further breakdown to far smaller particles when exposed to weathering at the earth's surface. Under special circumstances, however, other minerals, particularly feldspars, may also be important constituents of sandstones.

Shales are rocks made up of clay mineral particles that are too small to be distinguished except under the extreme magnification of an electron microscope. Clay minerals are the stable silicate phases formed at the earth's surface by weathering of most kinds of igneous silicate minerals, especially feldspars. Because feldspars are the most common minerals in igneous rocks, clay minerals resulting from the weathering of feldspars are

(a)   (b)   (c)

Figure 1.34 Typical exposures of sedimentary rocks.
(a) Sandstone exposed in a road cut in western Texas.
(b) Shale exposed along a stream bank in eastern New
York. (c) Evaporite exposed in a road cut in southern
Utah. (b: New York State Geological Survey; c: U.S.
Geological Survey)

also the most common sedimentary minerals. It has been estimated that about half of the earth's sedimentary rocks are clay-rich shales, about one-quarter sandstones, and about one-quarter carbonate rocks and evaporites.

Carbonate rocks, as the name implies, are those composed of carbonate minerals, principally calcite and dolomite. The two common carbonate rocks are **limestone,** composed predominantly of calcite, and **dolomite,** composed principally of the mineral dolomite (this is one of the few cases where the same name applies to a rock and a specific mineral). These carbonate rocks form when dissolved calcium and magnesium atoms, derived from the weathering of preexisting rocks, are precipitated either directly from the waters of lakes and oceans, or indirectly by the action of organisms in making shells and skeletons, many of which are composed of calcium carbonate. Such rocks precipitated from dissolved elements are called

Figure 1.35 Gradational relationship of evaporite to
the three more common sedimentary rock types shown
in Figure 1.33. Areas of photograph and enlargement
same as in Figure 1.33. (Photograph: Yale Peabody
Museum)

 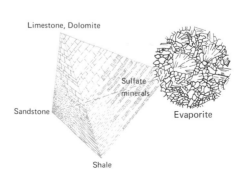

Evaporite

Limestone, Dolomite

Sulfate minerals

Sandstone

Evaporite

Shale

47

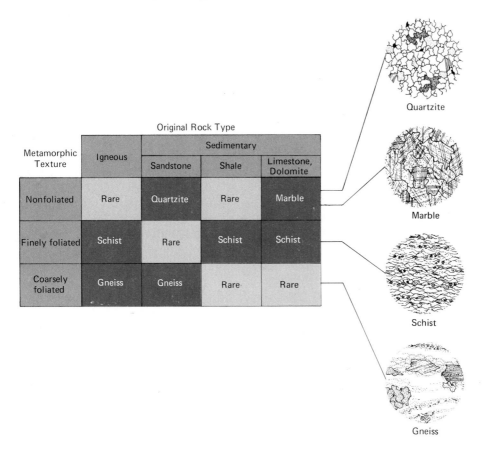

Original Rock Type

Metamorphic Texture	Igneous	Sedimentary		
		Sandstone	Shale	Limestone, Dolomite
Nonfoliated	Rare	Quartzite	Rare	Marble
Finely foliated	Schist	Rare	Schist	Schist
Coarsely foliated	Gneiss	Gneiss	Rare	Rare

Quartzite

Marble

Schist

Gneiss

Figure 1.36   Classification of the common metamorphic rocks by texture and original rock type. The enlarged drawings are of an area about 3 mm (⅛ in.) in diameter; the photographs show an area of rock at actual size. (Quartzite, marble, gneiss: B. M. Shaub; schist: Yale Peabody Museum)

chemical sedimentary rocks to distinguish them from sandstones and shales made up of preexisting solid particles, which are called **detrital sedimentary rocks.**

Evaporites are relatively uncommon chemical sedimentary rocks that form when the dissolved materials of lake or ocean waters become highly concentrated through evaporation. Under such circumstances dolomite, gypsum, anhydrite, and halite are the principal minerals deposited (Figure 1.27). Evaporite rocks usually contain two or more of these minerals.

Sedimentary rocks commonly occur as pure, unmixed sandstones, shales, carbonates, or evaporites, but all gradations exist among these major types. Sandstones, for example, may contain clay or carbonate minerals, giving rise, respectively, to ''muddy'' or ''calcareous'' sandstones. Some of these intermediate compositions can be conveniently described by the terms shown in Figure 1.33.

Quartzite

Marble

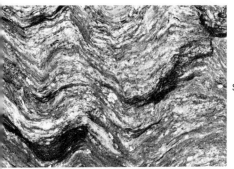

Schist

Gneiss

1.11 Metamorphic Rocks

This final class contains rocks formed when heat, pressure, and the chemical action of liquids and gases alter the minerals in preexisting igneous or sedimentary rocks.[3] Such alteration is one of the commonest processes in the solid earth; usually it involves temperatures high enough to melt the original rocks completely, thus forming a magma which eventually crystallizes as new igneous rock. When temperatures are too low for complete melting but high enough to disrupt the crystalline structure of some or all of the minerals in a rock, then the minerals may recrystallize—*without* completely melting—to become metamorphic rocks. In spite of the rather restricted conditions under which they form, metamorphic rocks make up about one-fourth of the earth's crust (Figure 1.28).

The most common minerals of metamorphic rocks are those that are also abundant in igneous rocks: feldspars, quartz, micas, and amphiboles (Figure 1.27). In addition, however, many minerals that are rare in igneous rocks are common products of metamorphism. Particularly important are the *aluminum silicates* and *garnets,* minerals which form from the metamorphism of the clay minerals in shales.

Metamorphic rocks are more varied and complex than normal igneous and sedimentary rocks, and many names and classification schemes have been proposed for them. Figure 1.36 shows a simple, widely used classification that is based not on mineral composition, as were the previous classifications of igneous and sedimentary rocks, but on the nature of the original rock before metamorphism and on the different textures of the resulting metamorphic rock. As the minerals of igneous or sedimentary rocks recrystallize during metamorphism, they normally became oriented in parallel, sheetlike planes. Such rocks are said to have a **foliated texture.** Finely foliated rocks in which the parallel planes of mineral orientation are closely spaced are called **schists;** more coarsely foliated rocks are called **gneisses** (Figures 1.36, 1.37). Either type of foliation can result from the metamorphosis of many

[3] In a sense the weathering of igneous rocks by the action of water and air at the earth's surface is a kind of metamorphism, but metamorphic rocks are normally considered to be those formed by chemical changes when buried below the zone of weathering.

(a)

(b)

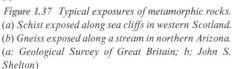

Figure 1.37 Typical exposures of metamorphic rocks.
(a) Schist exposed along sea cliffs in western Scotland.
(b) Gneiss exposed along a stream in northern Arizona.
(a: Geological Survey of Great Britain; b: John S.
Shelton)

kinds of igneous and sedimentary rocks. In addition, a few metamorphic rocks lack a foliated arrangement of the constituent minerals. The most common such rocks are **quartzites** and **marbles,** which sometimes form from the metamorphosis of sedimentary sandstones and carbonate rocks, respectively (Figure 1.36).

By far the most abundant metamorphic rocks are gneisses, most of which are similar to ordinary granite in composition and are thus called **granitic gneisses.** It has been estimated that such gneisses account for about 80

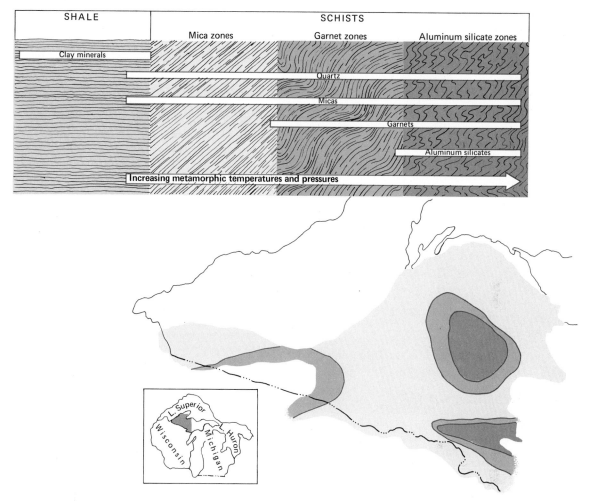

Figure 1.38 Changes in the mineral components of shale when metamorphosed to schist at increasing temperatures and pressures. The clay minerals are first altered to quartz and mica, some of which then converts to garnet and, finally, to aluminum silicate minerals. The presence of these key minerals can be used to identify "zones" of increasing metamorphism in the schists. The map shows the progressive metamorphic zones as exposed in schists of upper Michigan. (After Ernst, Earth Materials, 1969; generalized from James)

percent of all metamorphic rocks, schists for about 15 percent, and quartzites and marbles for only about 5 percent.

Classifications of metamorphic rocks based on mineral composition are more complex than the textural classification of Figure 1.36 because different minerals may be formed from the same original rock by differing metamorphic temperatures and pressures. Figure 1.38 shows how increasing temperatures and pressures lead to progressive changes in mineral components during metamorphosis of a sedimentary shale. Similar "zones" of mineral change have been defined for the progressive metamorphism of other common sedimentary and igneous rocks.

SUMMARY OUTLINE

Matter

1.1   *Elements and compounds:* the bulk of the solid earth is composed of only 10 elements chemically combined in compounds.

1.2   *Atoms and molecules:* the atomic properties of elements—the number and arrangement of electrons, protons, and neutrons—determine the elements' chemical and physical properties, and the manner in which elements combine to form compounds.

1.3   *Gases, liquids, and solids:* transformations among these three fundamental states of matter continuously occur among the elements and compounds making up the earth's rocks, water, and air.

1.4   *The solid state:* most of the compounds of the solid earth are aggregates of small crystals, each of which has a regular, orderly arrangement of the internal atoms.

Minerals

1.5   *Chemistry of the solid earth:* most of the earth's solid crust is composed of crystalline minerals dominated by oxygen-silicon atom groups; a small but significant proportion of crustal minerals lack silicon; only a very few common minerals lack oxygen.

1.6   *Silicate minerals:* adjacent silicon-oxygen tetrahedra share oxygen atoms in differing ratios to make four principal groups: framework, sheet, and chain silicates, and silicates with isolated tetrahedra.

1.7   *Nonsilicate minerals:* the most important mineral groups which lack silicon are carbonate, sulfate, and oxide minerals, all of which contain oxygen, and halide and sulfide minerals, which lack oxygen.

Rocks

1.8   *Igneous rocks:* about two-thirds of the solid crust is made of rocks that previously crystallized from molten magma. By far the most abundant types are silicon-rich granite and iron- and magnesium-rich basalt.

1.9   *Origin of igneous rocks:* laboratory experiments with artificial silicate melts have suggested that some granitic rocks originate from the chemical evolution of basaltic magmas.

1.10   *Sedimentary rocks:* when exposed to water and air at the earth's surface, igneous rocks are altered to particles of sand and clay, and to dissolved elements. These materials are then transported and accumulated by water, air, or ice to form sedimentary rocks, the most abundant of which are sandstones, shales, and carbonate rocks.

1.11   *Metamorphic rocks:* when exposed to high temperatures and pressures, preexisting igneous or sedimentary rocks may be altered without complete melting to form metamorphic rocks. The most abundant of these are gneisses and schists.

ADDITIONAL READINGS

*Ahrens, L. H.: *Distribution of the Elements in Our Planet,* McGraw-Hill Book Company, New York, 1965. A brief, intermediate survey of the distribution of the elements, stressing their occurrence in igneous rocks.

Desautels, P. E.: *The Mineral Kingdom,* Grosset & Dunlap, Inc., New York, 1968. A popular introduction to gems, minerals, and mineral collecting, with sumptuous illustrations.

*Ernst, W. G.: *Earth Materials,* Prentice-Hall, Inc., Englewood Cliffs, N.J., 1969. A comprehensive introduction to minerals and rocks.

Huang, W. T.: *Petrology,* McGraw-Hill Book Company, New York, 1962. A standard intermediate text on rocks.

Mason, B.: *Principles of Geochemistry,* John Wiley & Sons, Inc., New York, 1966. A clearly written and accessible introduction to all aspects of the chemistry of the earth.

Mason, B., and L. G. Berry: *Elements of Mineralogy,* W. H. Freeman and Co., San Francisco, 1968. An outstanding intermediate text on minerals.

*Skinner, B. J.: *Earth Resources,* Prentice-Hall, Inc., Englewood Cliffs, N.J., 1969. An authoritative and readable introduction to mineral resources.

*Turekian, K. K.: *Chemistry of the Earth,* Holt, Rinehart and Winston, Inc., New York, 1972. A brief but authoritative introduction.

Vanders, I., and P. F. Kerr: *Mineral Recognition,* John Wiley & Sons, Inc., New York, 1967. A guide to mineral identification, with many fine color photographs.

*Zim, H. S., and P. R. Shaffer: *Rocks and Minerals,* Golden Press, New York, 1957. An excellent elementary guide to mineral and rock identification.

* Available in paperback.

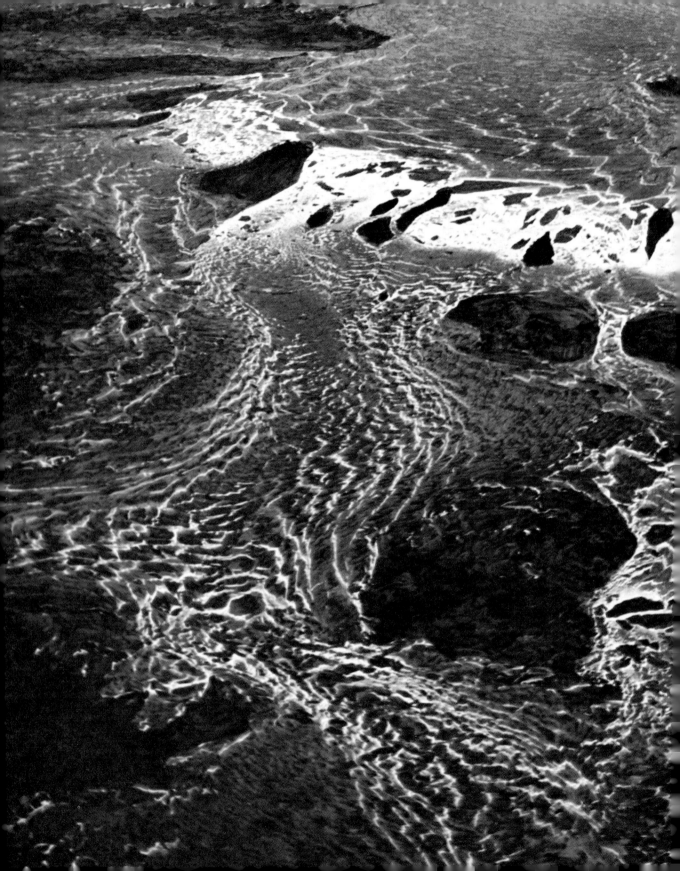

2

The Earth's Interior

Unfortunately, the only rocks and minerals that earth scientists can study *directly* are those exposed on the earth's surface and those which can be sampled in mines and bore holes. The deepest mines extend downward only about 3 km (2 mi), and the deepest bore hole drilled to date (an oil well in Texas) reached a depth of about 8 km (5 mi). Since the earth's radius is almost 6,400 km (4,000 mi), clearly only its most superficial skin can be directly sampled (Figure 2.1). For this reason, our understanding of the earth's interior depends on various indirect clues, most of which are provided by four physical properties: its *wave transmission, heat transmission, magnetism,* and *gravity*. Each of these can be measured at the surface and then used to infer characteristics of the materials below. By combining evidence from these properties, geophysicists have developed a model for the structure and composition of the earth's vast interior; the nature of this model and the supporting evidence for it will be the subject of this chapter.

EARTHQUAKES AND THE EARTH'S INTERIOR

The most important clues to the earth's interior are provided by the manner in which various shock waves are transmitted through the body of the earth. Such shock waves have two sources: some result from man-made explosions of dynamite or nuclear devices; others are caused by **earthquakes,** which are natural movements along ruptures bounding large blocks of the solid earth. In general, shock waves from explosions are most useful for determining relatively local properties of the earth's crust, whereas earthquakes may reveal both small- and large-scale patterns of the earth's internal structure. Indeed, the study of earthquake wave transmission has

Lava flow at night, Hawaii. (Richard S. Fiske, U.S.G.S., and Science)

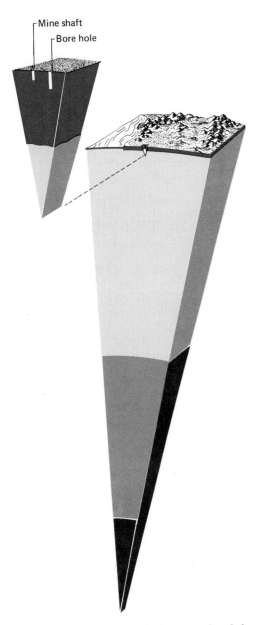

Figure 2.1 *Scale drawing of a deep mine, bore hole, and the earth's interior.*

Mine shaft

Bore hole

provided more information about the earth's interior than have clues from the additional properties of gravity, magnetism, and heat transmission combined.

2.1 Earthquake Waves

The rock movements responsible for earthquakes are not randomly distributed throughout the solid earth, but occur only in the outermost 700 km (400 mi), a little over 10 percent of the distance to the center of the earth (Figure 2.2). Only a few occur at depths as great as 700 km; most take place in the outermost 70 km representing only about 1 percent of the radius of the earth. In addition to these depth restrictions, earthquakes show a geographic concentration along two great world-encircling belts. Most large earthquakes, and virtually all of those occurring at depths below 300 km (180 mi), take place in a zone around the edges of the Pacific Ocean; a second zone of smaller earthquakes occurs along chains of submerged mountains that extend through the ocean basins (Figure 2.3).

Earthquakes arising in these regions cause various kinds of shock waves to move upon and through the earth in much the same way that waves and ripples are caused by dropping a pebble into quiet water. These waves, known collectively as **seismic waves,** are of three principal kinds: **body waves,** that travel from the source or **focus** of the earthquake through the main mass of the earth to emerge at the surface; **surface waves,** that travel only in the outermost layers of the earth; and **free oscillations,** vibrations of the earth as a whole, caused by only the largest earthquakes, which can be likened to the vibrations of a bell when struck by a hammer (Figure 2.4). Most of our knowledge of the earth's interior has come from study of body waves, but in recent years analyses of surface waves have provided additional information, particularly about the outermost layers of the earth. Free oscillations were first recorded by special instruments after the disastrous Chilean earthquake of 1960, although they had been predicted earlier. Like surface waves, they are useful in interpreting the structure of the earth's outermost layers, but they also provide some independent clues to the deep interior. All three wave types are measured by sensitive instruments called **seismographs** which, when anchored to solid rock, record even the slightest earth shocks (Figure 2.5). Universities and government research institutes have established around the world about 150 major *seismic*

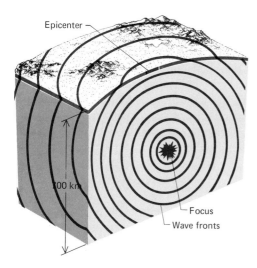

Figure 2.2 Relative locations of focus and epicenter of an earthquake. Earthquake foci do not occur below 700 km.

Figure 2.3 Locations of epicenters for all earthquakes recorded from 1961 through 1969. (From National Earthquake Information Center, U. S. Dept. of Commerce)

observatories containing such instruments, and these are the principal source of information on earthquake wave transmission.

Earthquake body waves, on which most of our knowledge of the earth's deep interior rests, can be subdivided into two types known as **primary** (or merely *P*) and **secondary** (*S*) waves. In *P* waves the individual particles of the material through which the wave travels are *pushed* back and forth in the direction of wave movement; in *S* waves the particles are *pulled* sideways, at right angles to the direction of wave motion (Figure 2.6). Because of these different motions, the two waves move at different speeds: in the same material, *S* waves travel only about half as fast as *P* waves. The names reflect this fact, for "primary" *P* waves arrive at distant seismic observatories before "secondary" *S* waves from the same earthquake. In addition to these inherent velocity differences, the speed at which *both* waves move depends on the physical properties, particularly the density and elasticity, of the materials they are passing through. Both types move faster as the density decreases and the elasticity increases. Because of this influence on velocity, *P* and *S* waves are bent or *refracted* at the boundaries of layers with differing densities or elasticities, just as light waves

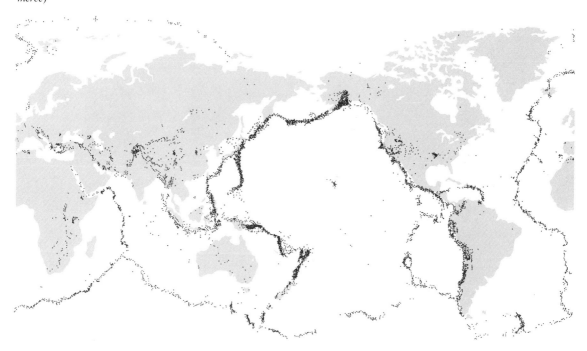

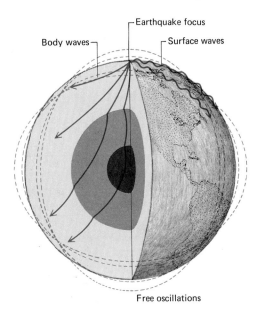

Figure 2.4 *The three principal types of seismic waves: body waves, surface waves, and free oscillations.*

Figure 2.5 *Seismographs. (a) Schematic diagram. Earthquake waves cause movements of the underlying rock and attached support; the weight tends to remain stationary while the pen records the motion. (b) A seismograph registering a large, distant earthquake. (United Kingdom Atomic Energy Authority)*

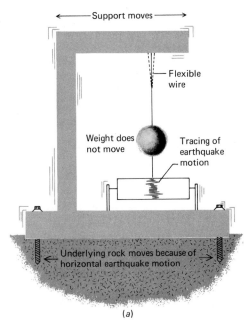

are bent in passing from less dense air into more dense glass (Figure 2.7). Like light waves, they may also be *reflected* from such boundaries without penetration. Because of both reflections and refractions, the earthquake body waves reaching a distant seismic observatory consist not only of simple *P* and *S* waves that have travelled by the most direct route, but also of dozens of additional waves that have reached the station by more circuitous reflected and refracted paths.

2.2 Seismic Discontinuities

Observations on the arrival time of such waves coming from thousands of different earthquakes recorded at many observatories have provided data for the calculation of **velocity-depth curves,** which show the change in speed of both *P* and *S* waves as they go deeper into the earth. These curves, shown in Figure 2.8, are the primary tools used for interpreting the earth's interior.

The velocity-depth curves of Figure 2.8 show two depths at which earthquake wave velocities abruptly change: one at the relatively shallow depth of about 16 km (10 mi) (marked "Moho"), the other much deeper, at about 2,900 km (1,800 mi) (marked "C-M"), or about halfway to the earth's center. Such abrupt velocity changes

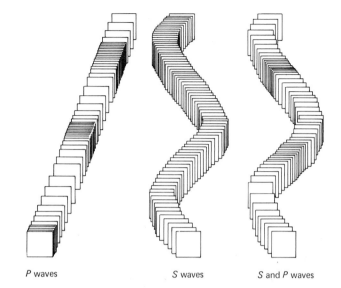

P waves   *S* waves   *S* and *P* waves

Figure 2.6 *Schematic representations of the motions of P and S waves as they pass through the body of the earth.*

Figure 2.7 *Direct, reflected, and refracted paths of P and S waves.*

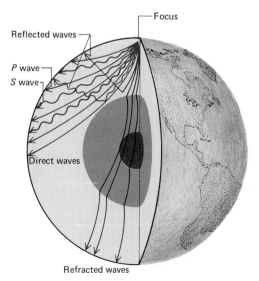

are called **seismic discontinuities;** they clearly show that the materials of the earth's interior are not uniform in density and elasticity. Instead, the two major discontinuities define three concentric layers with differing properties: a thin outer **crust,** a central **core,** and a thick **mantle** lying between the crust and core. The mantle makes up about 80 percent of the earth's total volume, the core about 19 percent and the crust only about 1 percent.

The outermost discontinuity, that separating crust and mantle, was discovered in 1909 by a Serbian geophysicist, A. Mohorovičić. It has since come to be known as the **Mohorovičić discontinuity,** which is commonly abbreviated simply as "the Moho." The Moho seismic discontinuity provides the principal evidence that the rocks and minerals exposed on the earth's surface are but a thin veneer or "crust" overlying a much larger volume of materials with differing properties—the mantle and core.

The second major discontinuity, first recognized in 1924, is commonly referred to as the **core-mantle discontinuity.** Note in Figure 2.8 that not only does the velocity of *P* waves decrease dramatically at this discontinuity, but *S* waves *stop completely* at that depth, that is, they do not penetrate the core (see also Figure 2.7). This fact is of prime importance because *S* waves, you will recall, result from a sidewise *pulling* movement of the transmitting particles. Unlike the *pushing* motions of *P* waves which can be transmitted through both solids and liquids,

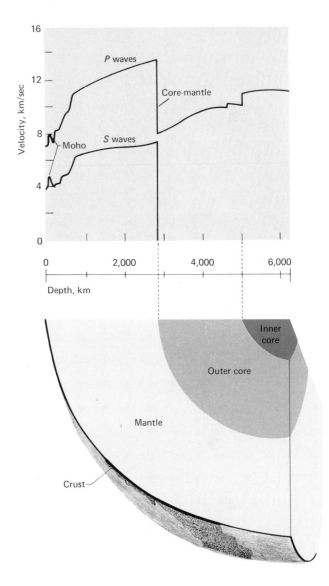

Figure 2.8 *Velocity-depth curves for P and S waves (above) and the inferred internal structure of the earth (below). Major velocity changes occur at the Moho and core-mantle discontinuities. (Data from Anderson et al.,* Science, *vol. 171, 1971)*

S waves can only be transmitted in solids, where the particles have enough adhesion to be pulled by one another. Liquids, of course, lack such adhesion and therefore do not transmit S waves. The lack of S wave transmission through the core provides strong evidence that the *core is liquid*, whereas the transmission of both P and S waves through the crust and mantle indicates that both the *crust and mantle are composed of solid materials*. To speak of "the solid earth" is therefore something of an oversimplification since only the outer four-fifths of the earth's volume is made up of solid materials.

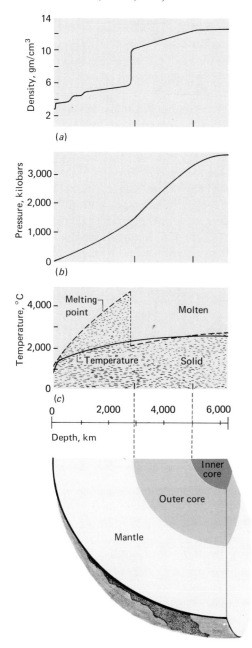

Figure 2.9 Inferred densities (a), pressures (b), and temperatures (c) of the earth's interior. (Data from Stacey, Physics of the Earth, *1969; Verhoogen,* American Scientist, *vol. 48, 1960)*

In addition to the two major seismic discontinuities defined by sharp breaks in the velocity-depth curves, minor discontinuities occur at several depths where the curves show less abrupt changes (Figure 2.8). These minor discontinuities provide additional clues to the structure of the core and mantle.

2.3 Internal Densities, Pressures, and Temperatures

We have noted that the velocity of P and S waves is dependent on both the *density* and the *elasticity* of the materials through which they pass. Neither of these properties can be directly measured for materials deep within the earth, and thus it is not completely certain whether the velocity changes seen in the velocity-depth curves result from changes in one or the other property, or both. Calculations based on the earth's gravity show, however, that the *average* density of the *entire* earth is about 5.5 grams per cubic centimeter (gm/cm^3). This figure puts a restriction on the densities of the material in the mantle and core (because the crust makes up only about 1 percent of the earth's volume, it can be effectively ignored), for they must combine to have this average density. Because of this restriction, and certain others caused by the manner in which the earth wobbles on its axis of rotation, it is possible to construct a reasonable model of the density distribution within the earth that is consistent with the velocity-depth curves. Such a model is shown in Figure 2.9a. Note that the densities of the mantle materials increase from the crustal value of 2.8 to about 5 at the base of the mantle. Then there is a sharp density increase, to about 10 at the boundary of the core, and thereafter an increase from 10 to about 12 at the center of the earth. The average of these values, based on the different volumes of mantle and core material, is 5.5.

From the density curves, it is then possible to calculate the probable increase in pressure downwards, which varies with both the amount and density of the overlying material. Such calculations show a steady pressure increase, from atmospheric pressure at the surface (1 bar), to pressures about 1,000 times as great at the base of the crust and over 3 *million* times as great at the earth's center (Figure 2.9b). These internal density and pressure curves are derived from the earthquake wave velocity-depth curves (Figure 2.8), but it should be stressed that these determinations also require educated guesses about

61

the composition and elastic properties of the earth's interior materials and are therefore less certain than are the velocity-depth curves themselves, which are based on direct records of earthquake wave velocities.

Earthquake wave data provide little information about still another fundamental property of the earth's interior, its range of temperatures. The most obvious and important temperature inference that can be made from the velocity-depth curves is that the materials of the mantle, as solids, must be cooler than their melting point, whereas the materials of the core, as liquids, must be heated beyond their melting point.

Additional clues to the earth's internal temperatures are provided by still another physical property that can be measured at the surface—the manner in which heat is transmitted from the earth's warm interior to its cooler surface. Observations in mines and bore holes show that temperatures always increase rapidly downward through the outer few miles of the earth's crust. The rate of this increase varies from place to place, giving rise to differing *temperature gradients;* these range from about 9°C for each kilometer of depth (25°F/mi) to about 52°C/km (150°F/mi), with an average value of about 28°C/km (80°F/mi). Thus at the bottom of the deepest bore holes (about 8 km) temperatures are often far above the boiling point of water. Below the very shallow zone of direct measurement in mines and bore holes, the earth's internal temperature gradient is poorly understood. The rapid rate of temperature increase seen in the outer crust probably does not persist downward, however, for if it did the materials making up much of the mantle would almost certainly be melted (Figure 2.9c shows one estimated temperature-depth curve which fits this requirement).

THE CORE AND EARTH MAGNETISM

Studies of earthquake wave transmission show that the earth's interior has a basic threefold division into a thin solid crust, a massive solid mantle, and a central liquid core. As we have seen, earthquake wave data also permit the calculation of the probable densities and pressures within these fundamental units. There are, however, many other things that we would like to know about the earth's interior. For example, the composition and structure of the minerals that comprise the mantle and core can only be inferred for, unlike the minerals of the outer crust, these

•

deeper materials cannot be directly sampled. Furthermore, the mantle and core are certainly not uniform, static masses but dynamic units that show variations in their composition and structure. Although much remains to be learned about these variations, geophysicists have combined various indirect clues to arrive at reasonable hypotheses about some of them. For the remainder of the chapter we shall review these clues and hypotheses, beginning with the earth's central core and progressing outward to the mantle and, finally, to the deeper, hidden parts of the crust.

2.4   Composition of the Core

Seismic evidence indicates that the core is largely composed of a liquid whose probable density ranges from about 10 gm/cm^3 at the outer core boundary to about 12 gm/cm^3 at the earth's center. Most geophysicists now believe that this liquid is *molten iron,* with small amounts of other molten elements, particularly nickel and either sulfur or silicon, mixed in. Three principal lines of evidence suggest this.

Molten iron has a density of about 7.5 gm/cm^3 at the earth's surface, but calculations show that at the extreme pressures encountered in the core, the iron atoms would be forced into a close-packed arrangement with a density slightly higher than that suggested for the core by seismic evidence. The slight discrepancy is believed to be due to small quantities of a lighter element, probably sulfur or silicon, in the liquid iron.

Additional evidence for an iron core is provided by the composition of **meteorites,** fragments of matter from interplanetary space that fall to the earth's surface. The materials in meteorites are believed to be similar to those that long ago consolidated to make the earth and other planets of our solar system (this will be discussed in more detail in Chapter 13). The composition of meteorites should, therefore, be similar to the bulk composition of the earth; that is, the most common elements in meteorites should also be the most common in the earth's core and mantle (recall that the highly variable rocks and minerals of the crust make up only 1 percent of the earth's volume and can be disregarded in considerations of the earth's bulk composition). Meteorites are of two principal kinds: **iron meteorites,** composed predominantly of iron, but containing between 5 and 20 percent nickel, and **stony meteorites,** composed of silicate minerals. The iron meteorites suggest that the earth's core may have a similar composition; in contrast, the stony meteorites may be similar in composition to the earth's mantle.

2.5 Earth Magnetism

The third suggestion that the core is composed of molten iron comes from an important source of indirect evidence about the earth's interior that we have not yet introduced—that provided by surface measurements of the earth's magnetic field. The most familiar demonstration of the earth's magnetism is the common compass, the needle of which is merely a small bar magnet mounted on a swivel so that it may align itself with the earth's magnetic field. Compass observations have long shown that the earth behaves much as if it were a gigantic bar magnet with poles near the axis of rotation (Figure 2.10). Because the earth's magnetic poles are near the rotational poles, the compass needle always aligns itself in approximately a north-south direction.

Figure 2.10 A generalized view of the earth's magnetic field. The magnetism is believed to be caused by motions within the liquid iron core.

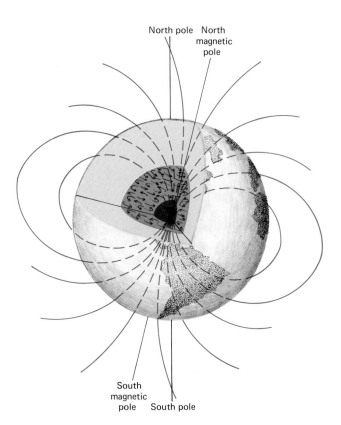

North pole   North magnetic pole

South magnetic pole   South pole

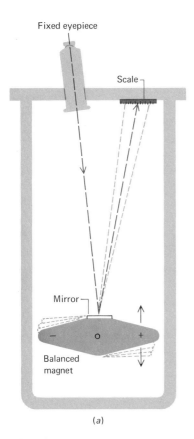

Fixed eyepiece

Scale

Mirror

Balanced
magnet

(a)

(b)

Figure 2.11 Magnetometers. (a) Schematic diagram. A balanced magnet moves either up or down with slight changes in the earth's magnetic field; the position of the magnet is read by means of the eyepiece, mirror, and scale. More complex electronic magnetometers are commonly used for surveying large areas. (b) A simple magnetometer in use. (Texas Instruments)

The exact strength and direction of the earth's magnetic field at different points on the surface is measured by sensitive instruments called **magnetometers,** which can be used on land, at sea, or even in moving airplanes for rapid surveys of large areas (Figure 2.11). Such measurements show that the earth's overall magnetic field consists of two components which can be analyzed separately to yield different kinds of information. The largest component, the **main magnetic field,** originates deep within the earth and is a principal source of information about the interior. The second, much smaller component originates from the distribution of magnetic minerals, primarily magnetite, in the earth's outer crust; it is of great value in understanding crustal composition and structure.

The origin of the main magnetic field is far more difficult to understand than is the simple magnetism caused by magnetic minerals in the outer crust. The main internal field cannot have such a source, for when magnetic minerals are heated they quickly reach a tempera-

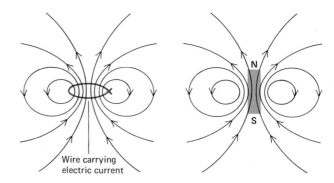

Figure 2.12 *The magnetic field produced by a loop of wire carrying an electric current is similar to that produced by a bar magnet.*

Wire carrying
electric current

ture, known as the **Curie temperature,** at which they lose their magnetism yet still remain solid. Curie temperatures are thus usually far lower than the melting temperatures of minerals, and are also far lower than the temperatures to be expected in the solid mantle. For these reasons, any solid magnetic minerals present below the cool outer crust have almost certainly lost their magnetism and cannot be a source of the main magnetic field. There remains, however, one other potential source of internal magnetism— that associated with electrical currents (Figure 2.12).

All electrical currents have associated magnetic fields, a fact first established in 1819 by H. C. Oersted who noticed that an electric current in a wire affected a compass needle placed near the wire. Now if there were some means by which electrical currents were generated deep within the earth, then the main magnetic field could be understood as a direct natural consequence. The silicate minerals that make up the crust and, probably, most of the mantle are poor electrical conductors and thus are unlikely sources of electric currents and electromagnetism. Iron, on the other hand, is an excellent conductor (in fact, the magnetic properties of iron can be traced to minute currents within its atoms) and thus motions in a core of fluid iron (see Figure 2.10) might well create both electrical currents and the main magnetic field. This possibility adds additional strength to the suggestion that the core is, indeed, composed of molten iron.

2.6   Core Structure

Surface measurements of the main magnetic field show that most of its force could be produced by core-generated electrical currents that cause an **axial dipole magnetic field,** that is, a field with two opposing poles at opposite ends of an axis, much as in a simple bar magnet (see Figure 2.10). After this large dipole field is subtracted from the main field there remains, however, a small **residual field** which, although originating in the core, is apparently not the result of the same current-producing core movements that cause the larger dipole field. Magnetic observations over the past hundred years show that the magnetic patterns of this residual field *move steadily westward* at a rate of about 19 km (12 mi) per year. The main dipole field, on the other hand, has remained rather constant in position. This westward drift of the residual field is of great importance for it suggests that the core, at least in its outermost parts, is rotating less rapidly than the mantle and crust; only such a difference in speed of rotation would account for the steady change in position of the residual field with respect to the earth's surface where it is measured (Figure 2.13). This differential movement further suggests that the residual field may result from turbulent eddy currents caused by the drag of the solid

Figure 2.13   Westward drift of the residual field from 1922 to 1942. Strong positive values of the residual field are shown in color; strong negative values in gray. Note the slight westward shift of the patterns during the 20-year interval. (From Vestine et al., Description of the Main Magnetic Field, *1947)*

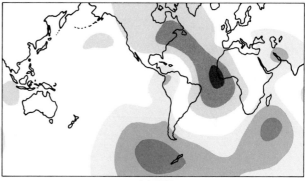

(a) 1922

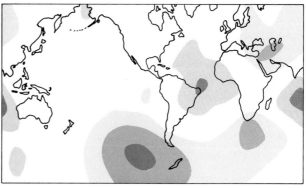

(b) 1942

mantle as it moves around the less rapidly rotating fluid core. Such eddy currents could produce small-scale electrical currents that might, in turn, give rise to the residual field.

Unfortunately, there is more difficulty in explaining the kinds of movements in the fluid core that might be responsible for the principal, dipole component of the main magnetic field. Not only are there theoretical problems in understanding how *any* motion in a fluid of uniform composition can produce magnetic fields, but there are also obstacles to explaining the underlying *causes* of the motions. Speculative suggestions for getting around these objections do exist, however, and most geophysicists now accept large-scale motions of the fluid core as the most probable source of the earth's main magnetic field. These internal motions are undoubtedly strongly influenced by the earth's overall rotational motion; this interaction probably causes the close coincidence of the magnetic and rotational poles.

One final bit of information about the structure of the core is provided not by magnetic data, but by earthquake wave transmission. The *P* wave velocity-depth curve of Figure 2.8 shows a relatively sharp velocity change within the core, at a depth of about 5,000 km (3,200 mi). This discontinuity indicates a change in physical properties in the innermost part of the core; the small spherical inner core defined by this discontinuity is generally considered to be under such intense pressures that the iron atoms are closely compressed and respond as a solid, rather than as a liquid as does the bulk of the core. Such a solid inner core would best account for the observed velocities and patterns of wave transmission, but the evidence is still far from conclusive.

THE MANTLE

The mantle makes up about 80 percent of the volume of the solid earth, but because the average density of mantle material is about half that of the underlying core, the mantle accounts for about 67 percent of the earth's total mass. Besides occupying most of the earth's interior volume and contributing two-thirds of its total mass, the mantle is of extraordinary importance because its properties have largely determined the nature of the crust which overlies it. Direct studies of crustal rocks thus provide some clues to the hidden mantle but, as with the core, most of our understanding of mantle composition and structure is based on indirect evidence.

2.7 Mantle Composition

Unlike the liquid iron of the core, the solid mantle is believed to be composed predominantly of silicate minerals, just as are the more familiar rocks of the crust. This composition is indicated by three lines of evidence. The principal evidence of mantle composition is provided by meteorites which, we have seen, are of two principal types: iron meteorites, whose composition is probably similar to that of the core, and stony meteorites, composed of silicate minerals. Because there are compelling reasons for believing that meteorites are similar to the materials that originally consolidated to make the earth (to be discussed in Chapters 11 and 13), stony meteorites may well reflect the general composition of the huge volume of material in the mantle. Stony meteorites do not in fact show the variety of compositions seen in crustal rocks; instead, they are composed largely of the silicate minerals *olivine* and *pyroxene* and are similar in composition to the relatively rare ultramafic rocks found in the crust (see Figure 1.29). This suggests that the bulk of the earth's mantle may have an ultramafic composition.

The second evidence of mantle composition is provided by laboratory experiments on the wave transmission characteristics of various silicate minerals at high temperatures and pressures. Such studies show that most crustal rocks transmit seismic waves too slowly to account for the velocities observed in the mantle (Figure 2.14a). There are, however, various mixtures of dense silicate minerals which have mantlelike wave transmission properties and among these are mixtures dominated by olivine and pyroxene. The properties of one such mixture, called *pyrolite*, are shown in Figure 2.14b. This experimental work therefore supports the inference, based on meteorites, that olivine and pyroxene make up much of the mantle, at least in its outermost parts.

The final evidence of mantle composition comes from rare occurrences at the earth's surface of olivine- and pyroxene-rich ultramafic rocks which are believed to have crystalized in the upper mantle (Figure 2.15). These rocks contain small amounts of certain minerals (diamond, for example) which can form only under extremely high pressures, pressures greater than those normally encountered in the crust. Such minerals suggest that the ultramafic

Figure 2.14 Mantle composition. (a) Seismic wave velocity (color) and inferred density (gray) of the mantle, at a depth of about 200 km, compared to those for the two principal crustal rocks—granite and basalt. (Data from Clark, Geological Society of America Memoir 97, 1966) *(b) Seismic wave velocity (color) and inferred density (gray) of the mantle, at a depth of about 200 km, are identical to those of laboratory mixtures of olivine, pyroxene, and garnet. Such mixtures, in the proportions shown, are called* pyrolite. *(Data from Anderson et al.,* Science, vol. 171, 1971)

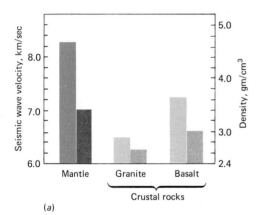

(a)

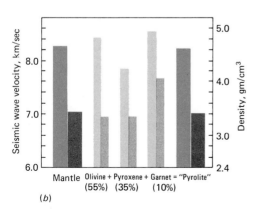

(b)

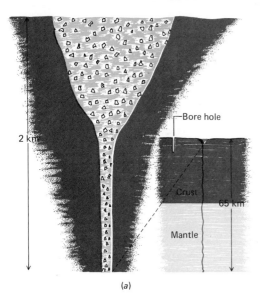

2 km

Bore hole

Crust

65 km

Mantle

(a)

(b)

(c)

Figure 2.15 "Diamond pipes," cone-shaped masses of ultramafic rock that are believed to represent materials of the upper mantle forced upward into the crust and exposed today at the earth's surface. (a) Schematic cross section. (b) Ultramafic rock being mined near the surface in South Africa. (c) A deep diamond pipe, the abandoned Kimberly diamond mine in South Africa. The ultramafic rocks below the thick, overlying soil were dug out, crushed, and screened to remove the diamonds. (b: De Beer Consolidated Mine; c: Bob Landry)

rocks originated in the upper mantle and were then somehow transported into the overlying crust, where they are exposed today on or near the earth's surface.

2.8 Mantle structure

Recent seismic studies indicate that the mantle is not uniform throughout but, instead, shows both horizontal and vertical changes in seismic properties. Figure 2.16

shows a modern survey of P wave velocities through the upper 800 km (500 mi) of the mantle, the region lying closest to the earth's surface which can be studied with high precision by seismic techniques. Note that in addition to the Moho marking the upper boundary of the mantle, there are three additional zones of very rapid increase in wave velocity bounded by regions of much less abrupt increase: these zones lie at depths of about 100, 400, and 650 km (approximately 60, 250, and 400 mi). There is also a single zone of sharp *decrease* in wave velocity lying not far below the crust, between depths of about 60 to 100 km (40 to 60 mi). These zones of rapid velocity change indicate levels at which the mantle materials undergo abrupt changes in their density, or elasticity, or both.

Some increase in density with depth is to be expected in the mantle because of the increasing pressures caused by the overlying materials, but calculations show this simple pressure increase to be both too small and too gradual to account for the sharp zones of velocity increase. Instead, it is relatively certain that the lower two zones, at least, mark regions in which there is an *abrupt* change in the crystal structure of the silicate minerals that make up the mantle. At the temperatures and pressures

Figure 2.16 Seismic wave velocities and inferred structure and composition of the outer 1,000 km of the mantle. (P wave data from Knopoff, Science, *vol. 163, 1969)*

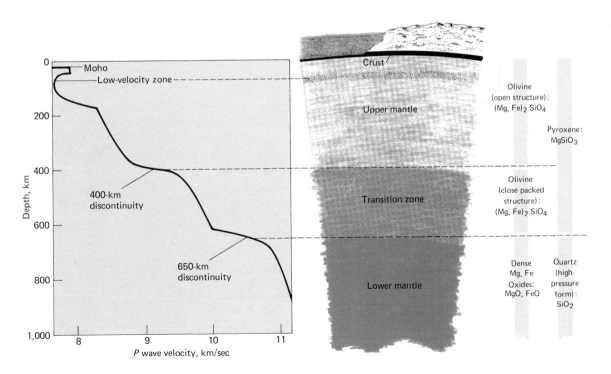

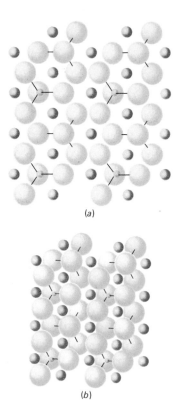

(a)

(b)

Figure 2.17 Structural change in olivine at high pressures. (a) Crystal structure of normal olivine as found in the crust and upper mantle. (b) More dense olivine with compacted crystal structure, inferred to be present in the transition zone of the mantle.

occurring near the earth's surface, olivine and pyroxene are relatively stable minerals. Theoretical calculations and laboratory experiments show, however, that at higher temperatures and pressures olivine converts abruptly to a more dense and closely packed crystal structure (Figure 2.17). At still higher temperatures and pressures the silicon is forced from both olivine and pyroxene to yield a kind of quartz, while the remaining magnesium and iron convert to oxide minerals with very compact crystal structures and high densities (Figure 2.16). These transitions occur at about the pressure-temperature conditions expected in the mantle at depths of 400 and 650 km, a finding that strongly suggests that the two lower zones of velocity increase are regions of abrupt structural changes in the minerals of the mantle. The region surrounding these changes is called the **transition zone** of the mantle; it separates the relatively thin and variable *upper mantle* from a much larger and probably more uniform mass of material comprising the underlying *lower mantle* (Figure 2.16).

The velocity increase in the upper mantle at 100 km depth is more complex because no transitions in the crystal structure of olivine and pyroxene should occur at such shallow depths. In addition, this zone of increase is closely associated with the overlying **low-velocity zone,** the single region of sharp velocity *decrease* in the mantle. Several lines of evidence suggest that the low-velocity zone is a region in which an *increase in elasticity*, rather than a sharp decrease in density, accounts for the low wave velocities; that is, it marks a zone where the minerals of the upper mantle are unusually warm, plastic, and mobile in contrast to the rocks above and below, which have the same composition but are cooler and more rigid (see Figure 2.16). Indeed, in some regions the low-velocity zone fails to transmit S waves, indicating that it is heated beyond the melting point to form local pockets of liquid magma. The velocity increase below the low-velocity zone apparently marks the zone in which the mantle materials become cooler and resume their normal rigidity. Because the warm low-velocity zone lies only slightly below the crust, it is probably the source for much of the molten magma that reaches the earth's surface to form volcanic rocks. In addition, the zone is intimately interrelated with movements of the crust. Both of these subjects will be developed more fully in the next chapter.

The velocity-depth relations shown in Figure 2.16 are based on detailed seismic data for the upper mantle lying

Man
and the Earth's
Interior

Over the long course of earth history, the deep interior has been largely responsible for the nature of the earth's surface on which man lives and from which he extracts most of his resources. On the much shorter time scale of human history, however, the earth's deep interior is of concern to man for one principal reason—as the source of destructive earthquakes, which can be among the most devastating of all natural catastrophes.

Although the center, or focus, of an earthquake always lies well within the earth's interior, the deep motions which cause the earthquake are commonly transmitted through the crust to the earth's surface. Such violent surface movements pose little threat to men standing in the open, but when they occur in urban regions they may cause the collapse of buildings with disastrous consequences. In addition, earthquakes occurring beneath the ocean may produce enormous *seismic sea waves* (often miscalled "tidal waves") which approach shore with heights as great as 100 feet to cause much death and destruction in low-lying coastal regions.

In an average year, 10 earthquakes causing widespread death and devastation occur somewhere on the earth; about 100 others cause serious local destruction; 1,000 do some damage; about 100,000 are strong enough to be felt as tremors; and about one million can be detected by seismographs. Although earthquakes may occur anywhere on the earth's surface, the most frequent and destructive are concentrated in the seismically active zone around the margins of the Pacific Ocean.

There is nothing that man can yet do to *prevent* earthquakes, but there is much that can be done to minimize their destructiveness. Buildings can be constructed to withstand almost any earthquake. Although such construction is unusually expensive, it is required by local building codes in many seismically active regions. Death and injury from seismic sea waves, although not property damage, have been minimized by establishing an elaborate warning system throughout the Pacific region. This system takes advantage of the fact that sea waves usually require many minutes or even several hours to travel from their source to populated islands and coastal regions. Thus when seismographs record strong oceanic earthquakes, seismic sea wave warnings are issued immediately and coastal inhabitants normally have time to evacuate to higher, inland regions. Much additional death and injury could be prevented if there were a means of predicting earthquakes themselves before they occur. Much research effort is currently being directed toward this problem, but no really reliable earthquake-predicting technique has yet been discovered.

Southern California earthquake damage, 1971. (Associated Press)

1867 sketch of a seismic sea wave striking a British steamer in the Virgin Islands. (Mansell Collection)

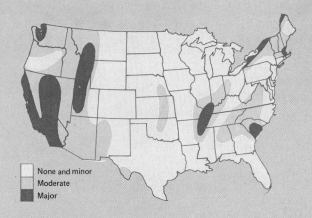

Seismic risk map, showing potential areas of earth-quake damage in the United States. (NOAA)

None and minor
Moderate
Major

beneath western North America. Generally similar patterns have been found in other regions, but these differ enough in detail to suggest that the low-velocity zone, and the deeper zones of change in crystal structure, are not simple, earth-encircling spheres but show lateral variations which, in part at least, reflect similar variations observed in the thin overlying crust.

THE DEEP CRUST AND EARTH GRAVITY

Only half of the crust's average 16-km (10-mi) thickness is accessible to direct sampling by even the deepest bore holes. The depths of the crust, like the mantle and core, are therefore known mostly from indirect physical measurements. As with the core and mantle, the principal evidence of deep crustal structure is provided by seismic wave velocities. Strong supporting evidence comes from an additional clue to the earth's interior that we have not yet introduced—that provided by gravity measurements.

2.9 Gravity

Gravity is a property of all bodies of matter by which they are attracted to other bodies by a force that increases as the mass (total amount of matter) of the bodies increases and as the distance between them grows smaller. This is the familiar force that causes unsupported objects to fall to the earth's surface; on a larger scale the gravitational attraction of the sun, whose total mass is very great, holds the planets in their orbits.

If the earth was a perfectly spherical body with a symmetrical internal structure, then the gravitational force acting at all points on its surface would be exactly the same. Neither of these conditions is true: the earth's subsurface materials are not distributed with perfect symmetry nor is the earth a perfect sphere, for its rotation causes it to bulge out slightly at the equator and become slightly flattened at the poles. Because of these variations in internal symmetry and external shape, measurements of the earth's gravitational force vary from place to place over its surface (Figure 2.18). These gravity measurements, made by sensitive instruments called **gravimeters** (Figure 2.19), can be considered to include three separate gravity components: a very large component resulting from the principal mass of the earth, and two smaller components, one resulting from local variations in the

Figure 2.18 Departures from uniform gravity caused by variation in the earth's external shape and internal symmetry. The earth is neither a perfect sphere, nor are its surface and internal layers uniform in thickness and composition; these variations lead to differing values of the earth's gravitational force when measured at different points on the surface.

Ellipsoid of earth

Perfect sphere

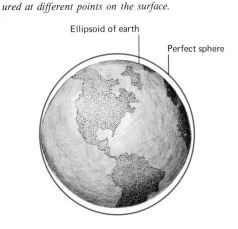

Earth's surface   Perfect ellipsoidal surface

Ocean

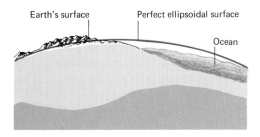

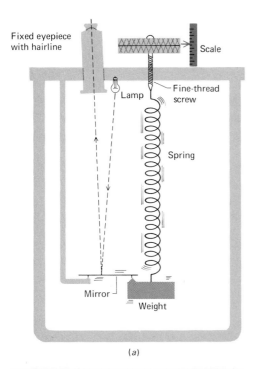

Figure 2.19 Gravimeters. (a) Schematic diagram. Gravity variations cause the suspended weight to move up or down. The screw is then adjusted to raise or lower the weight to a constant height (mirror horizontal, lamp centered in eyepiece); the gravitational force is read from the height of the screw. (b) A gravimeter in use. (Texas Instruments)

earth's shape, and the other from local differences in the thicknesses and densities of the underlying rocks. It is the two small components that make gravity measurements useful for interpreting the earth's interior, for they can be analyzed separately by first subtracting the large component contributed by the main mass of the earth, and by then making certain simplifying assumptions about either the local rock distribution or the general shape of the earth, depending on which parameter the geophysicist wishes to measure. Tens of thousands of such gravity measurements have been made on land and, more recently, from gravimeters on surface vessels at sea, and these have provided a wealth of information about both the earth's shape and the varying densities of rocks in the deep crust and upper mantle.

2.10 Isostasy

There is still another kind of information provided by gravity data. If reasonable values are assigned to each of the three gravity components, then it should be possible to *predict* the force of gravity at any point on the earth's surface. Such predictions have been made for many years, and their careful correlation with actual gravity measurements shows that the two fit *only* over continental areas of low relief. In mountainous regions the force of gravity is normally *lower* than would be predicted, whereas over the deep ocean basins it is normally *higher*. These **gravity anomalies,** as they are called, are caused by variations in the rocks of the crust: apparently crustal rocks are relatively less massive under mountain ranges and more massive under ocean basins. This systematic variation of gravity with surface topography was discovered over a hundred years ago and has since been confirmed by thousands of gravity measurements.

The discovery that mountain ranges are underlain by relatively less massive rocks quickly led to the suggestion that mountains stand high because they are "floating" on the underlying rocks in much the same way that a block of copper or iron will float in heavier liquid mercury. This concept of the "floating" of light crustal rocks, termed **isostasy,** had far-reaching implications for the study of the deep structure of the crust. In particular, it was hypothesized that such floating could come about in two different ways. If all the rocks of the crust had approximately the same density, then mountains would require a large buried *root* of similar material to support them on the "fluid," subcrustal material. To provide buoyancy, this

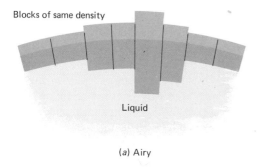

Blocks of same density

Liquid

(a) Airy

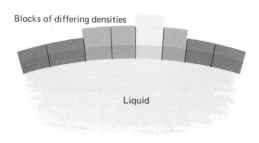

Blocks of differing densities

Liquid

(b) Pratt

Figure 2.20   The Airy and Pratt isostatic hypotheses. In the Airy scheme (a), mountains and lowlands are composed of the same materials and mountains float higher because they have large roots; in the Pratt scheme (b) they float higher because they are composed of lighter materials.

Figure 2.21   Moho depths and crustal thicknesses under continents and ocean basins.

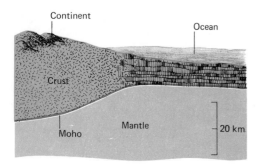

root would have to be much larger than the rocks exposed at the surface, just as the submerged parts of a log or iceberg are larger than the parts exposed above the surface (Figure 2.20a). But if there were density variations in the rocks of the crust, then mountains might "float" higher because they are composed of lighter materials than surrounding lowlands (Figure 2.20b). This second scheme was first proposed in 1854 by J. H. Pratt, an English geographer, and has come to be known as the *Pratt hypothesis;* the constant density scheme was proposed the following year by another Englishman, the astronomer G. B. Airy, after whom it is called the *Airy hypothesis*. These isostatic hypotheses were the first clues to the deep structure of the earth's crust because, under either hypothesis, simple calculations showed that the rocks exposed in continental mountain ranges must extend far beneath the earth's surface and "float" in rocks of higher density.

2.11   Seismic Evidence of Deep Crustal Structure

It was not until 1909, when Mohorovičić discovered the seismic discontinuity at the base of the crust, that there was any additional evidence of deep crustal structure. Subsequent observations showed that the depth of the Moho discontinuity averages about 35 km (22 mi) under the continents but, in striking contrast, *is much shallower under the ocean floors,* where it has an average depth of only 8 km (5 mi). These observations were extremely important for they showed that, on a continental scale, the base of the crust "mirrors" the topography of the earth's surface; that is, the crust is thickest under the high-standing continents and thinnest under the low-lying ocean basins (Figure 2.21) as predicted by the Airy scheme of isostasy.

This is not the whole story, however, for seismic evidence *also* indicated that the continents are composed of less dense rocks than are the ocean floors, a finding that supported the Pratt isostatic hypothesis. By making certain reasonable assumptions about the elasticity of crustal rocks, it is possible to infer crustal densities directly from seismic velocities—the faster the velocities, the more dense the rock. Early in the history of seismic studies it was noted that velocities in the continental crust were generally slower than in the oceanic crust, and these observations showed that continents are composed of lighter rocks than are the ocean floors. Seismic evidence thus

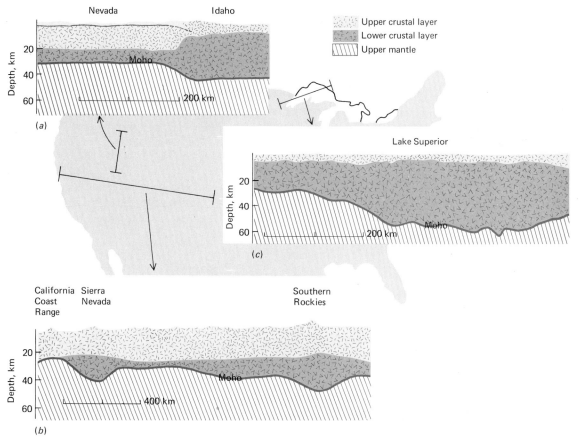

Figure 2.22 Variations in deep crustal structure and Moho depth, based on detailed seismic surveys of three U. S. Regions. Note that the crust thickens under the Sierra Nevada and Rocky Mountains, but not under the California Coast Ranges (b). It is also unusually thick under the Lake Superior region (c), where there is little surface relief. Most regions show two crustal layers, but Nevada appears to have three (a) and much of California only one (b). (Modified from Bott, The Interior of the Earth, *1971; after Hill and Pakiser)*

indicated that *both* the Airy and Pratt schemes of isostasy play a role in the deep structure of the crust.

Early seismic studies also suggested that the crust itself was composed of "layers" of rock, each having a relatively constant wave velocity (and, by inference, a constant density), and each separated by a minor discontinuity in wave velocity. Continents were considered to have two principal layers: an upper low-density layer and a lower layer of higher density. Oceans, on the other hand, showed evidence of only one principal layer whose density was about equal to the lower layer of the crust. These observations were interpreted to mean that the "oceanic crustal layer" was worldwide, extending under the lighter, upper continental layer as well as under the oceans.

Intensive modern seismic studies using precisely timed artificial explosions have shown that deep crustal structure, like that of the upper mantle, is more complex than this simple, layered model would suggest. Some continental regions show one or several small discontinuities within the crust, while in others the wave velocities increase uniformly with depth rather than changing abruptly (Figure 2.22). In all cases, however, there is a tendency for the velocities to increase downward, suggesting that although continental rocks may not necessarily be underlain by oceanic materials, they do become progressively more dense with depth. In oceanic areas, the presence of a single principal layer with a relatively high and constant velocity has been confirmed. Velocities and inferred rock densities found throughout the oceanic layer are about as high as the maximum velocities and rock densities found deep under the continents, confirming once again the old suggestion that the varying heights of the crust's continents and ocean basins are related to large-scale differences in the deep crustal rocks which underlie them.

Finally, recent seismic studies also show that the Moho discontinuity, marking the base of the crust, can be clearly recognized in most regions, but that its depth varies in unexpected ways (see Figure 2.22). In some regions the crust thickens to a maximum of about 80 km (50 mi) under mountains, but other mountain ranges are underlain by crust of normal thickness. Conversely, abnormally thick crust sometimes occurs in regions of low surface relief. It should be stressed, however, that even though there is no simple, universal relation between crustal thickness and surface relief in local regions, on a much larger scale the difference between the thick continental crust and the much thinner oceanic crust has been abundantly confirmed.

SUMMARY OUTLINE

**Earthquakes
and the Earth's
Interior**

2.1 *Earthquake waves:* seismic (earthquake shock) waves travel through the earth and provide the principal evidence of its internal structure; among the several types, body waves (primary and secondary) that pass deep within the earth are the most instructive.

2.2 *Seismic discontinuities:* the changing velocities of seismic body waves with depth indicate a basic threefold division of the earth's interior into a central core, a thick overlying mantle, and a thin surface crust; each is bounded by a sharp change or discontinuity in wave velocity.

2.3 *Internal densities, pressures, and temperatures:* the probable pressures and densities of the earth's internal materials can be inferred from seismic velocity-depth curves; internal temperatures, on the other hand, are more poorly understood.

The Core and Earth Magnetism	2.4	*Composition of the core:* seismic data indicates that the core is mostly liquid; this liquid is probably molten iron with small amounts of nickel and either sulfur or silicon.
	2.5	*Earth magnetism:* the earth's magnetic field is believed to originate from electrical currents caused by motions within the liquid iron core.
	2.6	*Core structure:* the changing patterns of the earth's magnetic field, as measured at the surface, provide clues to the structure of the hidden core.
The Mantle	2.7	*Mantle composition:* the mantle occupies 80 percent of the earth's volume; it is probably composed of magnesium- and iron-rich silicate minerals, which are principally olivine and pyroxene in its outer regions.
	2.8	*Mantle structure:* seismic discontinuities in the mantle indicate a warm, plastic zone just below the crust and two deeper, more rigid zones where olivine and pyroxene are believed to convert abruptly to more dense mineral structures.
The Deep Crust and Earth Gravity	2.9	*Gravity:* the force of gravity varies at the earth's surface due to variations in both the earth's shape and in the characteristics of the underlying rock.
	2.10	*Isostasy:* gravity measurements suggest that high-standing mountains and continents of the crust "float" on more dense underlying rock.
	2.11	*Seismic evidence of deep crustal structure:* the Moho seismic discontinuity, which separates the crust from the mantle, lies far deeper under continents than under ocean basins; the continental crust is therefore much thicker than the crust underlying the oceans.

ADDITIONAL READINGS

Bott, M. H. P.: *The Interior of the Earth,* St. Martin's Press, Inc., New York, 1971. An excellent intermediate text.

*Clark, S. P., Jr.: *Structure of the Earth,* Prentice-Hall, Inc., Englewood Cliffs, N.J., 1971. A brief, authoritative introduction, emphasizing the earth's interior.

Garland, G. D.: *Introduction to Geophysics,* W. B. Saunders Company, Philadelphia, 1971. An advanced text on the earth's interior.

Gaskell, T. F.: *Physics of the Earth,* Funk & Wagnalls, New York, 1970. A well-illustrated elementary introduction to the earth's interior.

*Hodgson, J. H.: *Earthquakes and Earth Structure,* Prentice-Hall, Inc., Englewood Cliffs, N.J., 1964. A brief, popular introduction stressing the relation of earthquakes to man.

Phillips, O. M.: *The Heart of the Earth,* Freeman, Cooper and Co., San Francisco, 1968. A readable introduction to all aspects of the earth's interior.

Stacey, F. D.: *Physics of the Earth,* John Wiley & Sons, Inc., New York, 1969. An intermediate-level survey of the earth's interior.

* Available in paperback.

3

The Earth's Crust

Rocks of the outermost crust can be directly collected, analyzed, and interpreted wherever they are exposed at the earth's surface or in mines and wells. Such study has produced both a wealth of facts and a host of unsolved problems. Continents and the ocean floor can be seen to be made up of different kinds of rocks, yet the origin of this difference cannot yet be explained. At the same time, rocks of the crust are known to be continuously created, moved, deformed, and destroyed, but the enormously energetic driving force behind these changes remains unknown. Such facts and problems that have grown from direct study of crustal rocks will be the subject of this chapter.

ROCKS OF THE CRUST

3.1 Continents and Ocean Basins

Imagine for a moment that the oceans could somehow be drained to leave the entire surface of the earth exposed as dry land. An explorer traveling in this strange world would discover that the earth's surface consists of two fundamentally different landscapes. The most obvious would be broad, flat plains covering millions of square miles, interrupted only by long, narrow chains of high mountains. These plains would contrast sharply with the second major feature, plateau areas—also relatively flat with linear mountain ranges—that would tower several miles above the surrounding plains and would be separated from them by the highest and most continuous scarps on earth. The lower plains would, of course, be the floors of today's oceans, and the higher plateaus the surface of today's continents. This division into low *ocean basins* and much higher *continents*, the most obvious

Selwyn Mountains, northwestern Canada. (National Film Board of Canada)

feature of the earth's crust, is caused by fundamental differences in the composition and structure of continental versus oceanic rocks.

In general, we know far more about continental rocks than about those making up the ocean floor. Continental rocks are more accessible at the earth's surface, and in mines and bore holes which reach maximum depths of 8 km (5 mi), or about one-quarter of the depth to the Moho discontinuity at the base of the continental crust. The Moho lies only about 6 km (4 mi) beneath the deep ocean floor, but because of the difficulties of drilling from floating platforms through several miles of water, only a relatively few ocean-bottom holes have so far been drilled, and the deepest extend only about a thousand meters into oceanic rocks. Rocks from the *surface* of the ocean floor have, however, been recovered by dredging along submarine cliffs and canyons in many parts of the oceans, and additional information about oceanic rocks is provided by islands that are far from the nearest continental land masses. These data on oceanic rock composition, combined with information from hundreds of thousands of surface rock exposures and tens of thousands of mines and wells from the continents, have shown that *continents are largely composed of silicic igneous rocks, particularly granite*, whereas *ocean basins are largely underlain by heavier, mafic igneous rocks, particularly basalt*. These direct observations of rock composition are further confirmed by seismic wave velocities which show values in the range expected for granitic rocks throughout the continental crust, but are generally higher in the oceanic crust, thus suggesting a predominance there of heavier, basaltic materials.

In summary, the differences between continents and ocean basins are primarily related to the concentration, in the continents, of the relatively light elements that are abundant in the minerals of granite—particularly silicon, sodium, and potassium; the rocks under the ocean floors, in contrast, have a higher concentration of heavier iron- magnesium- and calcium-rich minerals and are therefore more similar in composition to the inferred rocks of the underlying mantle than are continental rocks (Figure 3.1).

3.2 Continental Rocks

Although most of the surface rock of the continents is covered by a thin layer of soil, it is usually easy to determine the nature of the underlying ''bedrock,'' either by

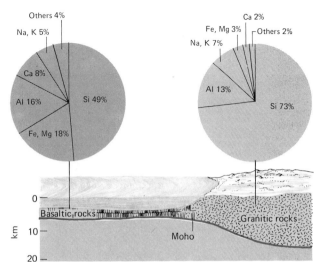

Figure 3.1 *Generalized chemical compositions of the basaltic rocks that underlie the ocean floor and the granitic rocks that underlie the continents.*

digging through the soil or, more commonly, by observing the rock where it is exposed in stream banks and cliffs or in man-made quarries, excavations, and road cuts (see Figures 1.30, 1.35, and 1.37). In addition, data from surface rock exposures may be supplemented by information on the "subsurface" rock distribution in areas that have been mined or explored for petroleum. In studying the rocks of a particular continental region, geologists normally seek out all of this evidence and use it to prepare a **geologic map** showing the distribution of different rocks as they would look if all the soil were removed from the surface (Figure 3.2). The degree of detail shown in geologic maps varies with the purpose of the map and the amount of field study that has gone into its preparation. At one extreme, areas with valuable mineral deposits may be mapped in such detail that 1 in. on the map equals only a few hundred feet on the ground (Figure 3.2d). At the other extreme, large areas are commonly mapped by rapid reconnaissance methods, usually with the help of aerial photographs, at scales in which 1 in. on the map equals 8 mi or more on the ground (Figure 3.2b). Most geologic mapping falls between these extremes (Figure 3.2c). In the United States, for example, most of the country has been geologically mapped at scales of 1 in. = 1 mi, and many local regions at scales as great as 1 in. = 1/4 mi. In preparing a geologic map each distinctive kind of rock, called a **formation,** is given a separate two-part name. The first

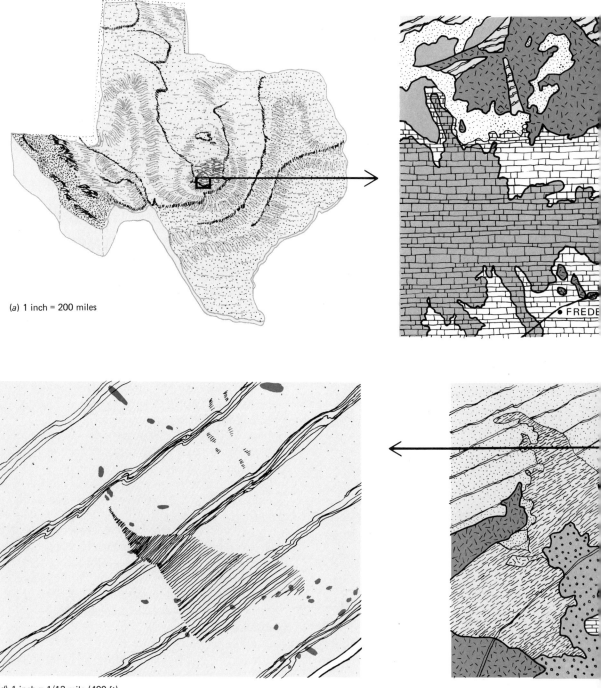

(a) 1 inch = 200 miles

FREDE

(d) 1 inch = 1/13 mile (400 ft)

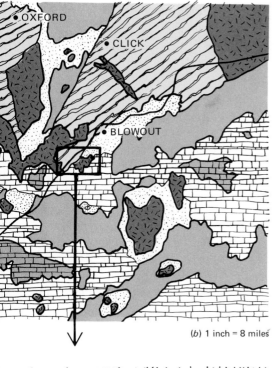

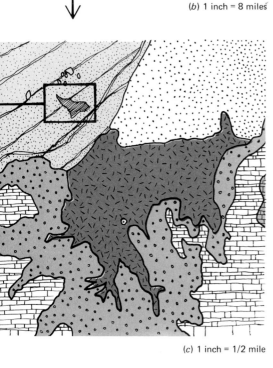

(b) 1 inch = 8 miles

(c) 1 inch = 1/2 mile

Figure 3.2 *Typical geologic maps at different scales. (a) Index map of Texas showing sedimentary veneer (black) and underlying basement complex (color); the latter is exposed at the surface over a roughly circular area in the center of the state. (b-d) Geologic maps of a portion of the southern basement-sediment boundary at progressively larger scales. Map d, which is exceptionally detailed, was prepared to locate small "soapstone" deposits (dark color; soapstone is a metamorphic rock made up of a fine-grained micalike mineral used in the paint, ceramic, paper, and rubber industries). (Compiled from various sources)*

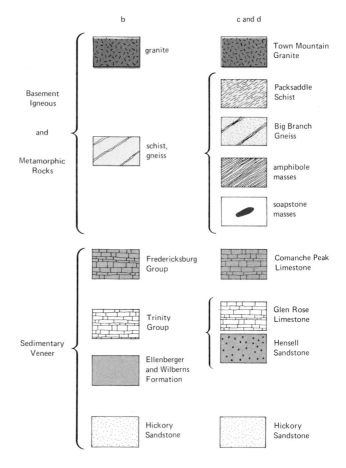

part of the name is a geographic locality where the rock is well exposed; the second part describes the general rock type. Thus Manhattan Schist, St. Louis Limestone, and Columbia River Basalt are typical formation names used in the preparation of local geologic maps. Other examples, from a single region, are shown in Figure 3.2.

Geologic mapping, combined with information on subsurface rock distribution, has shown that most of the earth's continents are covered with a relatively thin sheet of sedimentary rocks. This sheet is normally from a hundred to a few thousand meters thick, but in local areas it may thicken to greater than 15,000 m (50,000 ft). This sedimentary veneer is everywhere underlain by older igneous and metamorphic rocks which are collectively termed the **continental basement complex** (Figure 3.2). Although most of the land surface is covered by a veneer of sedimentary rocks, each continent has large areas in which the sedimentary cover is absent and the basement rocks are exposed at the surface. These regions are known as **shield areas;** their distribution is shown in Figure 3.3. Most of our knowledge of the basement rocks which make up the bulk of the continental crust has come from studies of these areas.

Geologic mapping shows that many kinds of rocks are found in shield areas, but granite, and metamorphic

Figure 3.3 The principal shield areas, which are regions where the igneous and metamorphic basement rocks of the continents are exposed at the earth's surface. Elsewhere on the continents the basement rocks are covered by a veneer of predominantly sedimentary rock. The arrow indicates the small area of basement exposure that is enlarged in Figure 3.2. (Modified from Kummel, History of the Earth, *1970, after various sources)*

Figure 3.4 *Extensive exposures of basement granite and granitic gneiss in the shield areas of northern Canada. (Canadian Dept. of Energy, Mines, and Resources)*

gneisses of granitic composition, are by far the most abundant. These granites, as plutonic rocks, originally crystallized deep within the crust but have subsequently been exposed by a wearing away of the crustal surface over the long course of earth history (Figure 3.4). Granite is also the dominant rock found in mines and wells that penetrate basement rocks in continental regions having a sedimentary cover. Below this surface zone that can be directly sampled, the principal evidence of crustal composition comes from gravity measurements and seismic wave velocities, both of which suggest a granitic composition for much of the deep crust.

Since granite is the dominant rock of the continental crust, geologists have long been concerned with trying to understand its origin. The most fundamental question is: How did the lighter elements and minerals characteristic of granite become concentrated into thick, separated continental masses lying adjacent to the heavier rocks of the ocean floor? No complete answer to this puzzle yet exists, but strong clues are provided by studies of *movements*

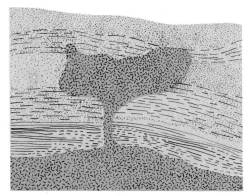

(a) Magmatic origin

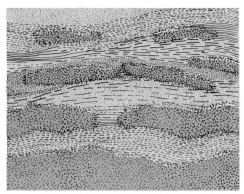

(b) In-place origin

Figure 3.5 Cross sections of shield area rocks suggesting different origins of granite. In (a) the granite apparently crystallized from a molten magma that moved upward from deeper levels. In (b) the granite appears to have formed by in-place melting of preexisting rocks without any appreciable movement of liquid magma.

of the crust which will be considered later in this chapter. A somewhat more restricted problem concerns the local processes that form granitic rocks.

Detailed mapping of the granitic rocks of the continental shield areas provides evidence of two different origins (Figure 3.5). In some areas the boundaries between large granite masses and the surrounding rocks are very sharp and distinct, suggesting that liquid granitic magma was squeezed into the surrounding rocks to make the granitic mass. In contrast, many areas show a complete gradation between granitic masses and surrounding metamorphic gneisses which suggests that the granite formed from a kind of "in-place melting"[1] of preexisting rocks without any appreciable movement of liquid magma. Typically such regions show a zone of rocks composed of thin bands of granite alternating with thin bands of gneiss which occurs between the areas of "pure" granite and "pure" metamorphic rock (Figure 3.6). Such mixed granitic rocks, called **migmatites,** indicate that the granites formed where the melting was most intense, but that in the surrounding areas temperatures did not exceed the melting range of preexisting minerals and only metamorphic rocks were produced.

Because of the abundance of migmatites and transitional granitic masses in the continental shield areas, most geologists now believe that much of the granite seen in the continents today formed by local melting of preexisting continental rocks, rather than by upward movement of liquid magma from deep levels of the earth's interior. Note, however, that if most of the continental crust has originated from melting of earlier crustal rocks, we are still faced with the fundamental question of *how* the silicon, sodium, potassium, and other elements characteristic of granite became concentrated in continental masses in the first place.

3.3 Oceanic Rocks

Most rocks recovered from drilling and dredging in the ocean basins, as well as those exposed on oceanic islands, are quite unlike the silicic granites that dominate the con-

---

[1] The question of the actual degree of "melting" that takes place in the formation of granites by this means has been much debated by geologists. Some believe that complete fusion takes place, but others feel that true granites may form from rearrangement of crystal structures in a solid or partially solid state. This problem has not yet been resolved, but it is less important for our purposes than the still more fundamental question of the *place* of origin of the granitic materials.

Figure 3.6 *Migmatite showing a small mass of granite (light color) surrounded by granitic gneiss (dark color). The hammer resting on the granite provides scale. (Geological Survey of Great Britain)*

tinents. Instead, oceanic rocks are largely mafic rather than silicic and, of these, by far the most abundant is basalt, the dark, fine-grained volcanic rock made up predominantly of pyroxenes and plagioclase feldspar. Such rocks are not confined to ocean basins, for they also occur as extensive basaltic lava flows on the continents. On the continents, however, basalt makes up only a small fraction of the total volume compared to that occupied by granites and granitic gneisses. Although basalt may occur among continental rocks, the converse is not true. The deep ocean basins have no granitic rocks, but are covered entirely by enormous volumes of basalt which must have formed as submarine volcanoes poured forth sheets of lava in first one area and then another. To understand oceanic rocks we must therefore examine the nature of volcanoes and their products.

Much of what is known about volcanoes comes from studies of oceanic islands and continental margins where volcanic rocks can be observed above sea level. Such studies have provided a wealth of data on the composition of both volcanic rocks themselves, and of the magmatic liquids and gases from which they form. In general, all volcanoes associated with oceanic islands as well as some found along the continental margins are composed of basalt just as are the volcanic rocks sampled from the ocean floor. In many regions, however, volcanic outpourings near continental margins are *not* mafic basalt but are, instead, more silicic andesites, the fine-grained volcanic equivalent of diorite (see Figures 1.29 and 3.9). Because of differences in their fluidity, basaltic and andesitic lavas produce somewhat different landscapes at the earth's surface. The fluidity of a magma is determined by its silicon content—the more silicon, the thicker and more viscous the magma. For this reason silicic andesites usually form steep-sided volcanic mountains and rugged terrains, whereas the more fluid mafic basalts tend to flow laterally to form extensive, flat-topped lava flows (Figures 3.7, 3.8).

Although studies of volcanoes exposed above sea level have provided much information, the most fundamental questions about volcanic rocks concern the enormous volume of ocean-floor basalts which cover about three-quarters of the earth's surface. At what depth within the earth's interior did these lavas form and how did they make their way to the surface? To answer these questions we must, once again, rely on indirect evidence, principally that provided by seismic waves.

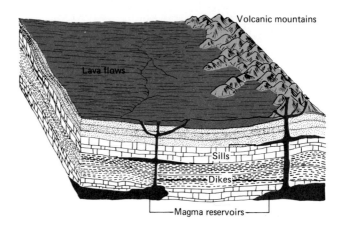

Figure 3.7 Volcanic landscapes. Basaltic lavas tend to form thin, widespread flows (darker color), whereas more viscous andesitic lavas form steep volcanic mountains (lighter color). Sills and dikes are, respectively, horizontal and vertical sheets of lava emplaced below the earth's surface. Dikes normally act as feeders through which lava passes from a deep reservoir to the surface.

Seismic waves penetrating the 8-km (5-mi) thickness of oceanic rocks lying above the Moho discontinuity show rather high velocities comparable to those found in laboratory experiments to be characteristic of basalt. For this reason, it appears most probable that the basaltic rocks exposed[2] on the ocean floor extend downward to the Moho discontinuity, where an abrupt increase in wave velocity indicates a change to the denser underlying rocks of the mantle (see Figure 2.21). It was long assumed that basaltic lavas originated from melting and upward movement of the lower parts of this basaltic crust, but it is now clear that they form not within the crust itself, but from a melting of the much deeper upper mantle. This is indicated by several lines of evidence.

First, regions of present-day volcanic activity are closely associated with the earth's deep earthquake zones (Figure 3.9). This suggests that volcanoes are related to events in the mantle, for all deep earthquake foci occur in the upper mantle far beneath the crust. Ruptures caused by deep earthquakes would provide ideal channels for the escape to the surface of molten magma formed deep within the earth. Second, recent seismological studies of individual volcanic eruptions in Hawaii and elsewhere have shown that the extruded basaltic magmas originate from fluid pockets in the upper mantle formed at depths ranging from 65 to 100 km (40 to 60 mi) beneath the

[2] The basaltic rocks of the sea floor are actually "exposed" only in rugged submarine cliffs, canyons, and mountains. Over most of the flat areas of the ocean floor they are buried beneath a thin blanket of sedimentary mud and sand. The thickness of this sedimentary cover can be easily determined from seismic evidence, for it transmits seismic waves much more slowly than does the underlying basalt; it has been found normally to be only about a hundred to a thousand meters thick.

(a)

(b)

Figure 3.8 *Volcanic landscapes. (a) A typical volcanic mountain built from viscous andesitic lava; off the western coast of Mexico. (b) A typical lava flow formed by a fluid, basaltic lava; in southern Idaho. (a: U.S. Navy; b: John S. Shelton)*

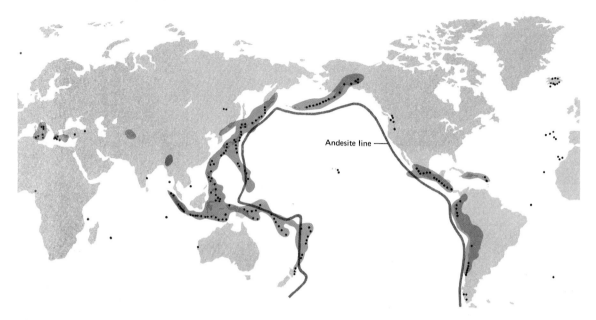

Andesite line ───

Figure 3.9 *Comparative positions of the earth's active volcanoes (*dark dots*) and deep earthquake zones (*light shading*). The "andesite line" marks the boundary between oceanic volcanoes discharging only basalt, and those near the continents that discharge mostly andesite.*

crust (Figure 3.10). In Section 2.8 we saw that this is precisely the depth range at which a plastic, *low-velocity mantle layer* (see Figure 2.16) is indicated by other seismic evidence. Local melting within this high-temperature plastic layer is therefore the most probable source for the basaltic lavas which have poured forth to form the oceanic crust.

If the basaltic rocks which make up most of the oceanic crust originate deep within the mantle, then we are faced with a fundamental problem because the increase in seismic wave velocity at the Moho discontinuity indicates that the mantle is composed of materials that are denser than the overlying basaltic crust. Thus the crustal basalts appear to have originated, in some manner, from *heavier* underlying rocks. Many lines of evidence suggest that the uppermost mantle consists largely of olivine and pyroxene and thus is similar in composition to the rare *ultramafic* rocks exposed at the earth's surface (see Section 2.7). If the upper mantle is composed of heavy, silicon-poor ultramafic materials, then how can their melting give rise to basaltic magmas which are lighter and contain more silicon bound up in the structure of feldspar minerals? The most probable answer relates to simple melting-point differences. Because feldspar melts at a lower temperature than either olivine or pyroxene, heating only slightly beyond this temperature might cause

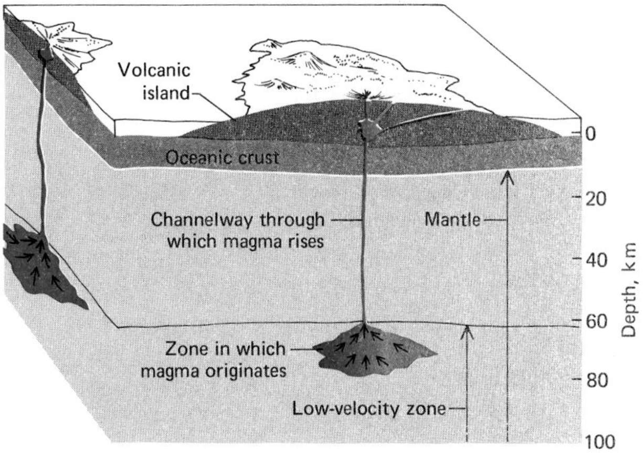

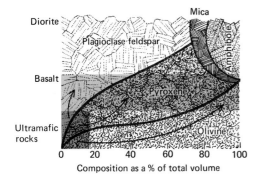

Figure 3.10 An idealized diagram of the origin of oceanic basalts, such as those making up the Hawaiian and other oceanic islands. Seismic evidence indicates that the basaltic magmas originate in the low-velocity zone of the mantle.

a **partial melting** of ultramafic mantle rocks which contain only small amounts of feldspar (Figure 3.11). This partial melting, in turn, could form local concentrations of silicon-enriched basaltic magmas surrounded by still-solid olivine and pyroxene. Most geologists now believe that basalts originate in this way even though it would require the partial melting of an enormous volume of silicon-poor mantle materials to account for the huge quantity of basaltic rocks found in the oceanic crust.

CRUSTAL MOVEMENTS: CONTINENTAL DEFORMATION

So far our discussion of continents and ocean basins may have incorrectly implied that they are static, changeless features of the earth's crust. There is abundant evidence, however, that crustal rocks are continually being moved and altered by dynamic forces originating beneath the crust. Flat-lying sedimentary rocks can be seen to have been crumpled into huge mountain ranges by enormous lateral pressures, while even larger blocks of crustal rocks have moved hundreds of miles along lengthy ruptures. These and other kinds of evidence indicate that most of the topographic features of the earth's surface—features such as mountain ranges, basins, and plateaus on the continents, and island chains, deep trenches, and submerged mountains on the ocean floor—have their origin in complex *crustal movements*. For the remainder of this chapter we shall consider the nature and possible causes of these fundamental movements.

Figure 3.11 Partial melting of ultramafic rocks to create basalt (modified from Figure 1.29). The small amount of feldspar in ultramafics has a lower melting point than either pyroxene or olivine. Heating sufficient only to melt this feldspar plus a small portion of the pyroxene and olivine (dark color) could produce a magma of basaltic composition (light color).

(a) Faults

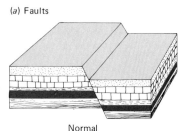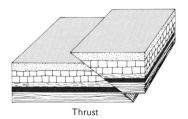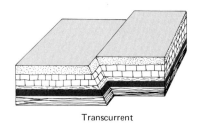

Normal   Thrust   Transcurrent

(b) Folds

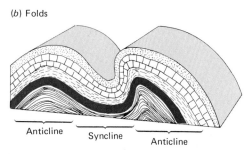

Anticline   Syncline
                    Anticline

Figure 3.12 Faults and folds. (a) *Three principal types of motion occur along fault ruptures: when there is extension along the rupture, one block usually moves vertically downward to produce* normal faults; *when there is compression, one block vertically overrides the other to produce* thrust faults; transcurrent faults *involve only horizontal motions between blocks.* (b) *Down-turned, troughlike folds are called* synclines *and upturned, archlike folds* anticlines.

Until the 1960s most evidence for crustal movements came from study of continental rocks which are, of course, far more accessible than are those of the ocean floor. Geologic mapping showed that intense movements had not taken place randomly over the surface of the continents but, instead, tended to be concentrated in long, narrow belts that usually parallel continental margins. Continental **mountain ranges,** formed in these belts by the compression and distortion of previously flat-lying and undeformed rocks, provide the principal evidence for movements of the continental crust.

3.4 Folds and Faults

When rocks of the crust are subjected to mountain-building forces they can respond in two ways: either they bend or they break. These two responses define the two major structural features seen in deformed crustal rocks: **folds,** the result of rock deformation with no breaking, and **faults,** rock fractures along which movement has occurred (Figure 3.12). Folds and faults occur in limitless variations of size and shape. Entire mountain ranges may be formed by a single huge fold or fault while, at the other extreme, pieces of rock small enough to be held in one's hand may be broken by tiny folds and faults that can be seen only under a microscope (Figures 3.13, 3.14).

Careful descriptions and geologic maps of the folds and faults seen in continental mountain ranges have been made since the nineteenth century, and these have provided a wealth of information about the geometry and distribution of various kinds of deformed rocks. In addition, these descriptive studies have raised additional questions about rock deformation that can best be attacked by theoretical or experimental studies. Most of the rocks exposed at the earth's surface are very *brittle*. If forces are applied to them, for example by striking with a hammer or squeezing with a vise, they yield only by breaking.

(a)

(b)

Figure 3.13 *Faults. (a) A small normal fault in shales of northern England. The dark layer is coal. (b) A normal fault in sandstones of western Colorado. The left-hand block has moved downward about 3 m (10 ft). (a: Geological Survey of Great Britain; b: U.S. Geological Survey)*

Because familiar rocks respond in this way, it is easy to visualize how rocks can deform by breaking along faults, but more difficult to understand how they can deform by *bending* to form folds. Obviously, folded rocks now exposed at the earth's surface must have originated under other conditions, conditions which made the rocks less brittle and more *ductile*, that is, able to flow and deform without fracture. Theoretical and laboratory studies have shown that ductility is not constant for a particular kind of rock, but varies with changing environmental conditions. Three environmental factors are particularly important in affecting rock ductility: pressure, temperature, and the rate at which deforming forces act on the rock.

The effect of variations in pressure and temperature upon rock deformation can be investigated by subjecting rock specimens, usually cut into the shape of cylinders about 3 to 15 cm (1 to 6 in.) long, to deforming forces at varying temperatures while they are held in special high-pressure presses (Figure 3.15). These presses simulate the increased pressures and temperatures to which rocks are subjected as they are buried deeper and deeper in the earth. Such presses can achieve pressures in excess of those occurring anywhere in the crust, and can be operated at temperatures high enough to melt most rocks. Studies using these presses show that most rocks tend

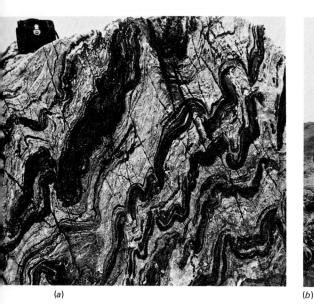
(a)

(b)

Figure 3.14 Folds. (a) Small-scale, in limestones of
northern Scotland; the camera case at top left provides scale. (b) Intermediate-scale, in limestones of
northern England. (a, b: Geological Survey of Great
Britain)

to become more ductile as pressures and temperatures
increase. Thus the brittle behavior we associate with rocks
at the earth's surface may be replaced by ductile deformation, and folding, when the same rocks are subjected
to deforming forces while deeply buried. By observing
the patterns of deformation seen in rocks now exposed
at the surface, it is sometimes possible to infer, from
laboratory experiments on similar materials, the approximate depths, temperatures, and pressures at which the
rocks were deformed.

The effects of temperature and pressure on rock
ductility are relatively easy to study in laboratory experiments and are now reasonably well understood. In contrast, the third major factor controlling rock ductility—the
rate at which the deforming forces act—is an area of great
uncertainty because very small forces, when applied over
very long times, can lead to ductile deformation in otherwise brittle materials. A familiar example is a wax candle
which may, over a long period, bend under its own weight,
but which can also be easily broken by a sharp blow.
Similarly, gravestones made of brittle natural rocks, such
as marble, have been observed to bend under their own
weight over many years. These observations suggest that
time is an extraordinarily important factor in rock ductility;
over long periods even brittle rock masses exposed at or
near the earth's surface might deform by ductile flow
under the action of relatively small forces, such as the
force of gravity. For larger forces of shorter duration, the

Increasing pressure

(b)

Piston

Cylindrical rock specimen

Confining fluid

Jacket enclosing specimen

Anvil

(a)

Figure 3.15 Experimental rock deformation. (a) Schematic drawing of a rock deformation press. (b) Cylinders of limestone shortened 15 percent at increasing pressures (left to right) in a deformation press. At low pressures the specimens break as in faulting; at higher pressures they deform without breaking as in folding. (Photos courtesy F. A. Donath; from American Scientist, *vol. 58, 1970)*

same rocks would break like brittle solids. Thus either folding or faulting might occur depending on the rate at which the deforming forces acted. Unfortunately, the effect of time on rock deformation has been little studied because laboratory experiments cannot duplicate the effects of small forces acting on rocks for thousands or millions of years.

3.5 Mountains and Geosynclines

On the largest scale, continental mountain ranges may be divided into two broad categories: those that originate largely from faulting, known as **fault-block mountains,** and those dominated by folding, which are called simply **fold mountains** [3] (Figures 3.16, 3.17). Most of the large mountain ranges seen on the earth today, such as the Rockies, Alps, and Himalayas, are fold mountains. Fault-block mountains are less conspicuous and are frequently associated with nearby fold mountains. The many small mountain ranges of the Great Basin region of western North America are typical of this type.

Detailed studies of many folded mountain ranges have shown them to have similar patterns of rock deformation. Viewed in cross section, the outer flanks typically consist of thick sequences of sedimentary rocks that become progressively less folded and deformed away from the central axis of the range. Toward the axial regions, these deformed but otherwise unaltered sedimentary rocks give way first to metamorphosed sedimentary rocks and,

[3] Still a third kind of mountain range originates not from deformation of preexisting solid rocks, but from the cooling of liquid volcanic magmas. Such *volcanic mountains* (see Figures 3.7 and 3.8) are common on the ocean floor but are far less important on the continents than either fold or fault-block mountains.

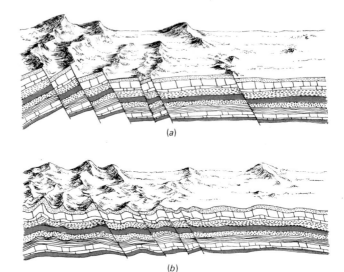

Figure 3.16 *The two principal types of continental mountain ranges: (a) fault-block; (b) fold.*

finally, to large elongate masses of granite which form the central core of the range (Figure 3.18).

In many regions the typical fold mountain rock sequence, grading from relatively undeformed sediments near the flanks of the range to granite at the core, can be shown to have formed by the progressive alteration of an unusually thick accumulation of sediments deposited

(b)

Figure 3.17 *Continental mountains. (a) Fault-block mountains in western Texas (on left page). (b) Fold mountains in southern Idaho. (a, b: John S. Shelton)*

near the margins of the continents. When the thin veneer of sedimentary rocks that covers much of the surface of the continents is traced laterally into the flanks of the mountain range, the total thickness of the original sediment can be seen to increase dramatically at just about the point at which the mountain-building deformation begins (Figure 3.18). Furthermore, this rapid thickening continues towards the axis of the mountain range where the metamorphosed sediments lose their identity by grading into granite. Such unusually thick sedimentary accumulations, known as **geosynclines,** are formed where large quantities of sediment, worn from the surface of the continents, are deposited by rivers and streams near the conti-

Figure 3.18 *Cross section of typical fold mountains. Sedimentary rocks (color) become progressively thicker and more deformed towards the axis of the mountain range where large masses of granite (gray) may be exposed.*

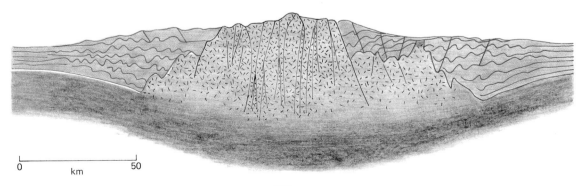

```
0                    50
     km
```

Figure 3.19 *The eroded roots of an ancient mountain range, exposed in the central Canadian shield area. (Canadian Dept. of Energy, Mines, and Resources)*

nental margins. Where this occurs, as in the Mississippi Delta region today, the sedimentary accumulation can reach thicknesses greater than 15,000 m (50,000 ft), in contrast to the 300 m (1000 ft) or so of sedimentary veneer which covers most of the continents. Pressures and temperatures become very high in the deeper, central portions of thick geosynclines and result in metamorphism and melting of some of the sedimentary rocks to form granite magmas. Ultimately, by processes that are not fully understood, these deeper zones move upward and, folding the buried sediments around them in the process, come to form the cores of fold mountain ranges.

The conspicuous folded mountain ranges seen on the continents today have all formed relatively recently in earth history. Most older fold mountain ranges no longer stand high above the continental surface but have been worn down to zones of little or no relief through millions of years of erosion by water and wind. Such ancient mountain ranges are best exposed in the continental shield areas where they can still be recognized, even though they now lack mountainous relief, by their characteristic pattern of elongate granite masses flanked by progressively less deformed sedimentary rocks (Figure 3.19). Such cores of former mountain ranges make up a large fraction of the granites forming the continental basement complexes.

The sedimentary debris resulting from the erosion of old mountain ranges typically accumulates in geosynclines near the continental margins where it ultimately becomes deformed to create new mountain ranges. Through this process the rocks of the continental crust are continually recycled, as granitic mountain cores are eroded to form sediments which are then melted deep within the crust to form new granites which become the cores of new mountain ranges. In this cycle, the steps involving the erosion of old mountains and the subsequent accumulation of sediments near the continental margins are more clearly understood (see Chapters 8 and 9) than are the forces that deform and uplift the geosynclinal sediments to produce new mountains. These complex forces are, however, beginning to be recognized as but a side effect of still more fundamental motions of the earth's crust.

3.6 Continental Drift

One of the first suggestions that continental mountains might result from larger-scale crustal motions was made in 1912 by Alfred Wegener, a German geophysicist. Wegener was impressed by the jigsaw puzzlelike fit between the shapes of some of the continental margins. The most obvious are those between western Africa and eastern North and South America but, with suitable juggling, rough fits can be made for other continental margins as well (Figure 3.20). This pattern led Wegener to suggest that the continents were originally joined in a single large land-

Figure 3.20 *A reconstruction of the probable relationships of the continents before "continental drift."* (*After Dietz and Holden,* Scientific American, *vol. 223, 1970*)

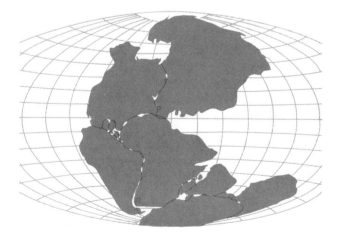

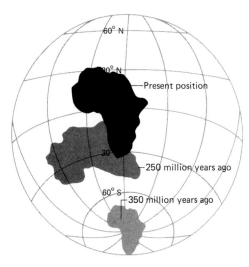

Figure 3.21 *Paleomagnetic reconstruction of Africa's position, relative to the south pole, from 350 million years ago to present.* (*After compilation by Clark,* Structure of the Earth, *1971, from various sources*)

mass that was subsequently broken apart to create the separate continents which we see today. If entire continents were moving relative to one another, then mountain-producing deformations along their margins might be explained as some sort of relatively minor frictional or drag effect at the edges of the moving mass. Wegener was unable to suggest a plausible energy source or mechanism for **continental drift,** as his idea came to be called, and it was largely dismissed until about 20 years ago, when new evidence derived from study of the earth's magnetic field began to suggest that large-scale continental motions had indeed taken place.

When igneous rocks first crystallize, any magnetic, iron-bearing mineral that they contain becomes magnetized in the direction of the earth's magnetic field. Any subsequent change in the orientation of the mineral with respect to the earth's magnetic field also affects the mineral's magnetism but, by using precise analytic techniques first developed in the 1950s, it is often possible to isolate the original magnetic component, which thus provides a record of the direction of the earth's magnetic field at the time the rock crystallized. By combining data from many such "fossil magnets" taken from rocks of different ages found on each continent, it is possible to get strong clues about the changing positions of the continents relative to the magnetic pole (which is assumed to have always remained relatively near the earth's rotational pole as it is today; see Section 2.6) (Figure 3.21). Such **paleomagnetic studies** of continental rocks provided additional evidence that the continents had moved relative to one another and led to a revival of interest in the idea of continental drift. General acceptance of the idea did not come, however, until the mid-1960s when it was discovered that still more fundamental crustal movements take place beneath the ocean floor.

CRUSTAL MOVEMENTS: OCEAN-FLOOR SPREADING AND PLATE TECTONICS

Among the most significant scientific advances of recent years was the discovery, first suggested as a hypothesis in the 1950s and then confirmed in 1966, that the oceanic crust is in continuous movement like a giant conveyor belt. This discovery of what has come to be called *ocean-floor spreading* was a result of two principal kinds of research: one of these concerned the topography of the ocean floor,

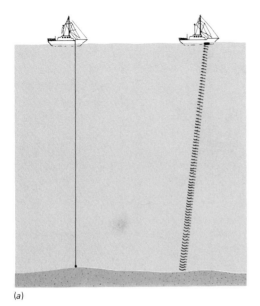

(a)

Figure 3.22 Methods of determining ocean-floor topography. (a) Before the 1920s the only method was the lowering of weighted lines (left); since then, increasingly sophisticated echo sounders (right) have been used to map the ocean floor. (b) A weighted line being used for sounding a shallow, uncharted area. (c) A recording echo-sounder showing a profile of the underlying topography. (b: U.S. Coast Guard; c: Ocean Science Laboratory)

(b)

(c)

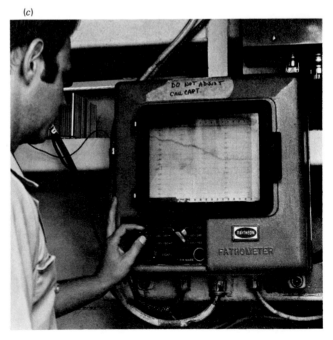

the other magnetism of the ocean floor. This section will first discuss these two research areas and their findings, and will then move on to consider how ocean-floor spreading has provided a unifying principle that also helps to explain continental deformation.

3.7 Topography of the Ocean Floor

Because of the ease with which we can observe mountains, hills, and valleys on the continents, most of us are not accustomed to thinking of the extraordinary problems involved in learning whether similar topography exists on the ocean floor. Before 1920 the only method of determining ocean depths and, conversely, sea-floor topography, was to laboriously lower a heavy weight, attached to thousands of feet of steel wire, over the side of a ship. Even though many such "soundings" were made, they were so scattered over the vast expanse of the oceans that they provided little evidence for ocean-bottom irregularities. The ocean floors were generally believed to be flat, featureless plains broken here and there by mountains that rose above the surface as islands.

In the 1920s the first *echo sounders* were developed for more rapid determination of ocean depths and bottom topography (Figure 3.22). These instruments measured

Figure 3.23 Idealized drawing of a portion of the mid-Pacific ocean floor with water removed, showing two flat-topped guyots in the distance. The canyon in the foreground is cut into a guyot, most of which does not show. (Painting by Chesley Bonestell, from Hamilton, Geological Society of America Memoir 64, 1956)

depths by noting the time required for sound waves to reach the bottom and be reflected back to a shipboard recording device. The first systematic surveys of ocean-bottom topography using these instruments were undertaken in parts of the Pacific Ocean in the early 1940s. The principal result of these early surveys was the discovery of a great many rounded, flat-topped mountains, called **guyots,** on the floor of the northern Pacific (Figure 3.23). The tops of the guyots are covered by many thousands of feet of water, yet the only known mechanism for producing their broad, flat tops is the wearing action of waves at the surface of the sea. Because their flat tops could have been produced only at sea level, either sea level had risen many thousands of feet since they were formed or, more probably, the ocean floor on which they stand was somehow lowered several thousand feet. Guyots thus provided the first suggestion of large-scale movements of the ocean floor.

Another significant result of early bottom surveys was the location and mapping of several **oceanic**

trenches—long, narrow zones where rocks of the ocean floor were puzzlingly depressed many thousands of meters *below* the surrounding terrain (Figure 3.24). Most of these trenches are parallel to continental margins and are usually bounded on their landward side by curving chains of offshore volcanic mountains that extend above the ocean surface to form **island arcs** (Figure 3.24). These unusually deep depressions immediately adjacent to high volcanic mountains also suggested that dynamic processes were taking place in the rocks of the ocean floor.

Since 1945 additional surveys of bottom topography have been undertaken in many areas: The Atlantic, Indian, Arctic, and eastern Pacific Oceans have been intensely

Figure 3.24 (a) *The principal oceanic trenches (dark color) and island arcs (light color).* (b) *Cross section through the Tonga Trench, with Mt. Everest superimposed for scale; the vertical scale is exaggerated 10 times in the upper sketch; the true scale is shown below.* (After Fisher and Revelle, Scientific American, *vol. 193, 1955*)

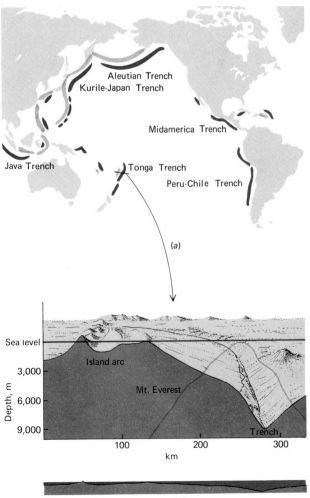

surveyed, though other oceans are known in less detail. These surveys have shown that the ocean floor, although relatively flat over large areas, also includes the longest, highest, and most continuous mountain ranges on earth. These ranges make up the **oceanic ridge-rise system,** a world-encircling belt of ridgelike mountains and broad rises that is of extraordinary importance for understanding movements of the ocean floor (Figure 3.25).

The most intensively studied part of the oceanic ridge-rise system is the *Midatlantic Ridge,* a chain of mountains lying in almost the exact center of the Atlantic Ocean (Figure 3.26). The existence and general position of the ridge has been known for many years because it comes to the surface in several places to make oceanic islands such as Iceland and The Azores. Furthermore, it has long been known that the ridge overlies a zone of rather shallow earthquakes. In the 1950s detailed studies of the topography of the ridge and the distribution of earthquakes under it revealed two important relationships: first, although the total width of the ridge averages several hundred kilometers, the earthquakes are concentrated in a narrow band under only the highest part or *crest* of the ridge. Second, the crest itself did not make up a single

Figure 3.25 *The oceanic ridge-rise system. (After Bullard,* Scientific American, *vol. 221, 1969)*

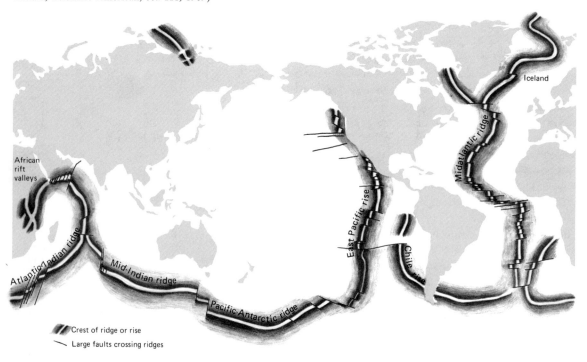

Crest of ridge or rise

Large faults crossing ridges

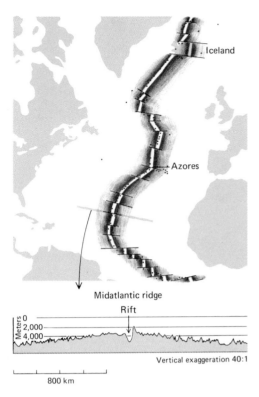

Midatlantic ridge

Rift

Meters
0
2,000
4,000

Vertical exaggeration 40:1

800 km

Figure 3.26 *A portion of the Midatlantic Ridge showing the location of earthquake epicenters (black dots). The cross section below shows the rift valley at the crest of the ridge.* (*After Heezen,* Scientific American, *vol. 203, 1960*)

long ridge but was, instead, composed of two steep-sided parallel ridges separated by a deep valley. Similar but smaller and less continuous valleys, called **rifts,** are also known from the surface of the continents where they can be shown to form from normal faulting caused by a tension or *pulling apart* of the faulted rocks on each side of the valley (Figure 3.27). This immediately suggested that the Midatlantic Ridge might represent a similar, but much larger, zone of crustal tension in which the Atlantic Ocean floor on both sides of the ridge was moving away from the rifted ridge; the rift valley at the crest of the ridge and the earthquakes underlying it would mark the line of separation. Other lines of evidence also pointed to the same conclusion. Where the tops of the ridge were exposed as islands, the rocks of the ridge can be sampled; they are everywhere basaltic or ultrabasic rocks, as would be expected if the ridge represents a fracture zone in which basaltic lavas poured out from the underlying mantle to build up the ridge. Furthermore, the entire width of the rift valley itself is exposed above the sea in Iceland where tension cracks and other evidence of movement can be directly observed (Figure 3.27a).

After the significance of the Midatlantic Ridge was discovered, interest quickly spread to similar features in other ocean basins. From topographic information, combined with data on the distribution of oceanic earthquakes, it was proposed in 1956 that the Midatlantic Ridge is only one segment of a much longer world-encircling system of oceanic mountains which represents a fundamental zone of fracture and movement of the earth's crust. This is the oceanic ridge-rise system which extends, with several side branches, down the center of the Atlantic Ocean, through the Indian Ocean, south of Australia and up the eastern Pacific Ocean. Although the system mostly occurs in ocean basins, it does extend into the continents in a few places such as the Red Sea, eastern Africa, and the Gulf of California.

3.8 Magnetism of the Ocean Floor

The discovery of the oceanic ridge-rise system, with its associated earthquakes and, in many areas, central rift valleys, strongly suggested that horizontal movements take place in the crustal rocks beneath the sea floor on a tremendous scale. Proof of this hypothesis of **ocean-floor spreading,** as it came to be called, was provided by magnetic studies of the rocks of the ocean floor.

(a)

(b)

Figure 3.27   Rifts. (a) A part of the rift valley of central Iceland. (b) A small rift in east Africa. (a: Sigurdur Thorarinsson; b: Sidney P. Clark, Jr.)

In Section 2.5 we saw that measurements of the strength of the magnetic field at the earth's surface are affected by two components: one, the earth's main magnetic field, originates from motions in the liquid iron of the core; the other originates from the distribution of magnetic minerals in rocks of the crust. There are no magnetic effects from mantle rocks lying between the crust and the core because the minerals making up the mantle are heated beyond the *Curie point* (the temperature at which they lose their magnetism). By subtracting the large component contributed by the main magnetic field, it is possible to study differences in the magnetic minerals of crustal rocks by making magnetic surveys of large areas with shipboard or airborne magnetometers. Such surveys have been conducted over land areas for many years and have shown complex magnetic patterns caused primarily by differences in the amount of the mineral magnetite present in the underlying crustal rocks. The first systematic magnetic surveys of the ocean floor were undertaken off the coast of California in 1955 and, surprisingly, showed patterns that were *far more regular than any from continental areas* (Figure 3.28). In particular, the oceanic magnetic patterns showed a series of extremely long, narrow bands running for hundreds of kilometers approximately parallel to the coastline. One additional feature of these banded magnetic patterns was of great significance. In certain areas the patterns were broken and offset for hundreds of kilometers along huge faults running at approximately right angles to the present coastline. Some of these faults were already known because they make large submarine cliffs that had been previously discovered by echo sounding. The fact that oceanic rocks had moved for hundreds of kilometers along these faults was first shown, however, by the offset patterns of magnetic measurements. Here again was strong evidence for large-scale movements of the ocean floor.

Once the linear oceanic magnetic pattern was discovered, the next question was, what caused it? In continental areas the much less regular magnetic pattern could be shown, by geologic mapping, to be caused by differing amounts of magnetic minerals in the underlying rocks. There was, however, another more likely cause for the regular oceanic patterns. This was related not to differing *amounts* of magnetic minerals but to differences in their *direction of magnetization*.

Rocks of the ocean floor are almost exclusively basalts containing a large and relatively constant amount of magnetite which makes them strongly magnetic. When lavas cool to form basalt, this magnetite crystallizes from

Figure 3.28 Contrasting linear magnetic patterns of oceanic rocks and irregular patterns of continental rocks. (After Raff, Journal of Geophysical Research, *vol. 71, 1966; Sauck and Sumner,* Aeromagnetic Map of Arizona, *1971)*

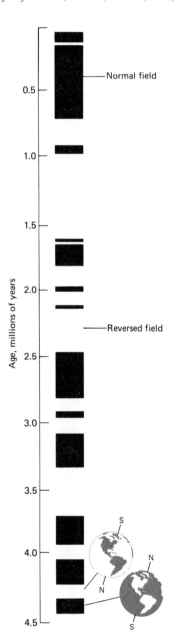

Figure 3.29 A time scale for magnetic reversals during the last 4.5 million years. Intervals of normal polarity are shown in color; reversed polarity in white. (*Modified from Cox,* Science, *vol. 168, 1969*)

the liquid lava and then eventually cools to the Curie temperature. As this temperature is reached the magnetic crystals become magnetized in the direction of the earth's magnetic field. Now suppose for a moment that the direction of the earth's magnetic field is not constant but shows regular changes. If this were the case, then lavas forming at different times would have different directions of magnetization reflecting the earth's magnetic field at the time the lavas cooled. (The original patterns would dominate because magnetic minerals permanently retain much of their original magnetic orientation gained on passing the Curie point and do not become completely reoriented to later changes in the earth's field.) Most significantly, at about the same time that the puzzlingly regular magnetic patterns were discovered in the oceans, additional studies of vertical sequences of magnetic mineral orientation in lava flows on land and in samples of sediments from the sea floor (fine sedimentary grains of magnetic minerals also become oriented to the earth's magnetic field when they are deposited by water) showed that the earth's main magnetic field has undergone frequent and rapid **reversals in polarity,** the north magnetic pole becoming the south pole and vice versa (Figure 3.29). The cause of these sudden magnetic reversals is still unknown, but they must somehow be related to changes in flow patterns in the liquid iron of the earth's core, where the main magnetic field originates.

Soon after the regular oceanic magnetic patterns were discovered, it was suggested that they might reflect sudden reversals in the earth's magnetic field in the following way: if basaltic lavas poured onto the ocean floor from a long crack in the crust, then cooled to form basalt, and were then somehow moved away from the crack before new lava was extruded and, furthermore, if a reversal in the earth's magnetic field took place between the lava outpourings, then the result would be *a series of parallel bands of basalt each having minerals with a different magnetic orientation* (Figure 3.30). These differences in magnetic direction might account for the regular, parallel patterns of ocean-floor magnetism. Unfortunately this idea could not be directly tested because it is impossible to determine the direction of magnetization without having carefully oriented samples of the magnetic rocks, and these are extraordinarily difficult to obtain for oceanic basalts covered by thousands of meters of water and hundreds of meters of sediments. Final proof that spreading basaltic bands *are* the cause of the oceanic magnetic pattern was obtained, however, from another kind of evi-

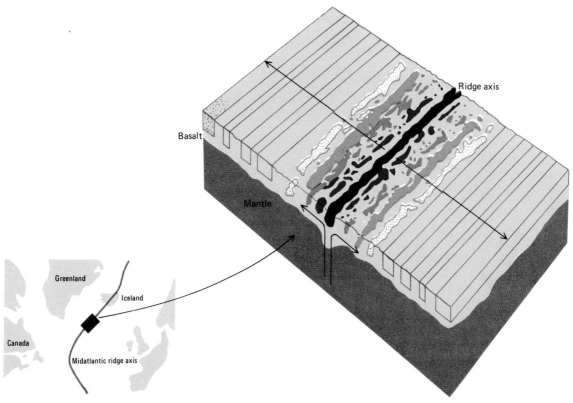

Basalt

Mantle

Ridge axis

Greenland

Iceland

Canada

Midatlantic ridge axis

Figure 3.30 *Interpretation of symmetrical magnetic patterns observed on a portion of the Midatlantic Ridge south of Iceland (color). Lavas from the mantle pour out along the ridge axis and then move laterally outward on both sides of the ridge forming symmetrical bands that are progressively older (lighter color) away from the ridge. Bands of reversed polarity are shown in white; normal polarity in color and gray. (Magnetic pattern from Vine, in* The History of the Earth's Crust, *1968)*

dence when, in 1966, the first detailed magnetic analysis was made of the Midatlantic Ridge.

This study showed parallel magnetic patterns similar to those previously found off California, but this time the bands were identical on each side of the Midatlantic Ridge, just as would be expected if lavas were pouring from the earthquake rift zone, solidifying into basalt, then being broken along the rift and moving away from it on both sides (Figure 3.30). Furthermore, the width of the parallel magnetic bands closely matched the relative times between reversals in the earth's magnetic field over the past several million years as established independently from magnetic studies of dated sedimentary rocks and lava flows. This discovery showed that the spreading basaltic rocks of the oceanic crust have behaved like a great, moving, magnetic tape which clearly records changes in the direction of the earth's magnetic field. Basaltic rocks of the oceanic crust must rise from the mantle along the oceanic ridge-rise system, solidify, and then be carried horizontally away from the ridges by powerful forces acting

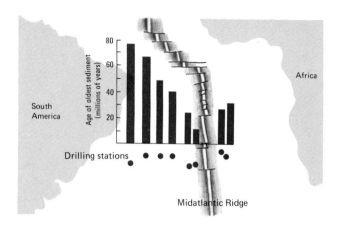

Figure 3.31 Confirmation of ocean-floor spreading from deep-sea drilling. Sediments lying just above the ocean-floor basalts become progressively older away from the axis of the Midatlantic Ridge. (After Maxwell et al., Initial Reports of the Deep Sea Drilling Project, *vol. 3, 1969*)

from below, within the hotter and less rigid rocks of the mantle. The oceanic crust thus becomes progressively older away from the ridge-rise system.

Following the dramatic indication of ocean-floor spreading from analyses of magnetic patterns, two other lines of evidence have added further confirmation of the motions involved. One, a careful study of earthquake motions, has shown that large-scale crustal movements take place with precisely the patterns suggested by the magnetic evidence. The other, an important program of drilling from shipboard into the deep ocean floor, has revealed that the thin sediments lying upon the oceanic crust show a regular increase in age away from the ridge-rise system, as would be expected if the underlying basalts similarly varied in age (Figure 3.31).

3.9 Plate Tectonics and Crustal Deformation

The confirmation of ocean-floor spreading quickly led to a reconsideration of *all* crustal motions, for the movement of large blocks of the ocean floor indicated dynamic crustal processes on a scale that was never obvious from studies of the more accessible continental crust. In particular, the position of earthquake belts and oceanic ridges showed the crust to be divided into seven major moving **plates,** each relatively rigid within itself but bounded by zones of profound crustal movement (Figure 3.32). Differential motion of these huge plates accounts for most of the dynamic features seen in the earth's crust and their study has come to be called **plate tectonics** ("tectonics" means "land movement" and is applied to mountain-forming crustal deformations).

Man and the Earth's Crust

Continental rocks supply most of the minerals used as resources by man. The only mineral removed in any quantity from the ocean floor is petroleum, and even that comes from wells drilled into rocks submerged under shallow waters along the continental margins, rather than from the deep ocean basins. This is partially caused by the inaccessibility of the ocean floor which has prevented large-scale exploration for mineral resources, but there are good reasons for believing that the basaltic rocks of the ocean floor generally lack the large concentrations of easily extracted minerals that occur in the more complex silicic rocks of the continents. If so, then the continents will *always* supply the bulk of man's nonrenewable resources.

On the continents, the thin veneer of sedimentary rocks contributes a disproportionate share of mineral resources. Most building materials (other than granite used as building stone), fuels, fertilizers, and minerals for industrial chemicals are extracted from sedimentary rocks, as are the principal supplies of iron, aluminum, and manganese. Sedimentary processes, to be discussed further in Chapters 8 and 9, have thus provided most of man's nonrenewable resources. The much greater volume of continental igneous and metamorphic rock is of primary importance in the distribution of only three major resources which are also unusually scarce: copper, used primarily by the electrical industry; lead, mostly used for automobile storage batteries; and zinc, used in various metallic mixtures and in anticorrosion treatment of iron and steel.

Copper, lead, and zinc occur as fractional trace elements in many different minerals, but the quantities involved are too small for profitable extraction. Under exceptional circumstances, however, the three elements have become locally concentrated in rare sulfide mineral deposits. These deposits are usually associated with present or former igneous activity, and appear to form as the gases, vapors, and liquids associated with molten magmas move through cracks in overlying solid rocks. There, the gases and liquids carrying volatile compounds of copper, lead, zinc, and sulfur cool and crystallize as solid sulfide minerals. Because both heat and water-rich gases and liquids appear to be involved in their origin, such occurrences are known as **hydrothermal ore deposits.**

Hydrothermal ore deposits do not occur at random on the continents, but tend to be concentrated in linear belts called **metallogenic provinces** (see map at right). The exact cause of the concentration is uncertain, but the recent discovery of lithosphere plate motions suggests that metal-rich vapors should be most abundant at plate margins where volcanic magmas from the mantle pour out onto the earth's surface. When such a zone occurs under or immediately adjacent to continental rocks, then conditions should be ideal for the escape of gases, vapors, and fluids into cool overlying rocks where they might crystallize as hydrothermal ore deposits. Although still highly speculative, such considerations may explain the puzzling concentration of otherwise rare metallic elements in metallogenic provinces.

Drilling into copper ore in an underground mine, Montana. (The Anaconda Co.)

An open pit copper mine in Arizona. (Cities Service Co.)

Distribution of principal world copper deposits (dots) and major earthquake zones (shaded). The close correspondence suggests that the copper originates in the mantle and is transported into the crust along plate margins. (Copper deposits from a compilation by Paul Eimon, 1970)

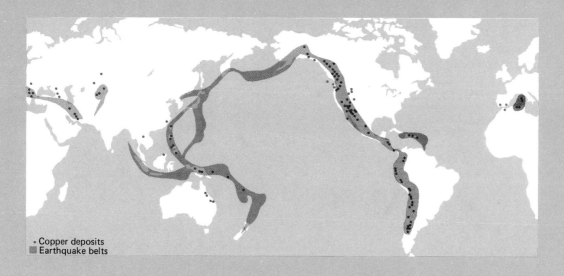

• Copper deposits
▮ Earthquake belts

The boundary zones between crustal plates are usually marked by earthquakes and volcanic activity, and are of three principal kinds. The first kind of plate boundary occurs along the oceanic ridge-rise system where basaltic lavas well up from within the mantle and pour out on the ocean floor to create new oceanic crust. This new oceanic crust tends to move horizontally away from both sides of the ridge-rise system and thus is added to separate moving plates on either side. This plate boundary can be called a **new crust zone** (Figure 3.33).

The creation of new crust along plate boundaries must be balanced by destruction of old crust at the other side of a moving plate unless the earth is continuously expanding in diameter, a prospect with no convincing evidence to support it. Instead, it is clear that many plates *are* bounded by zones in which rigid crustal rocks abruptly descend deep into the mantle where they are ultimately destroyed by melting. Such regions, called **subduction zones,** are usually marked by deep oceanic trenches, volcanic island arcs, and deep-focus earthquakes caused by the motions of the descending plate margin (Figure 3.33).

If crust is created at one side of a moving plate and destroyed at the other, then there must be additional

Figure 3.32 The major structural plates and their probable boundary types (the nature and position of many of the boundaries are still uncertain). The arrows show the inferred relative motions of the plates. Note that the plate boundaries generally follow volcanic and earthquake zones as shown in Figure 3.9. Several small plates are omitted. (After Morgan, Journal of Geophysical Research, vol. 73, 1968)

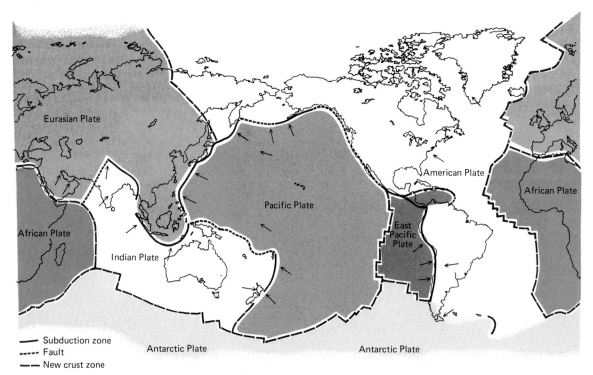

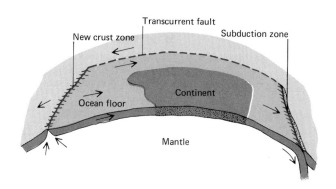

Ocean floor   Continent

Mantle

Figure 3.33 Idealized diagram of the three types of plate boundary: (1) new crust zones where volcanic crust is added to the plate along oceanic ridges; (2) subduction zones, usually marked by oceanic trenches and deep earthquakes, where the plate plunges deep into the mantle and is destroyed; and (3) transcurrent faults, where only horizontal motions between plates occur.

Figure 3.34 Separation and deformation of continental masses by plate motions. Formation of a new crust zone under a continent (a), leads to rupture and separation of the continent (b). Upon reaching a subduction zone the margins of the continent are deformed into mountain ranges (c). (After Dietz and Holden, Scientific American, *vol. 223, 1970)*

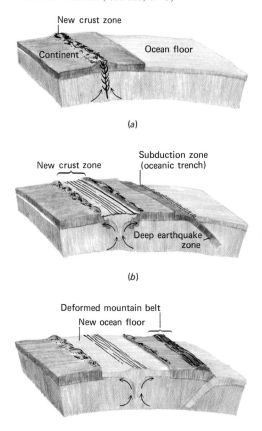

New crust zone

Continent   Ocean floor

(a)

New crust zone   Subduction zone (oceanic trench)

Deep earthquake zone

(b)

Deformed mountain belt
New ocean floor

(c)

boundaries where compensating motions take place without either plate creation or destruction. This third type of plate boundary is marked by long **transcurrent faults** caused by horizontal motions between adjacent plates (Figure 3.33; see also Figure 3.12).

A critical question involves the relation of continental masses to movements of these crustal plates. Because continents cover only about 30 percent of the earth's surface, much of the area of the seven major plates is occupied by oceanic crust. Each plate except the Pacific also contains, however, large areas of continental crust (Figures 3.32, 3.33). It appears that the thick continental masses are embedded in, and tend to move passively along with, the adjacent oceanic rocks of each plate. Furthermore, continental blocks can apparently be split apart and moved away from each other whenever they occur over a new crust zone. Such movement creates a new ocean basin with a midocean ridge and causes "continental drift," as had been previously postulated from other lines of evidence (Figure 3.34a, b).

When plate motions bring a continent in contact with a subduction zone, the results are more complex (Figure 3.34 b, c). Because continental rocks are significantly lighter than both the moving oceanic crust and the underlying mantle, continents are apparently seldom plunged downward at subduction zones but, instead, remain "floating" at the earth's surface while the adjacent heavier oceanic crust descends into the mantle. These movements of the adjacent oceanic crust may, however, cause deformation of sediments accumulated along the continental margins, and thus be responsible for the marginal fold mountain chains which are such a characteristic feature of the continents. The structure of continental rocks is therefore far more complex than that of the ocean floor because the continents persist indefinitely and are repeatedly moved and deformed at the earth's surface. In contrast, the rocks of the ocean floor are created, moved, and destroyed relatively quickly and uniformly and thus are both generally younger and less complex than continental rocks.

On a smaller scale, many of the complex structural features seen in continental deformation (Sections 3.4 and 3.5) can be related to plate motions. Thus many *compressional* features, such as folds, thrust faults, and folded mountain ranges, probably originate where a continent and an adjacent plate are pushed together at subduction zones (Figure 3.35a). On the other hand most *tensional* features, such as large-scale normal faults and fault-block mountains, form when continents are stretched and broken over new crust zones (Figure 3.35b).

The discovery of plate motions clearly indicates that the basaltic rocks of the oceanic crust originate from local melting of the upper mantle, as described in Section 3.3, but still leaves unresolved the question of the origin of the lighter, silicic rocks of the continents. Because volcanic rocks associated with subduction zones along continental margins are usually silicic andesites, rather than oceanic basalts (see Figures 1.29 and 3.9), some geologists have suggested that continents grow continuously along their margins by the addition of silicic materials that somehow separate from the mafic rocks of the descending plate or underlying mantle. An equally likely proposal suggests that the andesites result only from a deep melting of preexisting continental rocks, or of sediments derived from them, and thus shed no light on the *ultimate* source of the silicic continents. The question of continental origin thus remains a fundamental puzzle of the earth's crust.

3.10 Plate Tectonics and the Earth's Interior

Observation of crustal rocks exposed at the earth's surface, coupled with seismic data on the deep crust and upper mantle, have convincingly demonstrated that the earth's crust is formed and shaped by plate motions. A major uncertainty, however, concerns the underlying driving mechanism within the earth's interior which causes the creation, movement, and ultimate destruction of the plates.

Detailed studies of deep earthquakes caused by the descent of plates at subduction zones give an indication of the total thickness of the descending plates as well as the overall depth to which they extend into the earth's interior. Such studies show that the plates include not only

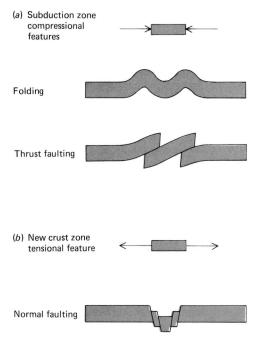

Figure 3.35 The relation of the most common structures of continental deformation to moving plate margins.

(a) Subduction zone compressional features

Folding

Thrust faulting

(b) New crust zone tensional feature

Normal faulting

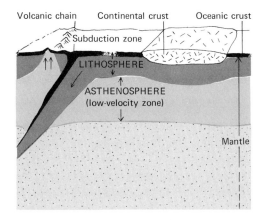

Figure 3.36 Lithosphere and asthenosphere. Moving plates consist not only of the crust (black), but also of a much thicker layer of the underlying mantle (dark gray); together these units make up the rigid "lithosphere" which moves on the hotter, plastic "asthenosphere" or low-velocity layer of the mantle (color). The figure shows a descending plate at a subduction zone.

the 8-km (5-mi) thickness of the oceanic crust, but also a much greater thickness, up to about 80 km (50 mi), of the underlying mantle as well. This conclusion is confirmed by calculations of the strength of crustal rocks, which show that plate-sized units made up of only the thin crustal layer would tend to crush rather than to move as rigid bodies. Apparently, then, both the crust and the cool, uppermost mantle combine to form the earth's rigid outer skin. Lying beneath this rigid **lithosphere,** as it has come to be called, is the plastic, low-velocity mantle zone which is far warmer and less rigid than the overlying mantle and crust (see Section 2.8). This plastic mantle zone, sometimes called the **asthenosphere,** evidently provides a lubricating layer over which the colder and more rigid lithosphere plates move (Figure 3.36).

Lithosphere plate motions must ultimately be driven by the heat energy concentrated in the underlying asthenosphere, but both the exact source of this heat energy and the way it is translated into horizontal plate movements remain uncertain. There are two likely sources for the earth's internal heat. One is *primordial heat* left over from the enormous frictional energy that is assumed to have been dissipated when the earth first consolidated from smaller masses of matter. Because of the insulating effect of the earth's great bulk, this initial heat would be lost very slowly and thus might still account for high internal temperatures even though the earth is very old. The second source, *radiogenic heat,* is provided continuously by the decay of radioactive nuclides within the silicate minerals of the crust and mantle. Although such nuclides make up only a tiny fraction of the earth's bulk, calculations show that their decay provides large quantities of heat to the earth's interior. Either primordial or radiogenic heat, or both, must in some fashion be concentrated in the warm asthenosphere, but there are many difficulties in understanding exactly how this takes place. Theoretical considerations indicate, for example, that a large fraction of the earth's heat-producing radioactive nuclides occur in the light silicic rocks of the continental crust rather than in the underlying mantle. Furthermore, there is no fully satisfactory explanation for the concentration of heat in a relatively narrow zone of the asthenosphere while the rocks both above and below remain cooler and more rigid.

Equally problematic is the question of exactly how the heat energy of the asthenosphere, whatever its origin, drives the regular horizontal plate motions of the overlying lithosphere. The most favored hypothesis relates plate motion to *convection currents* moving in either the underlying asthenosphere or, less probably, in a much greater thickness of the underlying mantle. Such convective movements, analogous to those seen in a pot of boiling water, occur whenever a fluid is heated from below in the earth's gravitational field. Convection currents always take the form of "cells" containing both rising and descending fluids; the rising areas might account for the upwelling of lavas along new crust zones, and the descending fluids might somehow pull down the oceanic crust at subduction zones. Although this hypothesis is perhaps the best yet advanced as the underlying cause of plate motion, it has nevertheless proved difficult to explain in detail how such regular fluid motions can account for the irregular outlines and changing configurations of the moving plates which shape the earth's crust.

SUMMARY OUTLINE

Rocks of the Crust

3.1 *Continents and ocean basins:* crustal rocks show a basic division into high-standing continents and low-lying ocean basins.

3.2 *Continental rocks:* continents are composed mostly of granite and granitic gneiss covered by a thin veneer of sedimentary rocks; much of this granitic material apparently originates from a melting of preexisting continental rocks.

3.3 *Oceanic rocks:* unlike the continents, the ocean floor is composed largely of basalt; these basalts apparently originate from a partial melting of ultramafic rocks in the underlying low-velocity zone of the mantle.

Crustal Movements: Continental Deformation

3.4 *Folds and faults:* deformation of continental margins by mountain-building forces may cause rocks to bend, producing folds, or break, producing faults; folding occurs only when rocks are made ductile by burial at high temperatures and pressures.

3.5 *Mountains and geosynclines:* most mountain ranges are made up largely of folded sedimentary rocks that originally accumulated along continental margins in thick sequences called geosynclines; near the axis of many mountain ranges the sediments have been melted to form elongate masses of granite.

3.6 *Continental drift:* the jigsaw puzzlelike fit of the continents, combined with studies of fossil rock magnetism, have suggested that the continents have moved or "drifted"; this phenomenon is now known to be related to still more fundamental movements involving the ocean floor.

Crustal Movements: Ocean-Floor Spreading and Plate Tectonics

3.7 *Topography of the ocean floor:* the ocean floor has the earth's longest and most continuous mountain chain; it is called the oceanic ridge-rise system.

3.8 *Magnetism of the ocean floor:* the regular magnetic patterns of ocean-floor rocks show that basaltic lavas have been continuously poured out at the ridge-rise system and then moved laterally to create parallel bands of new crustal rocks; this process is called ocean-floor spreading.

3.9 *Plate tectonics and crustal deformation:* new basaltic crust is added at the margins of large, rigid plates which move laterally as the opposite plate margin sinks into the mantle and is destroyed; these plates include both continental and oceanic crust; mountain-building deformations of continental margins are caused by plate motions.

3.10 *Plate tectonics and the earth's interior:* moving lithosphere plates extend downward to include the rigid, uppermost mantle; they move upon the warmer and more fluid low-velocity mantle zone; the ultimate cause of plate motions is uncertain, but they may be due to thermal convection in the underlying mantle.

ADDITIONAL READINGS

Billings, M. P.: *Structural Geology,* Prentice-Hall, Inc., Englewood Cliffs, N.J., 1972. A standard intermediate text on crustal folds, faults, and mountain building.

*Cailleaux, A.: *Anatomy of the Earth,* McGraw-Hill Book Company, New York, 1968. A popular introduction to the solid earth, emphasizing crustal rocks and structures.

*Clark, S. P., Jr.: *Structure of the Earth,* Prentice-Hall, Inc., Englewood Cliffs, N.J., 1971. Chapters 2 and 4 provide good summaries of crustal structures and plate tectonics.

Dickinson, W. R.: "Plate Tectonics in Geologic History," *Science,* vol. 174, pp. 107–13, 1971. A brief, authoritative review article.

*Keen, M. J.: *An Introduction to Marine Geology,* Pergamon Press, Oxford, 1968. An introductory survey with good discussions of ocean-floor topography, rocks, and structures.

Macdonald, G. A.: *Volcanoes,* Prentice-Hall, Inc., Englewood Cliffs, N.J., 1972. A clearly written intermediate text.

McKenzie, D. P.: "Plate Tectonics and Sea Floor Spreading," *American Scientist,* vol. 60, pp. 425–35, 1972. A popular review article.

*Takeuchi, H., S. Uyeda, and H. Kanamori: *Debate About the Earth,* Freeman, Cooper & Co., San Francisco, 1970. Excellent introduction to crustal magnetism, continental drift, and ocean-floor spreading.

Walton, M.: "Granite Problems," *Science,* vol. 131, pp. 635–45, 1960. Advanced but readable review of the controversy over the origin of granite.

*Wilson, J. T. (ed.): *Continents Adrift, Readings from Scientific American,* W. H. Freeman and Co., San Francisco, 1972. Popular review articles on earth structure, continental drift, and plate tectonics.

* Available in paperback.

PART 2

The Fluid Earth

Part 2 treats the two spherical shells of fluid that surround and enclose the solid earth. The innermost of these shells is made up of liquid—the waters occurring in the ocean and, in much smaller quantities, on the surface and in the soil of the land. The outermost shell is composed of the gases of the atmosphere.

Chapter 4 surveys the composition and structure of *the ocean* and considers the forces responsible for movements of ocean water.

Chapter 5 provides a similar survey of the composition and structure of *the atmosphere* whose thin, outermost gases extend for thousands of kilometers into interplanetary space.

Chapter 6 then considers motions of the lower atmosphere where interactions with waters of the ocean produce *weather*, the changing patterns of wind, temperature, and precipitation that occur near the earth's surface.

4

The Ocean

As land-dwelling creatures, we tend to think of the dry surfaces of the continents as the feature most characteristic of the earth. An observer in space, however, is most impressed not with the earth's land areas, but with the enormous volume of ocean water covering more than seven-tenths of the earth's surface. Among the planets of the solar system, only the earth appears to have such huge quantities of liquid water. Water is present on earth not only in the ocean basins, but in smaller amounts on land, where it occurs in lakes, rivers, streams, and as ground water saturating the uppermost levels of rock and soil. All of these waters make up a world-encircling sphere of liquid which is often referred to as the **hydrosphere** to distinguish it from the rocks of the underlying solid earth and the gases of the overlying atmosphere. About 97 percent of the water in the hydrosphere occurs in the ocean, which, through evaporation and rainfall, is also the ultimate source of the 3 percent of water occurring on land. In this chapter we shall be concerned with the composition and structure of this principal water reservoir, the ocean.

OCEAN WATER

4.1 Chemical Composition

In Part 1 we devoted an entire chapter to describing the chemical substances that make up the solid earth for, even though only a few elements are common in minerals and rocks, these elements occur in many different proportions and combinations. In contrast, the elements making up ocean water do *not* occur in widely varying proportions and combinations but are, instead, extremely limited in their variability. The most abundant elements of the ocean are shown in Figure 4.1. Hydrogen and oxygen, combined as water, are the most abundant for the ocean generally

Ocean waves. (U.S. Navy)

127

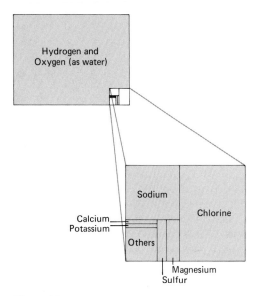

Figure 4.1 *The most abundant elements in ocean water. Hydrogen and oxygen, combined as water, account for 96.5 percent of normal ocean water. The remaining 3.5 percent consists largely of elements dissolved in the water, particularly chlorine and sodium.*

Figure 4.2 *The relative abundances of the dominant elements of ocean water (color) and crustal rocks (gray). Hydrogen, sulfur, and chlorine are enriched in the ocean whereas the relatively insoluble elements aluminum, silicon, iron, and nickel, although abundant in crustal rocks, are virtually absent from the ocean.*

consists of 96.5 percent pure water and only 3.5 percent of other elements, most of which are dissolved in the water to make a complex chemical solution. Although the 3.5 percent of dissolved material includes minute quantities of almost all the naturally occurring elements, chlorine and sodium predominate, accounting for 3 out of the total 3.5 percent. Magnesium and sulfur each make up about 0.1 percent while calcium and potassium account for only a few hundredths of a percent. All of these oceanic elements also occur in crustal rocks, but in differing proportions. The ocean is enriched in hydrogen, sulfur, and chlorine relative to crustal rocks, but contains proportionately less magnesium, potassium, and calcium (Figure 4.2). The abundant but relatively insoluble crustal elements aluminum, silicon, and iron occur only in trace amounts in ocean water.

For convenience in discussing variations in the composition of ocean water, it is customary to express the amount of dissolved matter not as *percent* ("parts per hundred") but as parts *per mille* ("per thousand"), abbreviated ‰, which merely moves the percentage decimal point one place to the right. Thus the normal 3.5 *percent* dissolved matter equals 35 per mille. The total amount of dissolved material, expressed as parts per mille, defines the **salinity** of ocean water, a property which is of extraordinary importance in understanding ocean structure. The waters of the open oceans (away from land areas) show little variation in salinity; thousands of salinity measurements made over the past hundred years from all the world's oceans show a total range of only about 33 to 37 ‰ with a mean of 35 ‰. Salinity is much more variable,

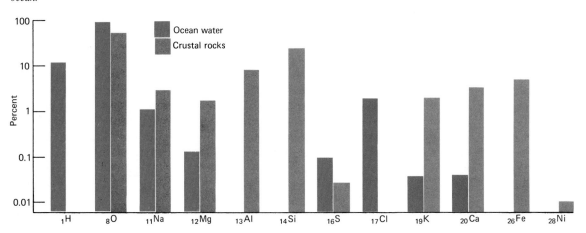

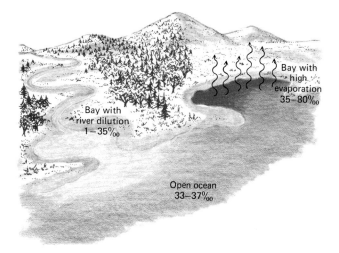

Figure 4.3 Variations in the salinity of ocean water caused by near-shore dilution and evaporation.

however, in shallow, near-shore ocean waters for two reasons (Figure 4.3). Lower than normal salinities occur because of dilution by "fresh water" running off the land where rivers and streams discharge into the sea. Rainwater and water on land normally have very little dissolved material giving them salinities of less than $1\ ^0/_{00}$. When such waters mix with ocean water, as they do in bays or estuaries, a full range of salinities from less than $1\ ^0/_{00}$ to the $35\ ^0/_{00}$ of normal sea water may occur. Similarly, when normal ocean water flows into shallow bays in warm regions with little or no rainfall or fresh-water runoff from land, the water in the bays may evaporate faster than it is replaced by normal ocean water, thus causing an increased concentration of dissolved material which gives *higher* than normal salinities. The highest salinities ever reached under such conditions are about $80\ ^0/_{00}$, for at this point certain elements, particularly calcium and sulfur, become so concentrated that they begin to crystallize from the ocean water as solid mineral compounds such as gypsum ($CaSO_4 \cdot 2H_2O$) or calcite ($CaCO_3$).

One of the most fundamental characteristics of ocean water was discovered in the late nineteenth century when the first chemical analyses began to be made of waters of different salinities from different parts of the world. These analyses showed that although the total *amount* of dissolved materials, expressed as salinity, may vary in the oceans, the *relative proportions of dissolved materials remain almost constant regardless of the salinity.* Thus, whether ocean water contains 1, 3.5, or 6 percent dissolved materials (representing salinities of 10, 35, or 60 ‰), the dissolved material will consist predominantly of chlorine and sodium, with the less abundant elements in a constant proportion to these two most abundant dissolved elements. These proportions are those shown in Figures 4.1 and 4.2. This discovery was of great significance for it indicated that there is, over long periods, an almost complete mixing of ocean waters; only through mixing could such a homogeneous distribution of the elements be produced.[1] Chemically, then, ocean water is a solution composed of a uniform mixture of many elements dissolved in water at concentrations ranging from less than 1 ‰ to 80 ‰ in near-shore waters, but having a relatively constant concentration of about 35 ‰ in the open oceans.

4.2   Physical Properties

The most important physical property of ocean water is its density (the relative weight of a given volume). Density is extremely significant because the structure and movements of masses of ocean water are strongly influenced by slight density differences. The density of ocean water, in turn, is primarily dependent on three properties: salinity, which we have already discussed, and two physical properties, *temperature* and *pressure*.

Because most of the dissolved elements in ocean water are heavier and more dense than the hydrogen and oxygen of pure water, the density of ocean water increases as the salinity increases. The effect of temperature on density is more complex. Most substances expand and become less dense when they are heated and, conversely, contract and become more dense when cooled. Pure water, when cooled, reacts in the normal way until it reaches a temperature of 4°C (39°F), just a few degrees above the freezing temperature. At that temperature, pure water has its maximum density; further cooling results in a *decrease* in density and an *expansion*, rather than contraction, of the water molecule. At the temperature of freezing, water (as ice) is about 10 percent less dense and occupies about 10 percent more space than at 4°C (Figure 4.4). This is why ice floats in water, rather than sinking, which is the fate of most substances that freeze from their own liquid.

[1] Precise modern chemical analyses show that very slight variations in the proportions of the rarer elements *do* occur in different parts of the oceans. These minute variations, which are sometimes useful in understanding the structure and movements of oceans, are usually measured in *parts per billion* and are so small that they do not invalidate the general concept of constancy of ocean water composition.

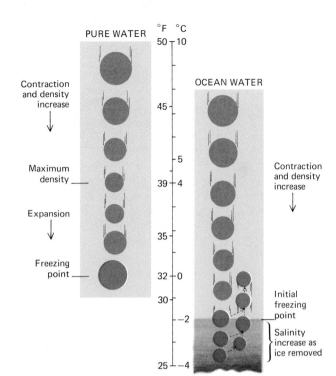

Figure 4.4 *Schematic comparison of density-freezing relationships in pure and ocean water (see text).*

Although *pure* water has its maximum density at 4°C, the dissolved elements in ocean water cause it to behave differently because they lower the freezing point. Instead of freezing at 0°C (32°F), as does pure water, normal ocean water (salinity 35 ‰) begins to freeze at −2°C (29°F). Furthermore, the density of ocean water is highest not at 4°C but at almost −4°C, which is, of course, below the freezing point. Thus ocean water increases in density down to the freezing temperature just as do most other substances. When ocean water begins to freeze at −2°C, however, the dissolved elements are *not* normally incorporated into the crystal structure of the ice. Ice crystals forming in ocean water are composed of almost pure water which is less dense than ocean water and floats just as in pure water. Because the dissolved elements are not incorporated into ice as ocean water freezes, freezing in a specific volume of ocean water *increases the salinity* as pure water is removed in the form of ice crystals. This salinity increase further depresses the freezing temperature so that new ice crystals will form only at lower temperatures. Thus, unlike pure water which freezes completely if held at or below 0°C, *ocean water has no fixed freezing point* for the freezing temperature

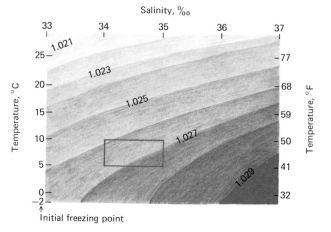

Salinity, ‰

Figure 4.5 Variation in the density of ocean water as a function of temperature and salinity. The highest densities occur in cold waters of high salinity (darker colors). Seventy-five percent of ocean waters fall within the range shown by the outlined rectangle.

drops as the salinity increases due to ice-crystal formation. The density of ocean water therefore has a complex relation to temperature, since low temperatures increase the density not only by the usual process of volume reduction, but also by increasing the salinity.

Pressure, the final factor affecting the density of ocean water, is the least important of the three. As with most substances, the density of water increases as it is compressed into a smaller space by increasing pressure. Water, however, is a relatively incompressible substance; that is, increased pressure leads to very small decreases in volume and consequently to very small increases in density. The pressure in a vertical column of ocean water increases with depth because of the weight of the overlying water. Bottom waters in the deepest parts of the sea are under pressures 1,000 times as great as surface waters, yet water is so incompressible that the density is increased only a few percent. For this reason, pressure is far less significant in determining ocean water density than are temperature and salinity.

The density of ocean water is usually expressed in relation to pure water at its temperature of maximum density (4°C or 39°F). Such pure water is assigned a density of 1. Ocean water at the same temperature has a slightly greater density of 1.027 because of the presence of dissolved elements. Changes in temperature and salinity affect this number only at and beyond the third decimal place. Because the density of ocean water is primarily determined by salinity and temperature, the three variables can be related in a useful chart which shows the differing densities resulting from various combinations of temperature and salinity found in waters of the open oceans (Figure 4.5).

In addition to density, temperature, and pressure, the three physical properties of ocean water that have been mentioned, certain other physical properties are sometimes studied. Among the most important are: optical properties, such as color and transparency; viscosity (resistance to flow); sound wave transmission; and electrical conductivity. While all of these properties are useful in specialized studies of ocean water, they are far less important than are density and temperature for understanding the structure and movements of the ocean.

OCEAN STRUCTURE

The constant ratio among dissolved elements in ocean water indicates that there is a general mixing of the oceans; such complete mixing occurs, however, only over hundreds or thousands of years. On a shorter time scale, the ocean is very *inhomogeneous*, being marked by horizontal and vertical differences in temperature and salinity which combine to create local differences in density. These variations give the waters of the oceans a characteristic *structure* which, although defined by relatively small variations in properties, is in many ways comparable to the structural differences that we have already seen in the core, mantle, and crust of the solid earth.

4.3 Layering

As with the materials of the solid earth, the oceans show a vertical separation into layers of differing density, with the heaviest and most dense layers on the bottom. Unlike the earth's inaccessible mantle and core, however, these layers of the ocean can be directly sampled at all depths; such sampling requires only that special water-collecting bottles and thermometers be lowered on a line from shipboard. A single line usually contains many bottles and thermometers for sampling at various depths (Figure 4.6). The water samples are analyzed for salinity; from this measurement and the temperature readings the water density at each level is computed from a chart similar to Figure 4.5. Such measurements require considerable time and effort, for at each station the ship must be stopped, the lines lowered and raised, and the water samples analyzed. Recently, electronic devices for measuring tempera-

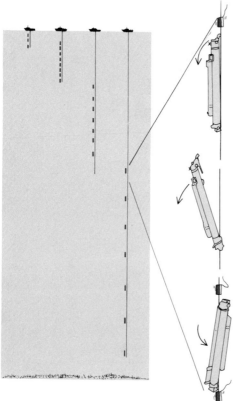
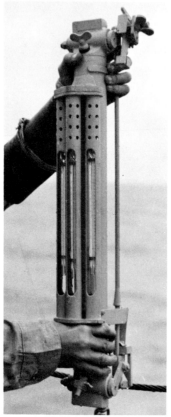

(a)

(b)   (c)

Figure 4.6 *"Nansen bottles," used to measure oceanic salinities and temperatures. (a) The bottles being lowered in various series. The enlargement at right shows their manner of operation. A small weight (color) descending the line releases the bottle and causes it to invert, trapping water inside and recording the temperature on special thermometers attached to the bottle. The trapped water sample is analyzed for salinity. (b) A nansen bottle, showing thermometers attached to the larger, water-collecting tube. (c) Attaching a bottle to the lowering cable. (b, c: U.S. Coast Guard)*

tures and salinities have been perfected, but these still require that the sensing devices be lowered on a line from a stationary ship. As yet, unfortunately, no more rapid means has been devised for determining deep water densities, although temperatures alone, but not salinities or densities, are now routinely recorded automatically at relatively shallow depths by a device known as a **bathythermograph** which may be towed from a slowly moving ship (Figure 4.7). In spite of these difficulties, many thousands of temperature-salinity-depth measurements have been made in all the earth's ocean basins, and these have provided enough information for a broad understanding of the vertical structure of the ocean.

Such studies show that all of the ocean, outside of cold polar regions, has the same characteristic pattern of temperature change with depth (Figure 4.8). In the upper hundred meters or so temperatures are quite variable but are usually relatively high, reflecting daily and seasonal warming of the overlying air and the effects of transport

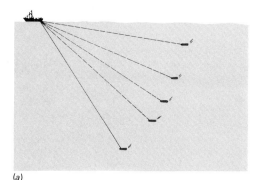

(a)

Figure 4.7 *Bathythermographs. (a) Boat towing the instrument, which continuously records depth and temperature as it descends to a maximum depth of about 300 m (1,000 ft). (b) Close-up of a bathythermograph. (c) The depth-temperature profile is etched on a smoked plate, shown being inserted into the instrument. (d) Retrieving the instrument after a dive. (b, d: Ocean Science Laboratory; c: Gar Lunney, Information Canada Photothèque)*

(b)

(c)

and mixing by winds. This thin zone of relatively high and variable temperatures is called the **surface layer.** Below the surface layer is a second layer, usually about a thousand meters thick, or about 10 times as thick as the surface layer, in which temperatures drop rapidly to very low values, normally only a few degrees above freezing. This zone of rapid temperature decrease is called the **thermocline layer.** The depth of its base varies in different regions from about 600 to 1,000 m (2,000 to 3,000 ft) below the surface. From the base of the thermocline layer to the sea floor, which lies at an average depth of 6,000 m (18,000 ft) below the water surface, is the **deep layer,** which contains the bulk of the waters of the oceans. Temperatures in the deep layer are uniformly cold, usually only a few degrees above freezing.

Salinity variations with depth have a much smaller range and a less regular pattern of change than do temperatures. The salinity of the surface water in the open ocean is largely determined by the ratio of evaporation to rainfall occurring in a particular region. In "dry" oceanic regions, surface salinities are increased by evaporation; in "wet" regions they are decreased because of dilution by rainfall. The normal range of surface salinities produced by these variations in the open sea is about 33 to 38 ‰. In general, "wet" oceanic regions, where precipitation exceeds evaporation resulting in lower than normal salinities, occur along the equator and in Arctic and Antarctic waters. In the temperate belts between, evaporation normally exceeds precipitation leading to higher than normal surface salinities (Figure 4.9). Deep waters, below about 2,000 m (6,000 ft), tend to be much less variable

(d)

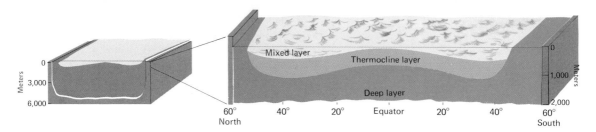

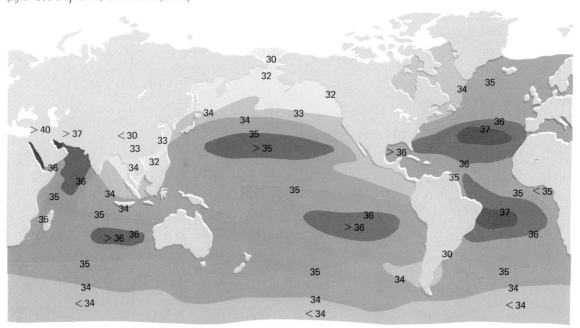

Figure 4.8 *A schematic representation of the three principal ocean layers. The thin, upper, "mixed" layer is relatively warm because of solar heating and atmosphere interactions. Beneath is the thicker "thermocline" layer where temperatures drop sharply; in the still thicker "deep" layer temperatures are relatively constant and average only slightly above freezing.*

Figure 4.9 *Average surface salinity of the ocean. (After Sverdrup et al.,* The Oceans, *1942)*

than surface waters; their salinities range only between 34.5 and 35 ‰. When waters with salinities beyond this narrow range occur on the surface, a **halocline layer** of increasing or decreasing salinity will be present between the surface and deeper waters, but the position and nature of such halocline layers are much more variable than is the universal thermocline layer.

The range of temperature variation with depth in the open ocean is far greater than that of salinity. For this reason, the density distribution of the oceans, although related to both temperature *and* salinity, is *primarily* controlled by temperature. As in all fluids, the most dense (and coldest) waters of the ocean tend to sink and the less dense (and warmest) tend to rise under the influence of gravity; this process is called **convection.** The vertical stability of the oceans is maintained by such convective movements leading to the great mass of high-density cold water in the deep layer, overlain by the warmer, less dense waters of the thermocline and surface layers. The thermocline layer, in particular, is of great importance in ocean structure because it forms a stable boundary of increasing density which tends to prevent interchange between the surface and deep waters.

4.4 Water Masses

Within the surface, thermocline, and deep layers it is usually possible to recognize smaller, geographically restricted bodies of water that can be distinguished from each other by relatively minor differences in temperature, salinity, or other properties. Such bodies, known as **water masses,** make up a second fundamental feature of ocean structure.

The origin of these distinctive water masses is controlled by an important fact of ocean structure: *all primary changes in temperature and salinity must take place near the ocean surface.* A moment's reflection will show why this is true. Such changes require energy, and all but an insignificant fraction of the energy reaching the oceans comes from the sun. Most of this solar energy is absorbed in the upper 30 m (100 ft) where it causes rises in temperature and increases in salinity due to evaporation. Similarly, both temperature decreases and salinity reductions take place only at the ocean surface (by, respectively, loss of heat to the atmosphere and rainwater dilution). Large-scale temperature and salinity change can occur, then, only where the water is exposed to solar energy and contact with the atmosphere. When a particular mass of water occurs far below the ocean surface it is effectively insulated from these effects and can change its temperature or salinity only by becoming mixed with other water masses having different properties. Because such mixing takes place very slowly, deep water masses tend to retain, for long periods, the characteristic temperature and salinity they obtained when last exposed at the surface. For this reason, water masses covering large areas of the oceans can often be traced to their source regions by their characteristic temperature-salinity patterns. Sometimes it is even possible to determine the degree of mixing between different water masses by observing intermediate temperature and salinity values.

In addition to temperature and salinity, there is one additional property of ocean water that is particularly useful in tracing water masses. Oxygen, we have seen, is the most abundant element in the oceans, making up 86 percent of ocean water. Most of this oxygen occurs combined with hydrogen in water molecules, but a very small portion (less than 0.01 percent of the total amount) occurs as gaseous oxygen dissolved in the ocean water. This dissolved oxygen is one of a small group of elements which are relatively rare in ocean water but are actively used in the life processes of ocean-dwelling animals and plants. These elements (various forms of carbon, nitrogen, and phosphorus are, besides oxygen, the principal ones) are exceptions to the general rule that the proportions of dissolved materials in ocean water are everywhere constant. Instead, these elements, all of which are present in very small quantities, tend to vary in abundance according to their utilization by marine animals and plants. For this reason these elements are called **nonconservative properties** of ocean water in contrast to the other elements which are **conservative properties.** In sunlit surface waters, tiny floating green plants are usually very abundant; these plants, like their larger land relatives, use solar energy to convert simple carbon, nitrogen, and phosphorus compounds into more complex organic molecules. Carbon, nitrogen, and phosphorus thus tend to be depleted in the upper waters of the seas where sunlight permits abundant plant life. Unlike green plants, which give off oxygen as a waste product, all animals require oxygen for their life process. For animals living in the sea, this oxygen comes from the small quantity dissolved in ocean waters. Furthermore, animals are not restricted by sunlight penetration to only the uppermost waters of the oceans, for they can live throughout the dark ocean depths, and on the sea floor, by eating plant and animal material moved downward from the sunlit waters above. Because oxygen-consuming animal life is present almost everywhere in the oceans, and because new oxygen is added to ocean water only near the surface by plant metabolism and contact with the oxygen in the atmosphere, it follows that the dissolved oxygen in deep ocean waters tends to become depleted by animal consumption. Herein lies the usefulness of dissolved oxygen in tracing ocean water masses, for the longer the water mass has been isolated from the ocean surface, the less dissolved oxygen it will have. Thus the amount of dissolved oxygen provides an estimate of length of time the water has been away from contact with the surface; for this reason, analyses of oxygen content are now routinely performed, along with temperature and salinity measurements, on water samples taken from the deeper parts of the oceans.

The most important result of water mass studies by analysis of patterns of temperatures, salinities, and oxygen contents has been to show the sources and general structure of the deep layer of cold water which lies beneath the thermocline and includes the bulk (about 80 percent) of the waters of the oceans. *Most of these waters originate on the surface in polar regions where there is no thermocline to prevent interchange between surface and deep waters.* Unlike the rest of the ocean, in polar regions the surface waters have temperatures only a few degrees above freezing just as do the deeper waters. In these regions there is thus no temperature change with depth and no threefold density subdivision of the waters; instead there is merely a single mass of cold, high-density water extending from the surface to the ocean floor (Figure 4.8). Furthermore, the density of polar waters may be increased by the formation of ice in the sea, which removes pure water and increases the salinity, and thus the density, of the remaining unfrozen water (see Figure 4.4). The highest-density waters of the ocean originate in this way, as cold, saline, polar water masses that tend to flow outward along and above the ocean floor and make up the deep ocean layer below the thermocline. These deep, cold waters may be further subdivided into smaller water masses both geographically, according to whether they originate in Arctic or Antarctic regions, and by small differences in temperature and salinity (Figure 4.10). As with the general deep layer of which they are a part, the

Figure 4.10 Schematic diagram of the principal water masses of the deep Atlantic Ocean. (*After Neumann and Pierson,* Principles of Physical Oceanography, *1966; from Wüst*)

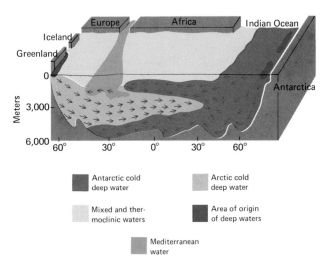

vertical position of these smaller water masses is always determined by density which, in turn, is primarily controlled by temperature and salinity. Normally, as we have seen, temperature is the primary factor in controlling density, and thus these smaller water masses tend to be slightly colder and more dense with depth. Sometimes, however, warm but highly saline masses may have high enough densities to occur below cooler, less saline water masses. Such a mass of warm, saline, high-density water originates in the Mediterranean Sea and flows outward into the deep Atlantic to form an anomalously warm, deep layer that can be recognized over thousands of miles (Figure 4.10).

Distinctive smaller water masses are less clearly defined in the intermediate thermocline layer, but, as in the deep layer, many can be recognized in the thin surface layer where temperature and salinity variations are at a maximum. These upper water masses are much less permanent than are masses of the deep layer because they are subject to rapid mixing and movement by winds and solar heating. These water masses of the surface layer, although they include only about 2 percent of all waters of the ocean, contribute disproportionally to large-scale movements and circulation of ocean water because of their interactions with solar energy and the atmosphere.

4.5 Ice in the Ocean

The formation of ice in sea water is a final important structural property of the ocean. We have seen that ice crystals forming in the ocean are made up of almost pure water; their formation increases the salinity of the remaining water so that further cooling is necessary for more freezing to take place. If this process continues, ice particles increase in abundance and freeze into a solid framework enclosing small cells of sea water which have too high a salinity to freeze at that temperature. The result is a frozen mixture consisting mostly of relatively pure ice crystals enclosing small quantities of salt and brine.

Complete freezing of ocean water, in the manner just described, sometimes takes place in shallow bays and lagoons in polar regions, but in areas of deeper water only a thin surface layer of **sea ice,** usually less than 3 m (10 ft) thick, forms in even the coldest polar seas (Figure 4.11). Such sea ice perpetually covers most of the Arctic Ocean and surrounds the ice-covered continent of Antarctica. It forms from the freezing of surface sea water

(a)

(b)

Figure 4.11 *Sea ice. (a) Off the coast of Antarctica;*
note the ship breaking through thin, transparent ice
near the center of the picture. (b) Close-up of Arctic
sea ice. (a: U.S. Navy; b: U.S. Coast Guard)

and the accumulation of fresh water, as snow, on the surface. Sea ice seldom grows thicker than 3 m because as the waters beneath the ice are cooled, they increase in density and sink to be replaced by warmer waters from below. Accumulation of snow and ice on the surface is also limited by the water temperature for, as snow accumulates on the surface, the ice below is depressed by the weight and an equivalent amount of ice is melted below by the warmer waters. Thus the thickness tends to remain constant even in areas of high snowfall.

In contrast to the thin sheets of sea ice that form on the oceans, ice on land, accumulating as snow on a solid surface with no water below to cause melting, can reach tremendous thicknesses. Most of Greenland and the continent of Antarctica are now covered by huge ice sheets that are more than a kilometer thick. In addition to these extensive ice sheets, most mountains of polar regions have valleys filled by smaller **glaciers,** which are thick rivers of ice that accumulate on land as snow. Where either ice sheets or glaciers touch the ocean, huge **icebergs** hundreds of meters thick may break away and float until they reach warmer water and melt. Icebergs (which along with sea ice form the two principal kinds of ice found in the ocean) therefore originate not from freezing of ocean water, but from the accumulation of fresh water, as snow, on the land surface. Most of the icebergs of the northern hemisphere occur in the North Atlantic and have their source in Greenland and arctic Canada. In the southern hemisphere, the huge Antarctic ice cap in some places flows beyond the boundaries of the continent as **shelf ice** floating on the surrounding seas. The largest icebergs of the ocean originate when huge masses of this ice break off and float northward (Figure 4.12). During the Antarctic winter, this thick, floating, land ice is bounded on the seaward side by a much wider belt of thin sea ice similar to that which continually covers the Arctic Ocean.

Figure 4.12 *Icebergs. (a) Surrounded by sea ice in the Arctic Ocean. (b) An enormous, tabular iceberg off the coast of Antarctica. (a: National Film Board of Canada; b: U.S. Navy)*

(a)

(b)

OCEAN MOVEMENTS: CURRENTS

So far we have considered only the relatively stable, large-scale structure of the ocean, and have said little about more rapid and dynamic movements of ocean water. Such movements are of two sorts: **currents,** which are motions that transport water from one part of the ocean to another, and **waves,** moving disturbances traveling on or through the water, like wind ripples crossing a field of wheat, which do not directly transport water from one place to another. Currents will be discussed first; the following section then treats ocean wave movements.

4.6   Patterns of Surface Currents

Current movements are closely related to the ocean's layered structure. The most obvious take place in the surface layer, the upper hundred meters or so of water above the thermocline where solar energy and interaction with the atmosphere are concentrated. Most knowledge of **surface currents,** as horizontal water movements occurring in this surface layer are called, has come from records of their effect on the movement of ships. From thousands of such observations, navigators have compiled **current charts** showing the general patterns of surface currents around the world. The first comprehensive charts of this kind were made in the mid-nineteenth century as aids to navigation; a simplified modern version is shown in Figure 4.13.

An examination of the surface current patterns in Figure 4.13 shows certain similarities in each ocean. The principal features are large circular patterns of flow called **gyres.** The Atlantic and Pacific Oceans each have three such gyres, a small gyre flowing to the left (counter clockwise) in their northernmost extensions, and two much larger gyres on each side of the equator, the northern one flowing to the right (clockwise) and the southern one to the left (counterclockwise). Flowing along the equator between these large gyres in the opposite direction is a simple, one-directional current called the **equatorial countercurrent.** The Indian Ocean extends only a little bit north of the equator and has only the southernmost, left-flowing gyre well developed. The Antarctic Ocean has no gyre but merely a continuous eastward current, circling the entire earth around the continent of Antarctica. All of the

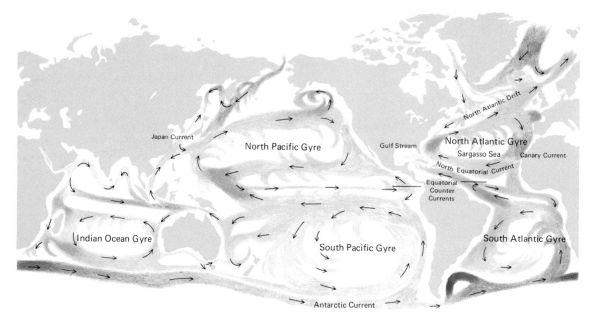

Figure 4.13 Principal surface currents of the ocean. Cold currents are shown in darker color, and areas of faster cold and warm currents by heavier shading.

major gyres persist in direction and undergo only relatively small-scale changes in position with the changing seasons; the smaller currents north of the equator in the Indian Ocean, however, completely reverse their directions in summer and winter.

The large gyres do not have common names, but can be informally referred to as the "North Atlantic," "South Pacific," etc. gyres (Figure 4.13). Certain segments of each large gyre have, however, been given separate geographic names. For example, the western part of the large North Atlantic gyre is called the *Gulf Stream*, the northern part the *North Atlantic Drift*, the eastern part the *Canary Current*, and the southern part the *North Equatorial Current*. Usually only the parts with relatively rapid water movements are called *currents*; parts with slower movements are called **drifts.** Within each gyre is a large region of relatively stable and immobile surface water called an **eddy.** The eddy within the North Atlantic gyre is called the *Sargasso Sea* because large quantities of the floating seaweed *Sargassum* accumulate within its relatively quiet and stable waters. The eddies within the other major gyres do not have common names. One other important fact concerning the ocean's surface currents is evident in Figure 4.13: all of the gyres tend to be asymmetrical to the west. Instead of having their centers in the middle of the ocean basins, the centers tend to be shifted to the left or westward in both the northern and southern hemisphere. As a result, the currents of each gyre tend to be narrowest and swiftest along their western sides. This is particularly evident in the Gulf Stream and Japan Current, the **western boundary currents,** as they are called, of the North Atlantic and North Pacific gyres.

Of all the currents making up the gyres of the oceans, the Gulf Stream at the western edge of the North Atlantic has been by far the most extensively studied. Like other western boundary currents, it is made up of warm water flowing away from the equator at relatively high speeds, 6 km/hr (3.7 mi/hr); it somewhat resembles a "river" of warm water moving in a "channel" of cooler water. The "river" varies greatly in size but is typically very large, averaging about 80 km (50 mi) in width and several hundred meters in depth. Its boundary is often sharply marked by changes in water color or transparency. The Gulf Stream was long thought to be relatively constant in structure and position, but recent surveys have shown that, in detail, it is not a single mass of moving water, like a large river, but more like a series of small, rapidly flowing interconnected "streams" separated by stable water (Figure 4.14). The position of the "streams" changes rapidly, sometimes as much as 150 km in only a few days. It is the average flow of many such filamentlike streams over the wide region shown in Figure 4.13 that makes up the rapid northward movement of water in the Gulf Stream. Limited observations on the western boundary currents in other gyres suggest that they have a similar structure.

4.7 Causes of Surface Currents

Probably no other single problem has attracted more study from oceanographers than the question of the *causes* of the regular patterns of flow seen in the ocean's gyres. Most of this study has involved estimating the nature and size of the forces acting on the surface waters of the oceans, followed by attempts to analyze these forces mathematically in an effort to see if they might produce the observed flow. Such analyses make use of the principles of **fluid dynamics,** a branch of physics concerned with the mathematical description of movements within fluids of all sorts, whether they be gases flowing in pipes, liquids running in channels, or water moving in the oceans. When applied to ocean movements, fluid dynamic analysis usually assumes that only four forces, acting in differing combinations and strengths, are responsible for most

Figure 4.14 Details of current flow in the northern Gulf Stream. The flow resembles a series of thin shifting "filaments" of north-flowing warm water. (After Munk, Scientific American, vol. 193, 1955)

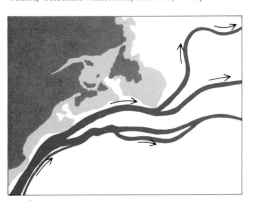

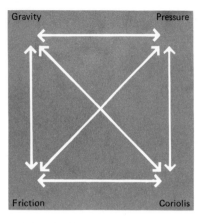

Figure 4.15 The four interacting forces responsible for most ocean movements.

water movements in the oceans. These forces are: the earth's *gravity*, which causes less dense water to rise and more dense waters to sink; *internal pressure gradients*, which cause waters to flow from regions of high pressure to regions of lower pressure; *friction*, which causes drag, turbulence, and changes in speed of flow at the boundaries between water layers and in water layers that come in contact with the solid earth below and the gaseous atmosphere above; and a final *Coriolis force*, which is related to the rotation of the earth around its axis (Figure 4.15). The first three of these forces are easy to visualize for they are observed in everyday life. The fact that oil floats on water is a familiar illustration of the effect of gravity on fluids of differing densities—the oil is less dense than water and rises to the surface. Pressure gradients are also familiar for they cause water to flow "uphill" in city water pipes. The water is merely kept at relatively high pressures by pumping it into overhead storage towers or locating reservoirs at higher elevations than the cities they serve. In this way the pressure on the water in the pipes is made higher than the surrounding air pressure caused by the weight of the atmosphere. When a faucet is opened, the water flows into the air, which is a fluid at lower pressure. Likewise, frictional effects in fluids are familiar; the whirlpools and eddies formed in rapidly flowing streams are caused by frictional drag on the water as it moves over the solid stream channel. The Coriolis force, the final force acting on the oceans, is less familiar for it normally acts on a scale too large to be observed in everyday life. Nevertheless, it is important to understand its effect, for it is one of the most critical forces contributing to movements in the oceans and, as we shall see later, in the atmosphere as well.

The Coriolis Force Coriolis movements are caused by the earth's rotation. An object on the earth's solid surface moves as the earth spins and the speed of movement is proportional to the latitudinal position of the object—it is greatest at the equator where the object moves 40,000 km (25,000 mi) through space each 24 hours as the earth completes one rotation, a speed of more than 1,600 km/hr (1,000 mi/hr). Conversely, the speed is zero at the poles where the object merely rotates in place but does not change position, in relation to objects in space, as the earth rotates. Speeds of rotational movement are thus at a maximum on the equator, a minimum at the poles, and have intermediate values between. This difference in speed

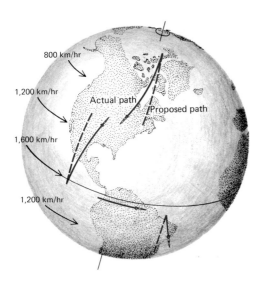

Figure 4.16 The Coriolis effect. Pole-to-equator differences in the earth's rotational speed cause moving objects that are not firmly attached to its surface to move to the right of their initial path in the northern hemisphere and to the left in the southern hemisphere.

800 km/hr

1,200 km/hr

Actual path

Proposed path

1,600 km/hr

1,200 km/hr

has no effect on objects, such as automobiles or trains, moving on the earth's solid surface for they are "attached" to the earth and rotate with it. However, objects moving in the fluid oceans and atmosphere are not attached to the solid earth; they tend to change speed relative to the earth's surface as they change latitude. This effect is illustrated in Figure 4.16. A rocket launched at the equator and moving north at a constant speed passes over land that is rotating eastward at progressively slower speeds than the rocket itself—the effect is that the rocket's path is *deflected to the right relative to the earth's surface*. If the rocket originates at the North Pole and moves southward, it will pass over more rapidly moving land and its path will still appear deflected to the right. If it moves along the equator there will be no deflection, but if it crosses into the southern hemisphere, the slower land movement will again cause deflection, but *this time to the left*. In a similar fashion, the earth's rotation causes moving masses of water in the oceans and air in the atmosphere to be deflected to the right in the northern hemisphere and to the left in the southern hemisphere. This is the Coriolis effect which, for convenience in calculations, is considered to be caused by a force, the **Coriolis force,** which varies in strength with latitude just as does the earth's speed of rotation.

The forces of gravity, pressure, friction, and the Coriolis effect, acting together, are the primary causes of ocean movements. Thus, in theory, with a knowledge of the magnitude and direction of each as they act on a mass of ocean water, one should be able to calculate the direction and amount of the resultant force and thus predict the exact speed and direction of movement of the water mass. This ideal is difficult to attain in practice, both because the values of some of the forces are poorly known in the oceans and also because the interaction of several forces is usually difficult to describe mathematically. For certain kinds of motions, however, one or more of the forces are insignificant and may be ignored, and the others then given values that appear to be reasonable based on actual observations of ocean movements. This kind of analysis has been particularly fruitful in attempts to understand the causes of surface currents.

Winds and Surface Currents There has long been reason to believe that winds play a large part in causing surface currents. Early current charts, compiled in the nineteenth century, showed a general similarity between sur-

147

face current direction and wind direction over most of the ocean's surface. This similarity led nineteenth-century scientists to the reasonable assumption that winds blowing across the ocean's surface were the primary cause of surface currents. Winds, however, tend to blow not in circular gyres but in parallel belts extending around the entire earth, and these wind belts generally correspond in position and direction to only the east-west portions of the gyres. The circular flow of the gyres was accounted for by the interference of the continental land masses which prevented development of world-encircling currents following the wind belts; the eastward round-the-world flow of the Antarctic Current, which occurs in the only ocean without continental barriers, further supported this idea. There were, however, several difficulties. Why were all the gyres asymmetrical with their swiftest flow along the western margins? And how, exactly, could wind flowing over water, a much denser fluid, "push" the water into such large and persistent movements?

To answer such questions as these, oceanographers began, around the turn of the century, to study the effect of winds on ocean movement by using mathematical analysis and the then recently discovered principles of fluid dynamics. The force of the moving wind was assumed to be transferred to the ocean's surface waters by friction. To make mathematical analysis possible, the effects of gravity and pressure gradients were omitted; only the effects of wind friction and the all-important Coriolis force were considered. The first generalizations to come from such analyses were rather surprising. In 1902 the Swedish oceanographer V. W. Ekman showed that the water movements caused by the wind should decrease very rapidly below the surface and, furthermore, would tend to move water not in the direction of the wind but to the right or left of the wind direction at an angle of 90 degrees. This conclusion resulted from a consideration of the Coriolis force (Figure 4.17). The wind was assumed to set in motion, by friction, a thin surface layer which then transferred its motion, by friction, to a thin layer below, and so on. With each transfer of motion between layers, energy would be lost so that the force would decrease rather rapidly with depth and would be virtually absent about 100 m (300 ft) below the surface (Figure 4.17a). In addition, each moving layer is moved slightly to the right (in the northern hemisphere) of the overlying layer by the Coriolis force. The direction of movement therefore shifts slightly at each lower level; at about the depth that the force would die

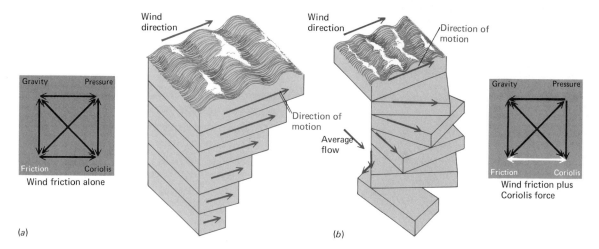

Figure 4.17 The "Ekman spiral." (a) Wind blowing across the water surface causes progressively less movement with depth. (b) At each level the direction of flow is shifted slightly to the right (in the northern hemisphere) by the Coriolis force. The net result is water flow at about 90° to the wind direction.

out, this shifting leads to movement exactly opposite to the wind direction (Figure 4.17b). The net result of all these changing, spiral-shaped movements would be water transport at right angles to the wind direction. These spirals of moving water proposed by Ekman have come to be called **Ekman spirals** and the relatively thin surface layer moved by the wind the *Ekman layer*.

Ekman's analyses have had a strong influence on thinking about ocean currents, but his "spirals" of movement have not been satisfactorily observed in the oceans, although water movement somewhat oblique to the wind direction is sometimes observed. Many oceanographers now feel that the assumptions in Ekman's calculations were oversimplified and that local water movements usually tend to more closely approximate the direction of the wind. His concept of a relatively thin "Ekman layer" where wind influence dominates has, however, been confirmed by many observations and is generally accepted. This Ekman layer is the *surface layer* that we have already discussed—the thin, relatively warm zone at the ocean's surface where mixing and movement are at a maximum.

Building on Ekman's studies, modern mathematical analysis of surface currents has been concentrated not on local effects of the wind on the water surface but on global patterns of wind-water interaction. Such studies have shown that the two forces first analyzed by Ekman—wind action and the Coriolis force—*can* reasonably account for observed worldwide patterns of surface currents. In particular, these studies have shown that the forces produced by the earth's parallel belts of eastward

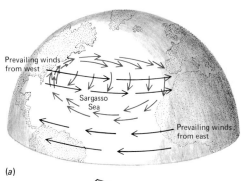

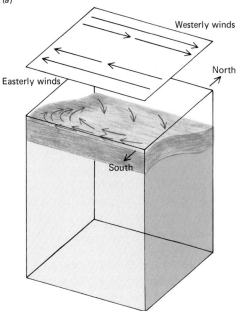

Figure 4.18 Origin of surface currents from wind action and the Coriolis effect. (a) The North Atlantic gyre, showing the prevailing wind belts. (b) Proposed origin of the circular flow and western boundary current (Gulf Stream). Water motions, initiated by the east and west wind belts, are moved to the right by the Coriolis effect (colored arrows), which is strongest to the north, away from the equator. This inequality causes a slight "pile up" of water along the equatorial flow which is balanced and returned northward by the swift western boundary flow.

and westward blowing winds are sufficient to drive the entire circular flow of the gyres; that is, the north-south flows at the sides of each gyre are a natural result of the opposing, wind-driven movements in the "top" and "bottom" parts of the gyres (Figure 4.18). Thus the "side" currents, such as the Gulf Stream, are not directly driven by local winds, but result from wind action in other parts of the gyre. Still more important, such analyses have shown that the asymmetrical shape and strong western boundary currents in each gyre can be explained by the Coriolis force (Figure 4.18). The correspondence of such models with the actual flow of the oceans is not perfect; in many places the ocean's eastern boundary currents move faster and the western currents, particularly in the southern hemisphere, more slowly than the models predict. Nevertheless, these theoretical results, based only on considerations of winds and the Coriolis force, strongly suggest that wind action is the principal driving force of the ocean's surface currents. The fact that the ocean's pattern of surface currents can be reasonably explained by wind movements still does not tell us the underlying *cause* of such movements. Instead, it merely shows that movements within the fluids of the earth's ocean and atmosphere are closely linked and must be considered together as one interconnected system of circulation.

4.8 Deep Circulation

Although the surface currents flowing in the ocean have been carefully observed and charted for over a hundred years, very little is known of current patterns in the underlying thermocline and deep layers. This is because the technical problems of measuring rapid movements in deep water are extraordinarily difficult, and have only recently begun to be solved by the design of "neutrally buoyant floats"—devices that can be adjusted to float at a preset depth and whose movements can then be followed by electronic tracking from ships or near-shore stations (Figure 4.19). These devices have shown unexpectedly rapid and complex movements of deep ocean waters in the few areas where observations have been made. Such results have stimulated the design of better equipment for measuring the deep currents which are a growing concern of oceanographic research. Nevertheless, the ocean is so large, and the number of oceanographic research vessels so few, that it will be a long time before knowledge of deep water movements even begins to approximate our understanding of surface currents.

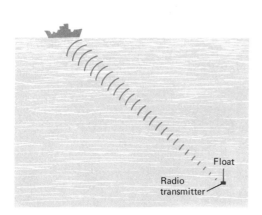

Figure 4.19 Measurement of deep current movements by use of a "float" designed to move with the current at a fixed depth. A transmitter reveals the position of the float to a shipboard receiver.

Because observational facts are so few, knowledge of the deep circulation of the oceans has come mostly from theoretical analyses of probable movements caused by density differences between deep water masses. The densities of deep water, you will recall, can be measured by special thermometers and water-sampling bottles, and are thus far easier to determine than are actual water movements at depth. Perhaps the most important result of such density determinations is the observation that vertical water movements throughout most of the deep oceans must be very slow indeed because the same patterns of layered water masses can be observed to persist without change over many years. If rapid vertical currents existed over large areas of the deep oceans, then mixing and disturbance of the structure, rather than long-term stability, would be expected.

In Section 4.4 we saw that most of the waters of the deep ocean originate from sinking of cold surface water in relatively small areas of the polar seas. This sinking must, of course, be balanced by rising water somewhere else; it is assumed, without strong evidence, that this rising is distributed over the entire ocean, and that it takes place too slowly to disturb the general layered structure of oceans. The deep water is assumed to move gradually upward through the thermocline layer and into the surface layer where it is moved rapidly by wind-driven surface currents until it eventually reaches the small polar areas where deep waters form and the process begins again.

From the information on deep density distribution it is also possible (as with surface currents) to apply the principles of fluid dynamics in an attempt to understand horizontal water movements. As always, such analysis requires simplifying assumptions. Wind action, of course, cannot directly affect deep waters. Instead, deep movements must be largely due to pressure gradients caused by horizontal differences in water density. The density differences, in turn, are caused by temperature and salinity variations; for this reason such movements are usually called **thermohaline circulations.**

In analyses of thermohaline movements, frictional forces are usually neglected and the local effects of gravity and the Coriolis force are considered to balance each other, as they would if no other forces were present. With these forces balanced, the principal water movements result only from density-induced pressure gradients. Calculations of such gradients from measured densities have shown that only very slow horizontal currents should be produced. It was for this reason that oceanographers were surprised by the recent discovery of rapid, deep currents as measured by neutrally buoyant floats. Slow movement and general stability over most of the deep ocean are still assumed to exist because of the long-term constancy of water mass characteristics. Locally, however, rapid movements of deep water by swiftly flowing currents now appear probable. The few observations so far available suggest that such currents, like the wind-driven surface currents, are strongest at the western boundaries of the ocean basins, indicating that deep currents are also influenced by the Coriolis force, although they are driven by pressure gradients rather than the wind.

OCEAN MOVEMENTS: WAVES AND TIDES

Currents transporting water from one part of the ocean to another are not the only movements that take place in the ocean. Other movements occur which do *not* involve large-scale displacements of water, but merely the passage of moving patterns of disturbance, called **waves,** on or through the water. In watching surface waves on the sea, it at first appears that the water itself *does* move rapidly in the direction of wave travel. Closer observation of small objects floating in the water, such as a cork or a piece of seaweed, shows that it is merely the *form of the wave*, rather than the water itself, that progresses. As a wave passes, the cork, and the individual water particles in which it floats, move not forward, but in a circular pattern that results in their ending up in approximately the same position as before the wave passed (Figure 4.20). Ocean wave movements therefore differ in a most fundamental way from the current movements we have already considered, for they do not involve large-scale transportation of water from one part of the ocean to another.

There are four principal ocean wave movements, each of which is caused by different dominant forces: the most familiar are *wind waves*, formed at the ocean surface by action of the wind, and *tides*, world-encircling waves caused by the gravitational attraction of the sun and moon. Somewhat less familiar are *seismic sea waves* (also called *tsunamis* and, incorrectly, "tidal waves") caused by earthquakes, volcanic eruptions, or other disturbances of the ocean floor. *Internal waves*, the final type, move beneath the surface of the ocean and are the most difficult to observe and interpret; their exact cause is still unknown.

Figure 4.20 Water motion in wave passage. An object floating in passing waves does not progress with the wave but inscribes an in-place circle (color) with each wave.

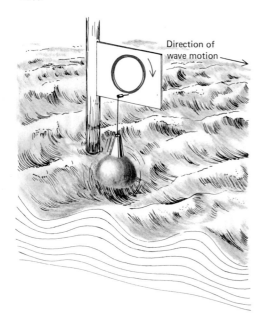

Direction of wave motion

Man
and the Ocean

The waters of the ocean contain enormous quantities of many valuable elements, but the concentrations of most are so low that huge volumes of sea water must be processed to extract them; this has not yet proved generally profitable. The only nonrenewable resources now obtained in quantity from ocean water are its three most abundant dissolved elements: sodium and chlorine, extracted by simple evaporation as common salt, and magnesium, extracted by adding dissolved calcium hydroxide to sea water which causes the magnesium to precipitate as magnesium hydroxide. The oceans are so large that the potential supplies of these elements, and any others that might be extracted in the future, are virtually inexhaustible. The ocean is of greatest significance to man not, however, as a reservoir of minerals, but for two other reasons—as an avenue of transportation and a source of food.

Ocean-going ships have always provided man's most efficient means of large-scale transport and, as a result, the largest and most influential cities throughout human history have generally been seaports. Even today, transportation by air competes successfully with ocean shipping only for objects of relatively small bulk and high value, the most important of which are human travelers! Much of our knowledge of the ocean's currents, waves, and tides has come from studies undertaken primarily to make ocean shipping safer and more efficient. As a corollary, the least-known parts of the ocean remain those that are seldom used as shipping routes.

As a source of food, the ocean is unique in providing man's last major food supply that is gathered from the wild, rather than being bred, nourished, and grown by man himself. Abundant and easily harvested fisheries are not present throughout the ocean, however, but are found only in relatively restricted regions where the patterns of oceanic circulation favor prolific marine life. Most of these occur near the margins of the continents where local current patterns cause nutrients from the deep layer of the ocean to be brought to the surface. Such motions are favored along the eastern and poleward sides of the major gyres, away from the swift water movements of the western boundary currents (see Figure 4.13). The self-renewing supply of fish and shellfish in such regions is large but not inexhaustible, and in some regions regulation is already required to prevent overharvesting which could ultimately eliminate many valuable species.

The final use man makes of the ocean is far less positive—as a receptacle for waste products. Coastal cities normally discharge sewage directly into the sea with little or no treatment; they also dump enormous quantities of solid waste and debris onto the ocean floor. Fortunately, the ocean is so large that these processes have as yet had little apparent effect on the waters and life of the open sea. In many partially enclosed near-shore waters, however, such waste disposal has led to destruction of marine life, including important fishery resources, as well as to pollution levels that restrict recreational swimming and boating. Hopefully, increased awareness of this condition may lead to corrective action before any large-scale damage is done to the overall balance of life in the ocean.

Ocean transport vessels entering New York harbor. (Port of New York Authority)

Herring harvest off the coast of Norway. (Norwegian Information Service)

Solid wastes being hauled to ocean dumping grounds. (J. Paul Kirouac)

Extraction of magnesium from ocean water, Freeport, Texas. From top: ocean water intake; large tanks in which calcium hydroxide is added to the ocean water to precipitate magnesium hydroxide; electrolytic cells which separate magnesium metal from other materials in solution; ingots of magnesium awaiting shipment. (Dow Chemical Co.)

4.9 Wind Waves

The waves produced on the ocean's surface by the action of winds, called **wind waves,** are among the most fascinating and easily observed phenomena of the ocean. Rather surprisingly though, they are still not fully understood. In the simplest case, any regular sequence of moving waves can be described by three properties: the length of the waves or *wavelength,* the *height* of the waves, and the *period* of the waves, which is merely the length of time it takes one complete wave to pass a fixed point (Figure 4.21a). From two of these parameters, wavelength and period, the *velocity* of the waves can easily be computed. Likewise, the ratio of wavelength to wave height determines the *steepness* of the wave (Figure 4.21b). Wind waves typically have wavelengths varying from less than 3 cm ($1\frac{1}{5}$ in.) to more than 600 m (2,000 ft). The maximum steepness is reached when the wavelength is about seven times as great as the wave height. At or below this ratio the wave form becomes unstable and the water plunges from the top to form a "whitecap" or "breaker."

Wavelength, height, and period are useful concepts for describing and discussing waves, but natural wind waves are usually far too complex and irregular to be understood by these properties alone. If you watch waves generated by a strong wind, you will see that it is impossible to follow single crests moving over long distances. Instead, the waves tend to grow and subside in complex patterns that defy simple recognition of wavelength and period. In considering such complex waves it is convenient to look at a completely smooth ocean surface over which a gentle wind has just begun to blow. The first effect is the production of very small surface irregularities less than 3 cm ($1\frac{1}{5}$ in.) across known as **capillary waves** (Figure 4.22). These capillary waves are so small that they are strongly influenced by the surface tension of the water and quickly disappear if the wind subsides. If the wind speed increases and the winds become more turbulent and gusty, much larger and more irregular waves, ranging in length and height from several centimeters to a meter or more, begin to form (Figure 4.22). Unlike small capillary waves, these larger waves will continue to move under the influence of gravity even after the wind subsides; for this reason they are called **gravity waves.** With further increases of wind velocity the larger gravity waves tend, for reasons that are poorly understood, to gain energy at the

Figure 4.21 Wave properties. (a) Wavelength and wave height. (b) The ratio of wavelength to wave height determines the steepness of a wave. When the ratio is less than 7, the crest becomes unstable and plunges to form a "whitecap" or "breaker."

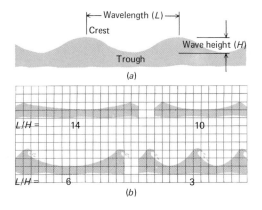

expense of the smaller with the result that the smaller waves are destroyed soon after they appear, whereas the larger waves continue to grow. The maximum height of these larger waves has been observed to depend on three characteristics of the generating wind: its *velocity*; its *duration*, or the length of time it blows; and its *fetch*, which is the distance over which it blows. The height increases as each parameter—wind velocity, duration, and fetch—increases. The maximum observed height of wind waves is rarely greater than 6 m (20 ft), although exceptional storm waves as high as 36 m (120 ft) have been reported (Figure 4.23). Even 6 m (20 ft) waves are unusual, for they require winds blowing at velocities greater than 80 km/hr (50 mi/hr), lasting for several days, and blowing over hundreds of kilometers of open water. It has been estimated that 45 percent of ocean waves are under 1 m (3.3 ft), 35 percent from 2 to 6 m ($6\frac{1}{2}$ to 20 ft) high, and only 10 percent have heights greater than 6 m (20 ft).

All that we have said so far has been based on direct observations of what happens on the ocean surface. Attempts to explain this observed sequence of wind wave development from basic principles of fluid dynamics have so far proved inadequate. Some workers have suggested that energy is transferred from wind to water largely by frictional forces; others have stressed the importance of small-scale pressure changes and turbulence created by gusty winds. In neither case has mathematical analysis yielded an accurate prediction of the sequence of origin and growth seen in natural ocean waves.

There is, however, another mathematical approach to wave analysis that has proved to be fruitful. This approach, devised in the early 1950s by two American oceanographers, G. Neumann and W. J. Pierson, Jr., makes use not of the classical hydrodynamic parameters of pressure, gravity, Coriolis effect, and friction, but, instead, of a statistical "averaging" of the energies of the many sizes and shapes of waves that form under winds of differing velocity, duration, and fetch. Complex natural waves are considered to result from the interaction of many super-

Figure 4.22 Sequence of wind wave formation. (*a*) Small capillary waves; wind velocity 4 km/hr. (*b–f*) Gravity waves at increasing wind velocities: (*b*) 10 km/hr; (*c*) 30 km/hr; (*d*) 45 km/hr; (*e*) 80 km/hr; (*f*) 95 km/hr. (*Photographs courtesy of Dr. F. Krügler, Hamburg*)

(a)

(b)

(c)

(d)

(e)

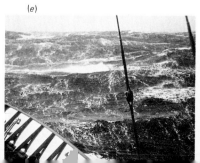

(f)

Figure 4.23 Unusually large waves in a storm off northern California. The ship, a Coast Guard Cutter on a rescue mission, is 100 m (327 ft) long. (U.S. Coast Guard)

imposed simple wave patterns (Figure 4.24). For a given set of wind conditions, a statistical combination of many such simple waves permits a calculation of the resulting distribution of energy in more complex waves of differing sizes. From this energy distribution, the general shape of the sea surface can then be estimated. These calculations agree rather closely with observations on natural waves and have formed the basis for an extremely useful scheme for predicting the maximum wave size to be expected under different conditions of wind velocity, duration, and fetch.

The complex wave patterns that we have been discussing occur in the **generating area,** the region where strong winds produce wave motions (Figure 4.25). Waves may, however, travel far beyond the generating area into calmer regions where large waves are no longer being produced by the wind. In this way, large storm waves often travel across hundreds or thousands of kilometers of open ocean with relatively little loss of energy. In the process of spreading from the generating area, the waves become more regular and are easier to describe by the simple properties of wavelength, height, and period. Such waves are called ocean **swell** to distinguish them from the more complex waves of the generating area which are known as a **sea.**

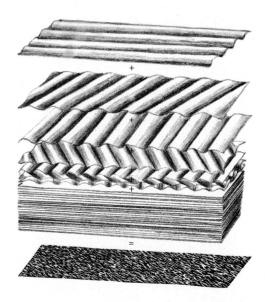

Figure 4.24 *Wave averaging. A series of simple wave forms are statistically "added" to simulate complex natural waves (bottom).*

Figure 4.25 *Development of waves in a "generating area" of strong winds. The waves may move far beyond the generating area; as they do they become more regular and are known as "swell."*

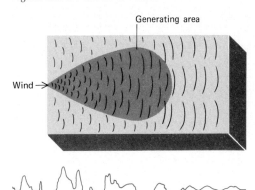

Generating area

Wind →

"Sea" wave profile in generating area

"Swell" wave profile beyond generating area

Waves traveling across the open ocean as swell continue to move with little loss of energy until they reach shallow costal waters where their energy is dissipated by interactions with the ocean floor. There is a simple relationship for calculating the water depth at which surface waves begin to "feel bottom" and lose energy (Figure 4.26). As a wave passes, the motion of water particles extends downward, with decreasing energy, to a depth about half as great as the wave length of the passing wave. Thus a wave 300 m (1,000 ft) long begins to "touch bottom" in water 150 m (500 ft) deep. When this happens the form of the wave begins to change because the bottom restricts the movement of water particles in the wave; the velocity decreases, the wavelength becomes shorter, and the wave height increases. This steepening of the wave continues as the water becomes shallower until the wave reaches an unstable height and "breaks" to form surf which dissipates most of its energy (Figure 4.27). Breaking normally occurs when the water depth is about $1\frac{1}{3}$ times the wave height. This wave dissipation in shallow water transfers energy from the ocean to the solid earth and results in a slow wearing away of coastal rocks.

4.10 Tides

A second fundamental ocean wave motion is seen in the **tides,** world-encircling ocean movements caused by the gravitational attraction of the sun and moon. The rhythmic rise and fall of the tides seen along the coastlines of the earth are the most regular and, next to surface waves, the most easily observed of all ocean movements. Since at least the seventeenth century, it has been understood that these movements are related to astronomical forces because regular tidal changes were observed to accompany the changing phases of the moon. Even though it has long been clear that tides are caused by the changing positions of the moon and sun, the exact manner in which the gravitational forces produce varying tidal patterns is still not completely understood.

To visualize the origin of ocean tides it is helpful to think first of the effect of the moon alone on an idealized earth completely covered by oceans. On such an earth, ocean water should bulge slightly outward in the direction of the moon because of the moon's gravitational attraction. At first it would seem that only a single bulge would exist directly under the moon and, indeed, this would be the case if gravity were the only force involved. The mutual

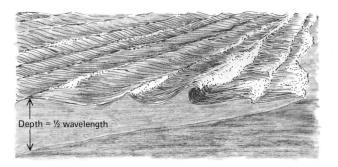

Depth = ½ wavelength

Figure 4.26 *Wave motions in shallow water. At water depths less than ½ wavelength, the wave motion intersects the ocean floor causing the wave to move more slowly and, ultimately, to steepen and "break."*

gravitational attractions of the earth and moon are, however, almost completely balanced by *centrifugal* forces caused by slight motions of the earth around the mutual center of gravity of the earth-moon system. This centrifugal motion creates a force in a direction *opposite* to the moon, and results in another tidal bulge on the side of the earth *farthest away* from the moon (Figure 4.28). The result of this balance of centrifugal and gravitational forces is to permanently deform the ocean surface into a slightly ovate (egg- or football-like) shape, with the long axis extending in the direction of the moon. This deformation can be thought of as a world-encircling wave with only two "crests," one at each tidal bulge, and two "troughs" in between. The forces involved in producing the tides are so small that the height of these waves is usually less than 1 m (3.3 ft) in the open ocean, but their wavelength is extraordinarily long for they extend halfway around the earth from one bulge to the other.

Figure 4.27 *Waves steepening and breaking in shallow water along the central California coast. (Josef Muench)*

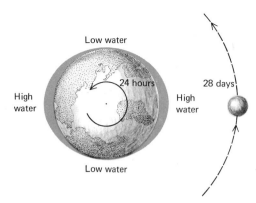

High water — Low water — 24 hours — 28 days — High water — Low water

Figure 4.28 Origin of the tides. The gravitational interactions of the earth and moon produce a low, world-encircling wave with two crests (high water) and two troughs (low water). The earth rotates "under" these waves each 24 hours, producing two intervals of high water and two of low water at each point on the ocean surface.

Figure 4.29 Spring and neap tides. Because of the gravitational attraction of the sun, tide ranges are greatest when the moon and sun are aligned (spring tides) and least when they are opposed (neap tides).

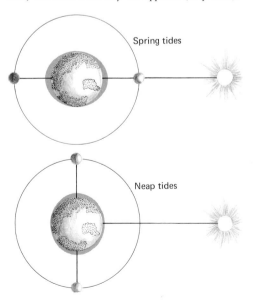

Spring tides

Neap tides

Daily tidal changes result from the fact that the tidal bulges remain in a relatively fixed position, that is, aligned with the moon, as the earth rotates on its axis every 24 hours. Thus, points on the earth's surface tend to rotate under both tidal bulges, and their intervening "troughs," every 24 hr. This is the reason that tidal records typically show two periods of high water and two intervening periods of low water each day. The tidal bulges themselves also move, however, as the moon rotates around the earth every 28 days. The effect of this movement is to make the tides, like the time of moonrise, come about 50 min later each day, thus completing one cycle and occurring at the same time once each 28 days (Figure 4.28).

This simple explanation of tides has neglected many complicating factors, the most important of which is the sun, whose gravitational attraction causes about *half* as much tidal deformation as does the moon. The principal effect of the sun is to cause tides to have their greatest range twice each month at **spring tides** (near the times of full and new moon) when the sun and moon are aligned and their forces act together; conversely, tidal ranges are lowest at **neap tides** (near the first and third quarter of the moon) when the solar and lunar forces are in opposition and act at right angles to each other (Figure 4.29).

There are still other factors that complicate natural tidal patterns. The earth is not, of course, completely covered by water, and thus the movement of the tidal "bulges" around the earth each day is strongly influenced by the shape and interconnections of the ocean basins. Furthermore, the extremely long wavelength of the tides causes them to "feel bottom" and behave as shallow water waves even in the deepest ocean basins. As a result, free movement of the tidal deformations are slowed down by frictional drag on the sea floor. Because of these and other complications, it has so far proved impossible to predict local tidal patterns from hydrodynamic calculations based on the forces of gravity, friction, pressure, and the Coriolis effect. Instead, tidal predictions are based on long observation of local patterns which tend to be repeated in regular ways along the coastlines of the earth.

Tidal observations taken over many years at thousands of stations around the world have shown that many regions, including most of the Atlantic Ocean, show a regular twice-daily tidal pattern with high and low tides of equal magnitude, just as predicted by our simple analysis. Such tides are called **semidiurnal** (Figure 4.30). In other areas, however, various complicating factors lead to two

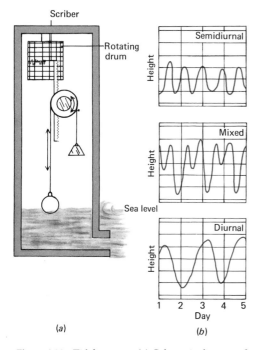

Rotating drum

Sea level

Semidiurnal

Mixed

Diurnal

1　2　3　4　5
Day

(a)

(b)

Figure 4.30 Tidal patterns. (a) Schematic diagram of a recording tide gauge. (b) The three principal types of tides, as recorded on a tide gauge. Semidiurnal tides have two high and two low waters of approximately equal magnitude each day. In mixed tides, some of the highs and lows are unequal. Diurnal tides show a single high and low each day.

additional tidal patterns. Most common are **mixed tides,** having two highs and two lows of unequal magnitude each day. Such tides dominate most of the shores of the Pacific Ocean. In a few places, such as the Gulf of Mexico and Southeast Asia, **diurnal tides,** with only one high and one low water per day, occur.

In the open ocean, the rise and fall of water with the changing tides is difficult to measure, but tidal observations on isolated oceanic islands suggest that the water seldom varies in height by more than 1 m (3.3 ft). In contrast, along the coasts of the continents much greater tidal ranges may occur when water movements become concentrated and focused by the shape of the land and by interaction with the shallow sea floor. The maximum tidal ranges so produced are found on the shores of the North Atlantic, particularly in Nova Scotia and around the English Channel, where the maximum range is about 15 m (50 ft) (Figure 4.31). On most shores, however, the maximum tidal range is less than 3 m (10 ft).

4.11　Seismic Sea Waves

All of the ocean movements we have discussed so far have been caused by either atmospheric forces (currents and wind waves) or external gravity forces (tides). Another ocean wave motion occurs that derives energy directly from movements taking place in the rocks of the solid earth. Such waves are commonly called **seismic sea waves,** but are also often referred to by their Japanese name, **tsunamis,** or by the misnomer "tidal waves"; they have no relation to tide-producing forces. Seismic sea waves do, however, resemble tides in having very low heights and very long wavelengths. In the open ocean they commonly have wavelengths of hundreds of *kilometers* but heights of less than 3 m (10 ft), so low that the crests pass unnoticed by ships in midocean. Unlike tides, however, the height of seismic sea waves may be greatly intensified as the waves move into shallow water. As with all surface waves, the effect of shallowing water is to increase the seismic seawave's height and reduce its wavelength. When they encounter a gradual shallowing extending over many kilometers, seismic sea waves, because of their long wavelength, may build up to heights as great as 30 m (100 ft) without exceeding the $\frac{1}{7}$ height-length ratio at which waves become unstable and break. Needless to say, such huge waves can lead to catastrophic destruction when they run ashore in populated areas where the height of the normal wind waves is much lower (see also p. 74).

Figure 4.31 *High and low tide in Nova Scotia, an area of extreme tide ranges. The wall in the right photo is about 8 m (24 ft) high. (G. Blouin, Information Canada Photothèque)*

Figure 4.32 *Schematic diagram of an internal wave. Such waves form at the boundary between deep water layers of differing temperatures and densities and have little or no reflection at the surface. Most have been discovered by noting systematic changes in temperature on continuously recording thermometers. (After Neumann and Pierson,* Principles of Physical Oceanography, *1966)*

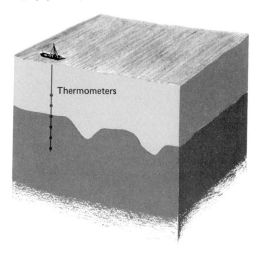

The cause of seismic sea waves is known to be related to crustal movements because individual sets of waves (there is never only a single wave but usually a series of about a dozen) can be observed to originate and spread from the sites of individual underwater earthquakes. It is still uncertain, however, exactly how the earthquake sets the water in motion. Because volcanic eruptions and submarine landslides often accompany earthquakes, some geologists feel that these may be the direct cause of the sea wave, rather than the movements of the earthquake itself. Large seismic sea waves occur almost exclusively in the Pacific Ocean for, as we have seen, most of the earth's earthquakes and active volcanoes are found in and around the Pacific.

4.12 Internal Waves

The final major ocean wave motion is the least understood and the most difficult to observe. These are **internal waves** moving *beneath* the surface of the sea. Such waves are produced along the boundary separating two fluids of differing density that move past one another. The surface wind waves we have already discussed are the result of such movements: a less dense fluid (air) moves over a more dense fluid (sea water) with the production of complex wave patterns. Similar wave patterns are also produced at the boundaries between the ocean *water* layers which, we have seen, often differ in density and have complex patterns of current movement. In particular, internal waves are especially prevalent at the boundaries of the thermocline layer which marks the major density discontinuity in the waters of the ocean.

Unfortunately, no really satisfactory direct means has yet been devised for recording and measuring deep internal waves; instead what little is known about them comes from indirect observations of two sorts. First, submarines often experience large wave movements when they are submerged far below the depths affected by surface wind waves. Second, continuous records of temperatures taken at depth by recording thermometers commonly show a rhythmic temperature rise and fall which can only be explained by internal waves moving along the boundary between layers of differing temperatures and densities (Figure 4.32). These limited observations suggest that internal waves occur throughout the oceans and may, indeed, be the largest of all ocean wave motions. Internal wave heights as great as 100 m (328 ft), almost 10 times that of ordinary surface waves, have been recorded.

SUMMARY OUTLINE

Ocean Water

4.1 *Chemical composition:* ocean water is a uniform solution of many dissolved elements, of which sodium, chlorine, magnesium, and sulfur are dominant; the concentration of dissolved elements defines the salinity of ocean water, which is normally 35 ‰ (3.5 percent).

4.2 *Physical properties:* density is the most important; it increases with increasing salinity and pressure and with decreasing temperature.

Ocean Structure

4.3 *Layering:* the ocean shows three principal vertical layers of differing density—a thin, warm, mixed layer at the surface; a thicker thermocline layer of decreasing temperature and increasing density below; and a still thicker mass of cold and very dense deep water which includes the bulk of the ocean.

4.4 *Water masses:* are smaller-scale water bodies characterized by distinctive temperatures, salinities, or oxygen content; they are especially useful for tracing the source of waters making up the deep layer.

4.5 *Ice in the ocean:* the dissolved matter in ocean water depresses the freezing point and prevents complete freezing; sea ice, formed from frozen sea water, is always relatively thin, even in polar regions.

Ocean Movements: Currents

4.6 *Patterns of surface currents:* currents are best known in the surface layer where they move in large, circular gyres; the swiftest motions occur in the assymetrical, western portions of the gyres.

4.7 *Causes of surface currents:* the similarity to wind patterns, coupled with fluid dynamic analysis of the four principal forces acting on the ocean (gravity, pressure, friction, and the Coriolis force), indicate that surface currents are driven primarily by atmospheric motions.

**Ocean Movements:
Waves and Tides**

4.8   *Deep circulation:* is poorly known because deep currents are far more difficult to observe than surface currents; most deep motions may be caused by temperature- and salinity-induced density differences.

4.9   *Wind waves:* are moving disturbances formed on the ocean surface by wind action; they increase in size as wind velocity, duration, and fetch increase and may move far beyond their generating areas to produce swell.

4.10   *Tides:* are world-encircling waves caused by the gravitational attraction of moon and sun; points on the ocean surface normally have two high and two low tides daily but this pattern is complicated by the shapes and limited interconnections of the ocean basins.

4.11   *Seismic sea waves:* large waves caused by undersea earthquakes.

4.12   *Internal waves:* form at density boundaries between ocean layers but are difficult to observe and are still poorly known.

ADDITIONAL READINGS

*Bascom, W.: *Waves and Beaches,* Doubleday & Company, Inc., Garden City, N.Y., 1964. A popular and entertaining introduction.

Ericson, D. B., and G. Wollin: *The Everchanging Sea,* Alfred A. Knopf, Inc., New York, 1967. A well-written popular survey of the study of the ocean.

Groen, P.: *The Waters of the Sea,* D. Van Nostrand Co., Ltd., London, 1967. A standard intermediate text covering ocean composition, structure, and motions.

*Gross, M. G.: *Oceanography,* Charles E. Merrill Publishing Co., Columbus, Ohio, 1971. A brief introductory survey.

Gross, M. G.: *Oceanography, A View of the Earth,* Prentice-Hall, Inc., Englewood Cliffs, N.J., 1972. A comprehensive and up-to-date intermediate text.

*Moore, J. R. (ed.): *Oceanography, Readings from Scientific American,* W. H. Freeman and Co., San Francisco, 1971. Many excellent popular articles.

Neumann, G., and W. J. Pierson, Jr.: *Principles of Physical Oceanography,* Prentice-Hall, Inc., Englewood Cliffs, N.J., 1966. A standard advanced text emphasizing mathematical formulation of ocean motions.

*Skinner, B. J., and K. K. Turekian: *Man and the Ocean,* Prentice-Hall, Inc., Englewood Cliffs, N.J., 1973. A good introduction to man's many uses of the ocean.

*Turekian, K. K.: *Oceans,* Prentice-Hall, Inc., Englewood Cliffs, N.J., 1968. An authoritative and readable introduction, emphasizing ocean chemistry.

Weyl, P. K.: *Oceanography, An Introduction to the Marine Environment,* John Wiley & Sons, Inc., New York, 1970. A clearly written intermediate text.

* Available in paperback.

5

The Atmosphere

Thin clouds high in the atmosphere, illuminated after sunset. (Benson Fogle, NCAR)

The gases of the earth's atmosphere are held in place by gravitational attraction, which concentrates them in a thin spherical shell near the earth's surface. The gases become progressively less dense away from the earth's surface until, at a height of several thousand kilometers, they pass into the thin, rarified gases of interplanetary space. By weight, about 99 percent of the total atmosphere occurs in the first 40 km (25 mi) upward, yet its thin, outermost gases extend so far into space that the total *volume* of the atmosphere is far larger than that of the solid earth and ocean combined. Its total *weight,* however, is far smaller.

Not only are the bulk of atmospheric gases concentrated in the lowermost atmosphere, but all of the earth's familiar patterns of changing weather take place in this dense layer adjacent to the surface of the earth. In addition, the lower atmosphere is far better understood than the upper atmosphere for it can be relatively easily observed during aircraft flights and with instruments carried aloft by balloons. Indeed, until the development of instrument-carrying rockets and satellites in the 1950s, direct observations of the atmosphere were limited to heights of about 32 km (20 mi), the upper operational limit of instrument-carrying balloons and jet aircraft. With the expansion of rocket and satellite technology, there is now an increasing amount of information being gathered about the upper parts of the atmosphere. In this chapter we shall survey the composition and structure of *all* the atmosphere and review recent discoveries concerning the upper atmosphere. Weather-producing motions of the lower atmosphere, many of which are controlled by interactions with the land and ocean surface, will be discussed in Chapter 6.

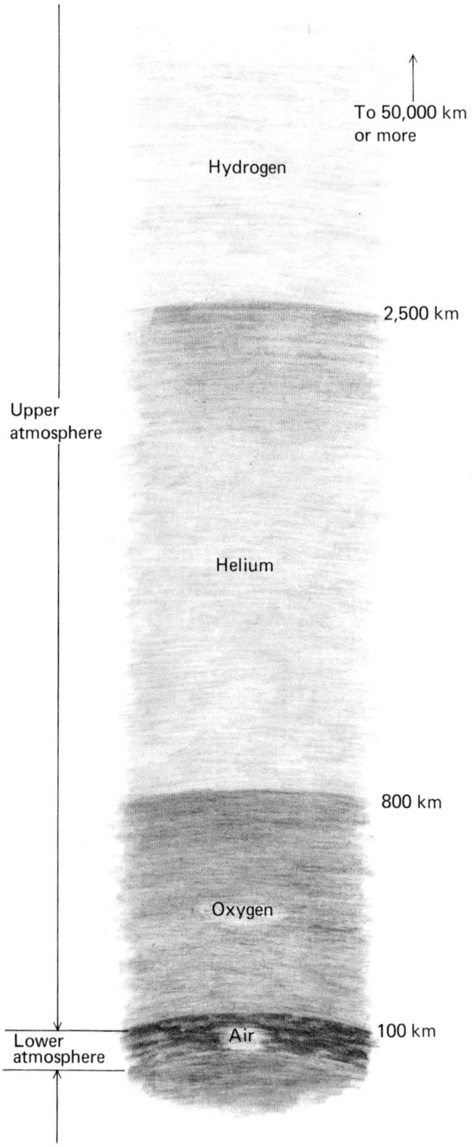

To 50,000 km
or more

Hydrogen

2,500 km

Upper
atmosphere

Helium

800 km

Oxygen

Lower
atmosphere

Air

100 km

Figure 5.1 Composition of the atmosphere. Air, a gaseous mixture dominated by nitrogen and oxygen, occurs in the lower atmosphere where solar heating causes turbulent mixing. Above this zone of turbulence, the thin gases of the upper atmosphere are stratified by weight. First oxygen, and then small quantities of lighter helium and hydrogen dominate progressively higher layers. Most atmospheric nitrogen, the heaviest abundant gas, is concentrated in the lower atmosphere.

We have mentioned the "lower" and "upper" atmosphere, but have not yet said precisely what separates the two. The principal difference is that the lower atmosphere is constantly being mixed and stirred by turbulent winds caused by the interaction of solar energy and the dense atmospheric gases near the earth's surface (Figure 5.1). In contrast, above a height of about 100 km (60 mi), turbulent mixing becomes minor and the gaseous particles in this much thinner upper atmosphere move principally by slow diffusion. This difference in turbulence is reflected by differences in the composition of the lower and upper atmospheres. The mixing of the lower atmosphere causes it to have a relatively uniform and constant composition, just as a similar mixing leads to the constant proportions of dissolved elements in the ocean. The upper atmosphere, however, does *not* have a uniform composition. Instead, gravitational forces lead to a general density stratification; the heavier gases move, by diffusion, to the lower levels and the lighter gases to higher levels. Because of this fundamental difference, it is useful to consider separately the compositions of the lower and upper regions of the atmosphere.

5.1 Composition of the Lower Atmosphere

The earth's atmosphere is composed of four principal chemical elements which may occur alone or combined in various compounds. These are: *argon, carbon, nitrogen,* and *oxygen*. In addition to these four principal elements, smaller quantities of *hydrogen* occur, as do traces of the "rare gas" elements, *helium, neon, krypton, xenon,* and *radon*, which, like argon, are extremely stable and never react to form natural compounds.

Because of the dominance of mixing in the lower atmosphere versus diffusion in the upper atmosphere, the two regions are characterized by differences in the concentration and mode of occurrence of these constituents. The bulk of the lower atmosphere is composed of only two elements, nitrogen (N_2)[1] and oxygen (O_2), which occur

---

[1]When gaseous elements such as hydrogen, nitrogen, or oxygen occur in high concentrations, the atoms' electrical properties cause them to unite in pairs to form molecules. These gaseous molecules made up of two atoms of a single element are indicated by the subscript "2." Such molecular gases dominate the lower atmosphere, but single, *uncombined* atoms occur in the upper atmosphere where concentrations are much lower.

mixed together, but uncombined with other elements, to make **air,** the gaseous mixture of the lower atmosphere (Figure 5.2). Nitrogen is by far the more abundant of the two, accounting for 78 percent of the volume of clean, dry air, while oxygen makes up only 21 percent. The remaining 1 percent is mostly stable argon (0.93 percent), but at least 15 other gases are present in trace amounts. By far the most important of these minor gases are carbon dioxide (CO_2) and ozone (O_3). So far we have been talking about *dry* air, but air may, as becomes all too obvious on a humid summer day, contain from a trace to as much as 4 percent of vaporized water (H_2O), which is another very important minor constituent.

Nitrogen Nitrogen, the dominant element in the lower atmosphere, is among the rarer elements in the rocks of the solid earth and waters of the ocean, occurring in rocks only as relatively uncommon nitrate minerals, and in the oceans only as relatively small quantities of dissolved nitrate (Figure 5.2*b*). Although nitrogen is a lesser component of crustal rocks and the oceans, it is a major constituent not only of the atmosphere, but also of the animals and plants of the living world, for it is a principal element in *proteins,* the basic structural compounds of all living organisms. Most animals and plants cannot directly utilize atmospheric nitrogen to manufacture these proteins. On the other hand, the small quantities of nitrogen found in soils and dissolved in the sea would be quickly depleted by organisms if there were no means of tapping the tremendous nitrogen supply of the atmosphere. This is done primarily by certain microscopic bacteria which have the

Figure 5.2 *Composition of air. (a) The most abundant elements in dry air. Uncombined nitrogen, oxygen, and argon dominate; carbon mostly occurs combined with oxygen as gaseous carbon dioxide. Air also contains varying amounts of water vapor, up to a maximum of about 4 percent. (b) The relative abundance of the dominant elements of air, ocean water, and crustal rocks. Of the principal elements of the atmosphere, only oxygen is also abundant in the ocean and solid earth.*

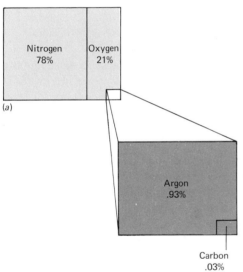

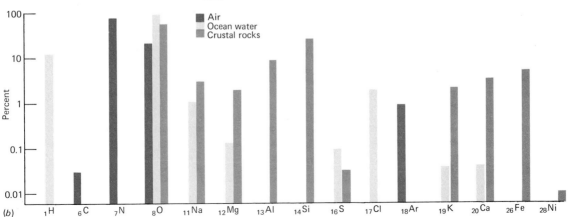

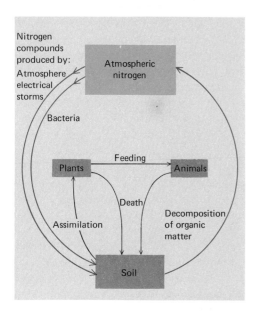

Nitrogen compounds produced by:
Atmosphere electrical storms

Bacteria

Figure 5.3 Interactions of atmospheric nitrogen and the living world. All of the nitrogen in the atmosphere cycles through plants and animals about once every 100 million years.

Figure 5.4 Interactions of atmospheric oxygen and the living world. All of the oxygen of the atmosphere cycles through plants and animals about once every 3,000 years.

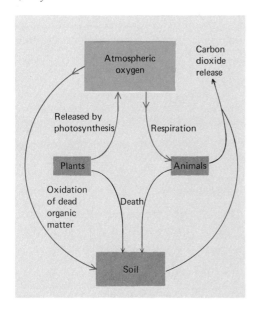

property of converting atmospheric nitrogen into water-soluble nitrate atom groups, the form in which nitrogen can be used by plants and animals for the manufacture of proteins. The nitrogen ultimately reenters the atmosphere as dead animals and plants are decomposed by other nitrogen-*releasing* bacteria (Figure 5.3). It has been estimated that it takes a very long time, about 100 million years, for all of the nitrogen in the atmosphere to pass once through this cycle.

Oxygen The other major constituent of the lower atmosphere, oxygen, is already familiar as the most abundant element in both crustal rocks and ocean water (Figure 5.2b). Unlike nitrogen, which is a relatively unreactive element, oxygen readily combines with almost all other elements to form compounds. Uncombined oxygen does not occur in the solid earth; it is found in the oceans only in relatively small amounts that dissolve in ocean water as a result of contact with the atmosphere. The immediate source of most of the uncombined gaseous oxygen of the atmosphere is therefore neither the solid earth nor the ocean. Instead, it is believed to have accumulated primarily as a result of photosynthesis by green plants. In this process sunlight is utilized to break down water into hydrogen and oxygen. The hydrogen, along with atmospheric carbon dioxide, is then used to produce complex organic compounds while much of the oxygen is given off into the atmosphere as a waste product. This free oxygen, in turn, is utilized by animals as an energy source, being ultimately released into the atmosphere combined with carbon as carbon dioxide which is taken up by plants to begin the cycle again (Figures 5.4, 5.6). It has been estimated that it takes only 3,000 years for all the oxygen in the atmosphere to pass once through this cycle. Thus, the free oxygen in the atmosphere, like the nitrogen, is closely interrelated with the life processes of organisms.

In the lower atmosphere most of the oxygen occurs in the two-atom molecular form O_2. In the upper regions of the lower atmosphere, between about 20 and 35 km (12 and 20 mi) high, small amounts of a three-atom molecule of oxygen called **ozone** (O_3) also occur. These molecules are produced by the action of the sun's ultraviolet radiation which is more intense at high altitudes than at the earth's surface. This radiation causes many O_2 molecules to separate or *dissociate* into single atoms of oxygen. Most of these quickly recombine to form O_2, but a very few atoms, about one in 10 million, attach themselves to

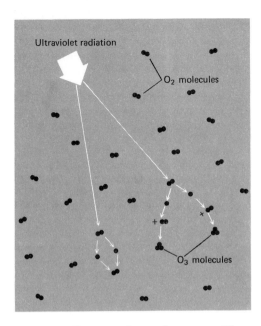

Figure 5.5 *The origin of atmospheric ozone. Ultraviolet radiation from the sun causes some O_2 molecules to split into separate oxygen atoms. Most of the atoms recombine to form O_2 again (left), but a few become attached to an O_2 molecule to form a molecule of ozone or O_3 (right).*

preexisting O_2 molecules to form ozone (O_3) molecules (Figure 5.5). These extremely small quantities of ozone in the higher parts of the lower atmosphere play, as we shall see later, a large role in the thermal structure and heat balance of the atmosphere. Because ozone is produced from O_2 primarily at higher altitudes where solar radiation is more intense, ozone distribution is an exception to the relatively uniform composition of the mixed lower atmosphere.

Carbon Dioxide   Carbon dioxide is among the most important minor constituents of the lower atmosphere, for, like ozone, it plays a fundamental role in atmospheric heat balance and is a major controlling factor in the earth's patterns of weather and climate. Like nitrogen and oxygen, atmospheric carbon dioxide is also intimately interrelated with the processes of living organisms, for it is the major source of the carbon which makes up a primary component of all life. Green plants directly use atmospheric carbon dioxide to synthesize more complex carbon compounds which, in turn, are the basic food for animals and nongreen plants. The carbon is ultimately returned to the atmosphere as a waste product of animal and plant respiration or decomposition just as free oxygen is contributed by green plant photosynthesis (Figure 5.6). The cycling of carbon dioxide through living organisms is far shorter than that of either nitrogen or oxygen; it has been estimated that it requires only 35 years for the relatively small quantity in the atmosphere to pass once through this cycle.

Water Vapor   Water vapor in the lower atmosphere is another extremely important minor constituent. The principal source of water vapor is the ocean; for this reason water vapor, like ozone, is not uniformly distributed in the lower atmosphere. Instead it tends to be concentrated near the oceans, that is, at lower altitudes and over large oceanic areas. The distribution of water vapor in the lower atmosphere is a primary factor in determining the earth's patterns of weather and climate, and will be a principal subject of Chapters 6 and 7.

Aerosols   In addition to gases, the lower atmosphere also contains varying amounts of tiny liquid and solid particles which are known, collectively, as **aerosols.** These include tiny particles of rocks, minerals, soil, salts from the sea, and also particles of biologic origin, such as tiny spores, pollen, and bacteria. Although they make up an insignifi-

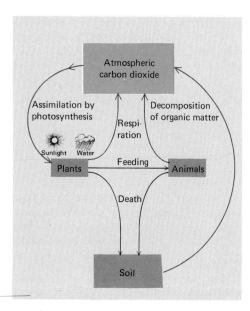

Figure 5.6 *Interactions of atmospheric carbon dioxide and the living world. All of the carbon dioxide of the atmosphere cycles through plants and animals about once every 35 years.*

cant fraction of the total volume of the lower atmosphere, these solid and liquid particles have a large influence on the earth's weather for they provide nuclei around which gaseous water vapor condenses to form fog, clouds, rain, and snow.

5.2 Composition of the Upper Atmosphere

The composition of the gases making up the upper atmosphere is still poorly understood by comparison with the lower atmosphere. Rather surprisingly, the *higher* parts of the upper atmosphere are known best, for this is the region in which artificial earth satellites normally orbit. Because of their extremely high velocities, satellites can orbit effectively only in the thin atmosphere at altitudes above 160 km (100 mi). At lower altitudes, the drag produced by the denser atmosphere causes them to rapidly decelerate and fall back to earth. The region between about 30 and 160 km (18 and 100 mi) high is the most poorly known because it lies above the range of balloons and aircraft but below the effective orbits of satellites. This critical region includes over half of the lower atmosphere as well as the transition zone between the lower and upper atmosphere and the lowermost 60 km (35 mi) of the upper atmosphere (Figure 5.1). In this important zone measurements must be made from rockets carrying instruments designed to analyze and sample the atmosphere as they ascend and descend. Unfortunately such rocket sampling presents many difficulties. To attain the desired height they must move at speeds that cause them to pass through the critical region in a very short time, usually less than 30 sec. Accurate observations and samples are very difficult to collect over such a short time from a rapidly moving rocket. The expense of such rockets, and the relative difficulty of recovering them, create additional problems. Furthermore, many of the elements and compounds of the upper atmosphere become unstable when returned to the earth's surface, and thus must be analyzed by instruments where they occur. Collecting such constituents by *any* means is ineffective. In spite of these difficulties, however, enough observations have been made to indicate the general composition and structure of the upper atmosphere.

As we have noted, at altitudes higher than about 100 km (60 mi) turbulent mixing of the atmosphere ceases to be a dominant process and the principal atmospheric elements become separated by their relative weights. In general, there appear to be three layers in the upper atmosphere, each dominated by progressively lighter elements (Figure 5.1).

The lowest layer, which extends from about 100 to 800 km (60 to 500 mi), is dominated by oxygen, although considerable nitrogen also occurs in its lowermost part where there is a zone of transition to the underlying turbulent air. Unlike the oxygen at lower levels, most of the oxygen in this upper zone occurs *not* as O_2 molecules but as single atoms of oxygen (O). These are produced by the same process that produces ozone in the lower atmosphere, that is, by solar energy which breaks the electrical charges holding oxygen atoms together as the O_2 molecule. When this process takes place in the lower atmosphere, the oxygen concentration is so high that most of the single oxygen atoms rapidly collide with each other and recombine to form O_2 or, much less frequently, O_3 (ozone) molecules. In contrast, at altitudes above 100 km (60 mi) the atmosphere is so thin that individual atoms may move great distances, even thousands of kilometers, without striking another particle and recombining. For this reason atomic oxygen, rather than molecular oxygen, occurs in the thin upper atmosphere. Unlike oxygen, nitrogen molecules (N_2) are not easily dissociated into nitrogen atoms by the sun's energy. Thus *atomic* nitrogen is a relatively rare constituent throughout the atmosphere. Because nitrogen and oxygen molecules (N_2 and O_2) are made up of two atoms, they are much heavier than single oxygen atoms and are concentrated in the lower atmosphere below the zone of lighter atomic oxygen.

Two additional gaseous layers occur above the zone of atomic oxygen. The lower is composed principally of widely spaced atoms of helium, the next to the lightest element, and the upper of atoms of hydrogen, the lightest element (Figure 5.1). Both of these elements occur in very small quantities in the lower atmosphere. Hydrogen is produced there primarily by solar energy which breaks down a few water molecules to release free oxygen and hydrogen. Helium is formed in the decay of various radioactive minerals of the solid earth, from which it escapes into the lower atmosphere. These two gases are so light that they tend to move steadily upward through the denser gases of the lower atmosphere and oxygen layer, and thus to become concentrated in the outermost atmosphere. The thickness of these outer zones is extremely variable because the thin gases expand greatly when heated by the sun, and contract when they are in the earth's cool shadow. As an average, the helium layer extends from about 800 to 2,500 km (500 to 1,500 mi) high, and hydrogen predominates above 2,500 km. The hydrogen layer is similar in composition to interplanetary space which also contains widely separated hydrogen particles. Indeed, there is a continuous interchange of particles at the boundary of the atmosphere as rapidly moving hydrogen particles escape the earth's gravitational field and interplanetary particles are trapped by it. Hydrogen is the only gas light enough to escape readily from the earth's gravitational field.

In addition to the major constituents (O_2, N_2, O, He, and H), the upper atmosphere, like the lower, also contains many minor components. A few nitrogen molecules are dissociated by solar energy to give atomic nitrogen (N) or nitrous oxide (NO), which is formed by combination of atomic nitrogen and atomic oxygen. Small amounts of carbon are also present in the upper atmosphere. Some occurs as carbon dioxide, just as in the lower atmosphere, but carbon in the upper atmosphere also combines with single oxygen atoms to give carbon monoxide (CO) and with hydrogen to give methane gas (CH_4). As in the lower atmosphere, ozone (O_3) also occurs in small amounts in the upper atmosphere, as does water vapor (H_2O).

5.3   Atmospheric Density and Pressure

In discussing atmospheric composition we have referred to the "dense" and "thick" gases of the lower atmosphere versus the "thin" gases of the upper atmosphere. The general upward decrease in atmospheric pressure and density is, along with composition, among the most fundamental properties of the atmosphere and it will be useful to consider its nature and causes in more detail before turning to other aspects of atmospheric structure.

In Chapter 4 we noted that the weight or density of ocean water depends on three primary factors: composition (which, in the ocean, is expressed as salinity), temperature, and pressure. The same three factors control the density of the atmosphere, but their relative importance is much different. Water, we have noted, is a very incompressible fluid; that is, great increases in pressure cause a given amount of water to reduce only slightly in volume and thus to increase only slightly in density. The result is that the pressure of overlying water has a relatively small effect on ocean density, which is controlled, instead, largely by temperature and composition (salinity).

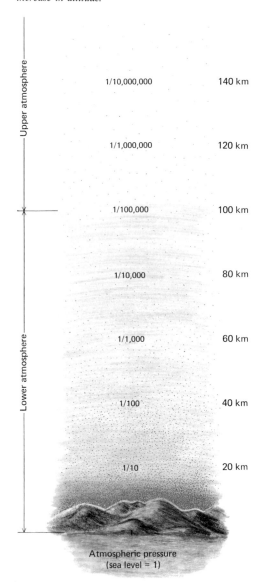

Figure 5.7 Decrease in atmospheric pressure (and density) away from the earth's surface. The pressure decreases by a factor of 10 for each 20-km (12-mi) increase in altitude.

Upper atmosphere

1/10,000,000	140 km
1/1,000,000	120 km
1/100,000	100 km
1/10,000	80 km

Lower atmosphere

1/1,000	60 km
1/100	40 km
1/10	20 km

Atmospheric pressure
(sea level = 1)

Gases, on the other hand, are extraordinarily compressible; that is, small changes in pressure lead to large changes in volume and density. This compressibility is so great that for unconfined gases, such as those of the atmosphere, pressure far exceeds temperature and composition as a primary factor controlling density. Indeed, the density is normally *directly proportional to the pressure.*

The principal factor determining atmospheric pressure, in turn, is the earth's gravity, for the readily compressible gases expand or contract until the weight of the overlying gas *just balances* the force of gravity at that altitude. Thus atmospheric pressures, and densities, decrease rapidly upward. Stated another way, most of the gaseous atoms and molecules of the atmosphere are compressed into a thin zone near the earth's surface by the force of the earth's gravitational attraction. If we ignore, for a moment, any complicating effects caused by temperature changes, then the pressure and density of normal air decreases by a factor of 10 with each 20-km (12-mi) increase in altitude. Thus pressures 20 km high are only $\frac{1}{10}$ as great as at the surface; at 40 km they are $\frac{1}{100}$ as great, at 60 km, $\frac{1}{1000}$, etc. (Figure 5.7). This is the reason that passengers in aircraft flying at altitudes above about 5 km, and mountain climbers in ranges such as the Himalayas which extend upward to heights of almost 10 km, require supplementary oxygen. At altitudes of only 7 km the atmospheric pressure and density, and thus the amount of oxygen available for breathing, are already reduced below the level required to sustain human life.

Although pressure effects caused by gravity are the primary control on atmospheric density, temperature also plays an important local role. As with liquids and solids, high temperatures *decrease* the density of gases by causing an expansion and volume increase (in gases which can move freely, this expansion also decreases the pressure); low temperatures, on the other hand, *increase* the density (and pressure) by contracting the volume. Winds in the lower atmosphere, for example, are caused by temperature differences which locally expand and contract the gases of the air.

Finally, compositional differences have a far less significant effect than pressure and temperature on atmospheric densities. There are, of course, no major compositional differences in the mixed lower atmosphere. The gravitational separation of gases in the upper atmosphere, with the heavier gases below and the lighter above, causes a small density decrease with altitude, but pressures are already so low at upper atmosphere altitudes that the decrease is minor compared with that caused by gravity in the lower atmosphere.

ATMOSPHERIC TEMPERATURES

5.4   Energy in the Atmosphere

High temperatures within the solid earth, we have seen, result largely from heating caused either by the decay of radioactive elements or by the retention, deep within the interior, of some of the original heat energy produced when the earth first consolidated from cosmic dust. In contrast to this *internal energy*, the energy of the fluid earth, including both the ocean and atmosphere, comes ultimately from an *external* source—the sun.

Most solar energy reaches the earth in the form of **electromagnetic radiation,** moving waves of combined electrical and magnetic fields which are familiar because visible light is made up of such waves, as are invisible X rays, gamma rays, and the waves used in radio and television transmission (Figure 5.8). Like all wave motions, electromagnetic waves have characteristic wavelengths; in the case of electromagnetic waves these vary from radio waves with lengths of hundreds of meters to gamma rays less than one *billionth* of a centimeter long. The energy of electromagnetic waves is *inversely* proportional to the wavelength; that is, the shorter waves contain vastly more energy than do the longer. The sun radiates electro-

Figure 5.8   Solar energy output and atmospheric absorption. Solar energy is concentrated in the ultraviolet, visible light, and infrared portions of the spectrum of electromagnetic radiation. Most of the ultraviolet, as well as some of the visible and infrared radiation, are absorbed by atmospheric gases and do not reach the earth's surface.

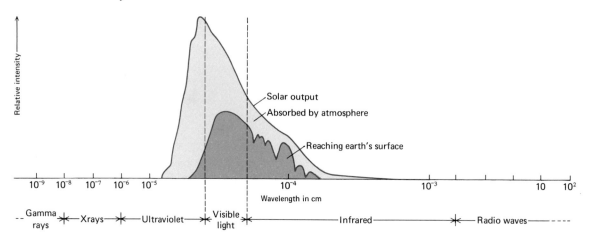

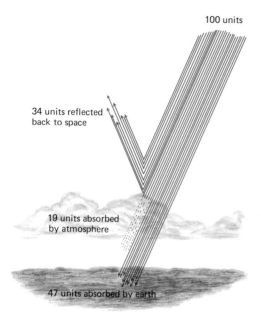

100 units

34 units reflected
back to space

19 units absorbed
by atmosphere

47 units absorbed by earth

Figure 5.9 *The fate of solar energy reaching the earth's atmosphere.*

magnetic waves with a wide spectrum of wavelengths, but most of its energy is concentrated in the relatively narrow range which includes visible light and invisible *ultraviolet* and *infrared* radiation, which are adjacent to visible light in the total electromagnetic spectrum (Figure 5.8). Because most solar energy is concentrated in these wavelengths, they are the principal source of atmospheric energy.

Solar electromagnetic energy interacts with the earth's atmosphere in a complex pattern that is determined both by the wavelength of the radiation and by the composition and concentration of atmospheric gases. In general, two things may happen to radiation as it passes through the atmosphere. It may be *reflected* back into space without change; or it may be *absorbed* either by atmospheric gases or by the surface of the land and ocean, a process which usually converts the electromagnetic energy into heat energy and raises the temperature of the atmosphere and the earth's surface. As an average, about 34 percent of the solar radiation reaching the earth is reflected into space without change; 19 percent is absorbed by the atmosphere, and the remaining 47 percent reaches the earth's surface to be absorbed by rock, water, and green plants (Figure 5.9).

The 19 percent of solar radiation absorbed by the atmosphere is not uniformly distributed throughout but, instead, different wavelengths tend to be absorbed by different atmospheric gases at different altitudes. In general, the high-energy, short wavelengths of the ultraviolet spectrum are absorbed by the upper atmosphere while some of the longer wavelength infrared energy is absorbed by the lower atmosphere. The remaining infrared energy as well as much of the energy in the visible spectrum penetrates to the earth's surface. As a result, heating of the upper atmosphere depends largely on ultraviolet absorption and heating of the lower atmosphere on infrared absorption.

The sequence of energy exchange in the atmosphere is further complicated by the fact that, ultimately, *all* of the energy imparted to the atmosphere and surface of the earth must escape through the atmosphere back into space. If it did not, the atmosphere and earth's surface would continually gain energy and would become progressively hotter. This escape of energy is possible because the gases of the atmosphere, and rocks and water of the earth's surface, behave on a very small scale like the sun itself when they become heated by energy absorption; that

177

Figure 5.10 *Temperatures of the lower atmosphere.*
Warm zones occur at the earth's surface and at about
50 km (30 mi); cold zones at about 15 and 100 km
(10 and 60 mi). These serve to divide the lower at-
mosphere into three zones: the troposphere and meso-
sphere, in which temperatures decrease with altitude,
separated by the stratosphere in which temperatures
increase with altitude.

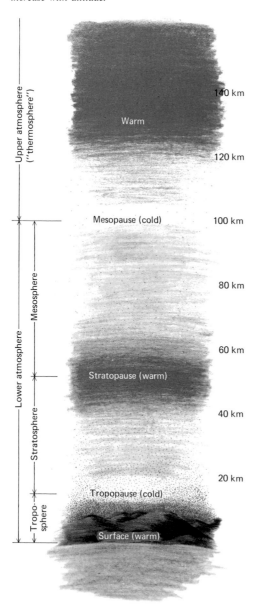

is, *they give off electromagnetic radiation* which removes energy and cools them. This energy may, in turn, be absorbed and reradiated by different atmospheric gases before it finally escapes into space. The temperature at any point in the atmosphere (or on the earth's surface) therefore depends not simply on the amount of solar energy absorbed, but on the balance between energy absorbed and energy reradiated or, in other words, on the relative gain or loss of energy at that point. In general, each gas in the atmosphere has different properties of absorbing and radiating energy. Understanding how these properties lead to the observed temperature distribution in the atmosphere is a primary goal of atmospheric research.

In addition to electromagnetic energy, a much smaller amount of solar energy reaches the earth in the form of rapidly moving particles of hydrogen gas which extend outward from the sun into interplanetary space as a sort of huge, but very thin, solar atmosphere called the **solar wind.** Unlike electromagnetic energy, which interacts with the earth's atmosphere at all levels, the energy from the solar wind affects primarily the uppermost, hydrogen-rich layer of the earth's atmosphere.

5.5 Thermal Structure of the Lower Atmosphere

The first step in understanding atmospheric temperatures is to measure them, and many thousands of observations from aircraft, balloons, rockets, and satellites have shown a consistent pattern of temperature variation with altitude. Starting at the earth's surface and going upward, temperatures first fall to about −50°C (−60°F), then rise to about 15°C (60°F), then fall again to −73°C (−100°F), and then finally rise to a maximum of approximately 1000°C (1832°F) that remains relatively constant into interplanetary space. Thus there are three levels of warmer temperatures separated by two levels of cooler temperatures (Figure 5.10). The two colder levels occur at heights of about 15 and 100 km (10 and 60 mi); the three warmest levels are at the earth's surface and at altitudes of about 50 km (30 mi) and above 120 km (75 mi). Between these levels temperatures have intermediate values. Most of this temperature variation takes place below 100 km (60 mi), in the mixed lower atmosphere, and temperature thus provides a convenient and widely used scheme for subdividing the lower atmosphere into three zones: the **troposphere** is the zone of decreasing temperatures between the surface and the 15 km temperature minimum; the

stratosphere is the zone of temperature increase between about 15 and 50 km; and the **mesosphere** is the second zone of decreasing temperatures between 50 and 100 km. (The regions of maximum or minimum temperatures bounding the zones are usually referred to by the name of the underlying zone, but with the word "pause" substituted for "sphere" as the suffix: *tropopause, stratopause,* and *mesopause*). The cold mesopause occurs at 100 km, the approximate altitude that separates the mixed lower atmosphere from the stratified upper atmosphere. Above the mesopause, temperatures in the upper atmosphere again increase to a maximum value that extends into interplanetary space. For this reason the upper atmosphere is sometimes called the *thermosphere,* although we shall continue to use the more general term "upper atmosphere." These high temperatures of the thin gases of the upper atmosphere will be considered after we examine the more variable temperature structure of the lower atmosphere.

Many thousands of measurements show that the three-level temperature stratification of the lower atmosphere is remarkably constant even though the precise altitudes and magnitudes of the temperature variations change with latitude and season. The *fact* of this variation is thus well established; the difficulties arise in attempting to understand its precise *causes,* that is, in determining just how the balance between gain and loss of solar energy leads to high energy values at the surface and at altitudes of 50 km and above 120 km, and to low values at altitudes of 15 and 100 km.

The high-temperature zone at the earth's surface is the most readily explained since about half of the incoming solar energy penetrates the entire atmosphere to be absorbed by heating the surface of the land and oceans. Most of this energy is radiated back into the atmosphere where it tends to be concentrated in a thin zone near the earth's surface by a phenomenon known as the **greenhouse effect,** so named because it superficially resembles the trapping of solar energy inside glass greenhouses (Figure 5.11). This effect results from a change in the wavelength of solar energy as it is absorbed and reradiated by the earth's surface. Incoming solar radiation is concentrated in the visible and short infrared wavelengths because none of the gases in the atmosphere strongly absorb these wavelengths (Figure 5.8). These wavelengths, however, are produced only by a body, such as the sun, that is so hot that it "glows" to produce visible light waves.

Figure 5.11   Atmospheric heating near the earth's surface by the "greenhouse effect." Short-wave solar radiation penetrates the atmosphere and warms the surface, which radiates at longer wavelengths that are more readily absorbed by atmospheric gases. The result is a concentration of heat near the earth's surface.

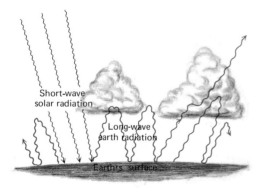

When cooler bodies, such as the earth's rocks and ocean, give off radiation it is in the invisible, longer wavelength infrared spectrum. Unlike visible and short infrared wavelengths, long infrared wavelengths *are* absorbed by atmospheric gases, particularly those such as water vapor (H_2O), carbon dioxide (CO_2), or ozone (O_3) with large molecules made up of three atoms. Carbon dioxide and water vapor are among the most concentrated *minor* constituents of the lower atmosphere where they absorb and reradiate much of the long wavelength infrared radiation given off by rocks and ocean water. By this process they tend to concentrate solar energy near the earth's surface. This concentration leads not only to relatively high temperatures but also to atmospheric turbulence and instability which, in turn, largely determine the earth's patterns of weather and climate.

The temperature maximum 50 km (30 mi) high at the stratopause has a somewhat different origin. Here energy is again absorbed by large molecules of atmospheric gases, but in this instance the gas is *ozone* rather than carbon dioxide and water vapor. Also, the energy absorbed is not radiated from the earth's surface in the long wavelength infrared spectrum; instead it is ultraviolet radiation coming directly from the sun.

The processes leading to the production of ozone in the atmosphere have already been mentioned (see Figure 5.5). Solar radiation, in this case short wavelength ultraviolet, tends to break down atmospheric oxygen molecules (O_2) into single atoms, whose fate depends on the concentration of oxygen which, in turn, increases rapidly toward the earth's surface. In the thin upper atmosphere, the oxygen atoms produced by breakdown of O_2 molecules seldom collide and recombine with other oxygen atoms to reform O_2; as a result atomic oxygen predominates in the upper atmosphere. However, as oxygen concentration increases downward, more and more oxygen atoms recombine and molecular oxygen begins to predominate at an altitude of about 100 km (60 mi) in the zone of transition between the lower and upper atmosphere. High-energy ultraviolet radiation penetrates, however, below this level into the O_2-rich lower atmosphere, where oxygen molecules continue to be separated into oxygen atoms. Most of these quickly recombine into O_2 molecules but a few become attached to a preexisting O_2 molecule to make ozone (O_3).

The rate of ozone production depends on the balance between the oxygen concentration and the availability of ultraviolet radiation of the proper wavelength. It

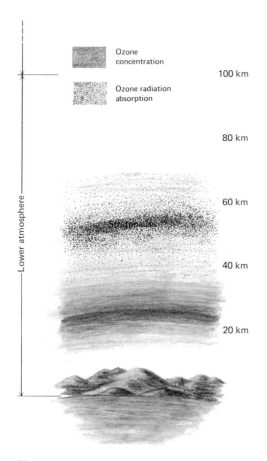

100 km

Ozone concentration

Ozone radiation absorption

80 km

Lower atmosphere

60 km

Stratopause

40 km

20 km

Figure 5.12   Ozone concentration and stratopause heating. Short-wave ultraviolet radiation produces ozone in the lower atmosphere with a maximum concentration (gray) around 25 km (15 mi). The less dense, upper part of the ozone layer absorbs longer-wave ultraviolet and is heated to form the warm stratopause (color).

at first increases downward as the oxygen concentration increases. However, as the radiation is absorbed the amount of ozone passes through a maximum and then decreases. Both direct observation and theoretical calculations show that this maximum concentration occurs at an altitude of about 25 km (15 mi) (Figure 5.12). Below this level most of the high-energy radiation has already been absorbed; above this level the ozone decreases as the oxygen concentration becomes smaller.

The level of maximum ozone concentration at 25 km has, however, almost no direct effect on atmospheric temperatures. This is because little heat is produced when high-energy radiation is absorbed by O_2 with the production of ozone. Like O_2, however, ozone itself also absorbs ultraviolet radiation, but of longer, less energetic wavelengths and this absorption produces a great deal of heat energy. Like the level of maximum ozone concentration, this ozone *absorption* maximum depends on the balance between concentration and the amount of radiation of the proper wavelengths present. Calculations show that it should occur at about 50 km (30 mi), the height of the temperature maximum of the stratopause. This second zone of high atmospheric temperatures is therefore caused by a complex sequence of interactions between atmospheric oxygen and solar ultraviolet absorption (Figure 5.12).

The cold zone at the tropopause is believed to be merely a zone of minimum energy absorption between the heated surface and the ozone absorption layer. Likewise, the minimum temperatures at the mesopause, although less well understood, appear to represent a zone of minimum energy absorption between the stratopause and the high temperatures of the upper atmosphere.

5.6   Upper Atmosphere Temperatures

In the extremely thin gases of the upper atmosphere temperature has a somewhat different meaning than in the more dense lower atmosphere. Because the gas particles are so widely spaced, the gases would not feel warm if touched, and a spacecraft passing through the upper atmosphere is only slightly warmed even though the "temperature" of individual gas particles is very high. Temperature in such thin gases is therefore not a property of the entire mass of gas, but reflects only the speed of movement, or energy level, of individual atoms and molecules. Because there are relatively few atoms and molecules in the upper atmosphere the absorption of small amounts of solar energy leads to high "temperatures."

It is very difficult to measure the energy of individual gas particles and thus direct temperature measurements are impossible in the upper atmosphere. Instead, temperatures are inferred from observations on gas density. In very thin gases the temperature is proportional to the density and composition and with a knowledge of any two of these three properties, the third is easily computed. Densities throughout much of the upper atmosphere are well known because they can be computed by observations on "satellite drag," that is, by noting how rapidly the orbital motions of satellites are slowed by atmospheric friction at different levels. This friction is a function of density and thus gives a direct measure of upper atmosphere densities. Unfortunately, the precise composition of the upper atmosphere, particularly the concentration of minor gases, is less certain, but by making reasonable assumptions it is possible to calculate temperatures from the density data. Such calculations show that gas particles in the upper atmosphere have extremely high temperatures at all altitudes above about 120 km (75 mi). Below this level temperatures drop rapidly to the cold mesopause at the base of the upper atmosphere.

The thin gases of the upper atmosphere are extraordinarily sensitive to slight energy changes; that is, small changes in the amount of energy absorbed lead to large changes in temperature and, as a result, in volume and density. Satellite drag density measurements show that upper atmosphere temperatures decrease sharply at night when the atmosphere is not directly heated by the sun (Figure 5.13). The temperatures also vary with slight fluctuations in the output of solar radiation.

Figure 5.13   Range of temperature variation in the upper atmosphere.

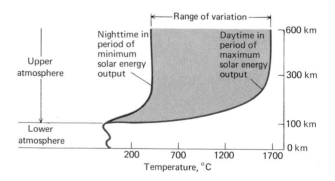

Man
and the
Atmosphere

Much of man's direct interaction with the atmosphere involves weather phenomena and will be considered in Chapter 6. In addition, however, there is a generally negative interaction involving foreign materials that man releases into the atmosphere which, at least temporarily, change its composition.

The most remarked upon and obvious of these man-made atmospheric additions are gases and aerosol particles that are toxic to animals and plants when concentrated by local weather conditions. The more abundant of these pollutants, and their principal sources, are summarized in the table on page 185. Such a summary shows very clearly that the principal contributors of toxic materials are automotive exhausts and sulfur-rich coal and petroleum burned for power and heating. It is for this reason that much effort is now being expended to restrict automotive emissions and to require the use of low-sulfur fuels.

Fortunately, most toxic pollutants are rather quickly removed from the atmosphere by natural processes so that, in spite of the local discomfort and even danger that they can cause, they have no long-term effect on atmospheric composition. Potentially far more serious, however, are possible man-induced changes in the proportions of *natural* atmospheric constituents. The burning of coal and oil as fuels, for example, has significantly raised the level of atmospheric carbon dioxide during the past hundred years. Since carbon dioxide is one of the large gas molecules that traps long-wave radiation to warm the lower atmosphere by the "greenhouse effect," some scientists have suggested that this increased carbon dioxide might be causing a general warming of the earth's climates. As another example, laboratory studies have shown that many kinds of phytoplankton—the microscopic floating plant life of the sea—are easily killed by certain long-lasting insecticides. Rainwater carries such insecticides from plants to rivers and ultimately to the sea where they may cause a reduction in phytoplankton. Because photosynthesis by phytoplankton accounts for a large fraction of the oxygen released into the atmosphere each year by green plants, this reduction might ultimately change the concentration of atmospheric oxygen with potentially disastrous results.

Fortunately, the most recent studies of both the carbon dioxide–climatic change possibility and the phytoplankton–oxygen reduction hypothesis have indicated that neither as yet poses a serious threat to the balance of atmospheric processes. These suggestions have served, however, to focus attention on the complex interactions between man's activities and the atmosphere that surrounds them, and thus may prevent still more serious problems from arising in the future.

Atmospheric pollution from industry and power plants. (NOAA)

Atmospheric pollution from automotive exhausts. (Dave Repp, National Audubon Society)

The results of atmospheric pollution. (United Press International)

Principal atmospheric pollutants emitted in the United States in 1968, in millions of tons. (National Air Pollution Control Administration)

	Gases			Liquid and solid particles
	Carbon monoxide	Sulfur oxides	Hydrocarbons	
Automotive	58	1	15	1
Power and Heating	2	22	1	8
Industry, Trash disposal	16	7	10	8

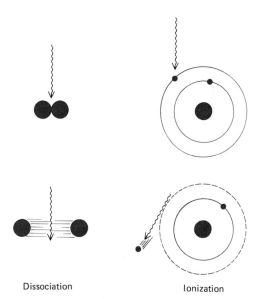

Dissociation   Ionization

Figure 5.14 *Disturbances of atmospheric gas particles by high-energy solar radiation. In* dissociation *the atoms of a molecule are separated. In* ionization *an electron is separated from a neutral atom, producing a charged ion.*

Figure 5.15 *Production of charged particles (ions) in the atmosphere. The ion concentration is a function of the upward increase in radiation and decrease in gas concentration.*

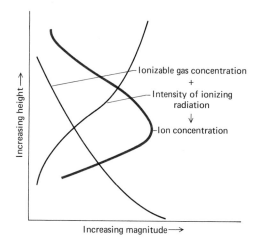

ELECTRICAL STRUCTURE OF THE ATMOSPHERE

In addition to composition and temperature, there is still a third basic structural property of the atmosphere to be considered. This property results from the presence of small quantities of *charged* gas particles among the much more abundant uncharged or *neutral* gas particles of the atmosphere. Because these charged particles respond to electrical and magnetic forces, they interact with the sun's electromagnetic radiation and the earth's magnetic field to give the atmosphere a characteristic *electrical structure.*

5.7 Charged Particles in the Atmosphere

We have already seen how high-energy solar ultraviolet radiation can break the bonds holding oxygen atoms together in molecules, a process known as **dissociation** (Figure 5.14). Extremely high-energy solar ultraviolet radiation of still shorter wavelength can have an even more profound effect on atmospheric gases. Instead of merely separating the atoms making up a molecule of gas, the radiation can separate an *electron* from either an atom or molecule of gas. This loss of an electron unbalances the normal electrical neutrality of the atom or molecule and leads to two separate *charged* particles: the negatively charged electron itself, and the remaining, positively charged atom or molecule, which is called an **ion** (Figure 5.14). This process of **ionization** of atmospheric gases by high-energy radiation depends, like simple dissociation, on the balance between the concentration of radiation and the concentration of atmospheric gases (Figure 5.15). Just as in the production of ozone by oxygen dissociation, the production of charged particles by ionization increases as high-energy solar radiation penetrates downward into the more concentrated gases of the lower atmosphere. As with ozone, there is a level of maximum ion concentration; below this level most of the ionizing radiation has been absorbed; above it, the decrease in concentration of atmospheric gases leads to decreased ionization. This layer of maximum ion concentration occurs far higher than the ozone maximum; it has an average altitude of about 300 km (180 mi), that is, in the lower part of the upper atmosphere far above the mesopause (Figure 5.16).

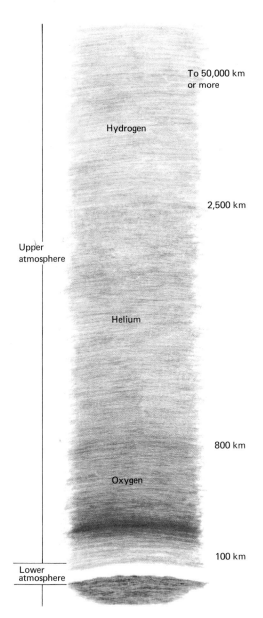

To 50,000 km
or more

Hydrogen

2,500 km

Upper
atmosphere

Helium

800 km

Oxygen

100 km

Lower
atmosphere

Figure 5.16 *Distribution of charged particles (ions)*
in the atmosphere. The maximum concentration occurs
at about 300 km (180 mi) in the lower part of the
upper atmosphere.

The exact chemical reactions leading to the production of atmospheric electrons and ions are far more complex and poorly understood than are the processes of ozone production. When high-energy radiation ionizes small quantities of various atmospheric gases, the charged ions and electrons often react with one another in *new* combinations to give entirely different ions or neutral gas molecules. These ions and molecules may, themselves, then be further ionized, leading to complex series of chemical reactions and end products. These processes are not fully understood because it is extraordinarily difficult to design rocket-borne instruments to measure the concentration of the many different ions in the atmosphere. In contrast, it is very easy to measure the concentrations of electrons, which are identical regardless of the composition of the gas particle from which they arose.

Since the production of an ion from *any* atmospheric gas produces an electron, measurements of electron concentration provide a precise measure of the *total* ionization taking place at a particular altitude, but tell nothing of the different kinds of atmospheric gases being ionized. For this reason, the total distribution of charged particles throughout the atmosphere is far better understood than are the exact processes leading to the observed distribution. It should also be emphasized that even at the altitude of maximum concentration, charged particles make up only an extremely small fraction of all atmospheric gas particles. It has been estimated that at 300 km, the maximum concentration altitude, only two out of each thousand atoms are ionized. These relatively few ionized atoms are responsible for the atmosphere's electrical properties.

5.8 The Ionosphere

Because of the occurrence of small numbers of ionized particles at all altitudes higher than about 80 km (48 mi), the atmosphere above that altitude is often referred to as the **ionosphere.** Note that the ionosphere includes all of the upper atmosphere as well as the uppermost part of the lower atmosphere (Figure 5.16).

Before the relatively recent development of rocket and satellite technology, the charged particles of the ionosphere were the only structural aspect of the high atmosphere that had been intensively studied. Such study was based on the fact that these charged particles reflect and absorb radio waves transmitted from, and received back at, the earth's surface. By observing the patterns of these

187

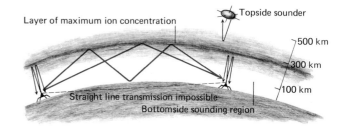

Topside sounder

Layer of maximum ion concentration

500 km

300 km

100 km

Straight line transmission impossible

Bottomside sounding region

Figure 5.17 Ionospheric "sounding" with reflected radio waves. The concentration of charged particles at various altitudes can be determined by ground stations and satellites. Reflection from the ionosphere also permits long-distance reception of radio waves (color).

radio wave reflections and absorptions, it was possible to make many important inferences about the nature of the ionosphere from radio equipment located on the earth's surface (Figure 5.17).

The discovery of the ionosphere, and the history of its study, is closely linked to the development of radio communications. In 1901 G. Marconi, an Italian inventor, showed that radio waves transmitted in England could be received in Canada, one-quarter of the distance around the earth. Because radio waves, like all electromagnetic radiation, were known to travel only in straight lines, this result was very surprising; instead of radiating out into space from the point of transmission, the waves were somehow kept near the curved surface of the earth. In 1902, physicists in the United States and England independently suggested that the cause might be the reflection of the radio waves in straight lines back and forth between the earth's surface and a "conducting" layer of charged particles in the atmosphere. The first direct proof that such a layer existed did not come until the mid-1920s when researchers, utilizing a sort of primitive radar which beamed radio waves directly upward and received back directly reflected waves, showed the presence of a strong reflecting layer at an altitude of about 300 km (180 mi) (Figure 5.17). This technique of ionospheric "sounding" with ground-based radar was perfected in the 1930s and has yielded a wealth of information about the distribution of charged particles in the ionosphere. Such techniques are still the principal means of studying the ionosphere *below* the charged particle maximum at 300 km. Ionization *above* this level cannot be routinely studied by this technique, however, because most radio waves are absorbed or reflected by the layer of maximum density and thus cannot penetrate to the other side. In recent years the "topside ionosphere," as the region above the maximum is called, has been intensively studied by means of radar-carrying satellites which orbit far above the maximum layer and send waves *downward* to be reflected back to the

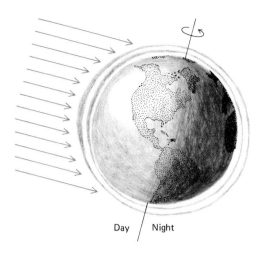

Day / Night

Figure 5.18 *Day-night differences in atmospheric ion concentration. During daylight hours increased solar energy forms an ionized layer beneath the 300-km (180-mi) layer of maximum ion concentration. This lower layer quickly decays at sunset whereas the maximum concentration layer persists through the night.*

satellite (Figure 5.17). These observations have produced a large body of knowledge about the distribution of charged particles in the atmosphere.

A principal result of ionospheric studies has been to show the close dependence of the distribution of charged particles on fluctuations in solar energy. These fluctuations are of two main sorts: the first results from regular day-night differences in energy; the second from less regular variations in the sun's energy caused by **solar flares,** explosive outbursts of energy lasting from a few minutes to a few hours, which take place, on the average, once every 3 days.

The ionosphere of the sunlit half of the atmosphere shows a different structure than does that experiencing the nighttime shading of half the earth (Figure 5.18). As might be expected, the amount of ionization increases during daylight hours, but in a complex pattern. The principal changes occur in the lower regions between about 100 km (60 mi) and the maximum ion density at 300 km (180 mi). At night there is far less ionization in these lower levels than during daylight (Figure 5.18). This is the reason for the general improvement in long-range radio communications at night; the radio waves travel farther because they are reflected from much higher ionized layers. The maximum ion concentration at 300 km, however, changes very little during the 24-hr cycle. Evidently, even though *ions* are most abundant at this level, the total gas concentration is so low that electrons, once separated, do not readily collide with ionic particles to reform neutral gases. In the denser gases at lower altitudes, such recombinations reduce the total ionization as soon as the driving energy from the sun is lost.

In contrast to these regular, daily changes in the ionosphere are the more dramatic and irregular effects of solar flares which, every few days, greatly disturb the structure of the ionosphere. During flare periods the sun gives off additional amounts of high-energy ultraviolet radiation which substantially increase the number of charged particles at all altitudes for periods ranging from a few minutes to several days. Certain kinds of long-range radio communications become difficult or impossible during such periods, which are known as **ionospheric storms.**

5.9 The Magnetosphere

The concentration of charged particles in the atmosphere decreases steadily above the 300-km (180-mi) level of maximum density but, at the same time, the total number of gas particles, both charged and neutral, decreases at a still more rapid rate. The result is that the *proportion* of charged to neutral particles increases upward even though the total *abundance* of charged particles decreases (Figure 5.19). This is readily understandable since high-energy ionizing radiation increases in the thin upper atmosphere where there are fewer particles to absorb the radiation. In the layer of maximum concentration at 300 km, less than 1 percent of the particles are ionized. At this altitude atomic oxygen is the principal atmospheric gas; the maximum ionization at this level probably represents removal of electrons from a small proportion of these oxygen atoms. At altitudes above 800 km, first helium and then hydrogen become the dominant gases; in these layers the gas concentrations are so low, and high-energy radiation so intense, that *most* of the gas particles are ionized.

In addition to having much higher proportions of charged particles, the outermost zones of the atmosphere differ from lower levels in another fundamental way. In the denser gases at lower levels, little movement can take place among the small number of *charged* particles because they quickly collide with other particles of opposite charge to reform neutral gases. In contrast, the ions and electrons of the much thinner helium and hydrogen layers

Figure 5.19 Schematic representation of the proportional increase in charged particles in the thin gases of the upper atmosphere.

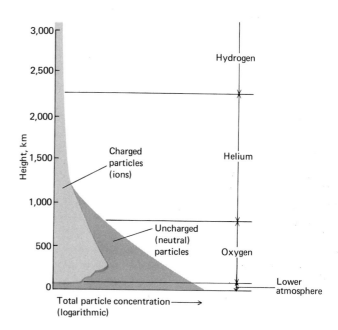

are free to move great distances without colliding with other particles and losing their electrical charges. The result is that charged particles in the highest levels of the atmosphere are profoundly influenced by the earth's magnetic field, which causes them to move over great distances and assume characteristic patterns of distribution. Because of this strong interaction with the magnetic field produced deep within the earth's interior, the outer ionosphere above about 500 km (300 mi) is commonly referred to as the **magnetosphere.**

The shape of the outer boundary of the magnetosphere which marks the upper limit of the earth's atmosphere is largely determined by the shape of the earth's magnetic field. It is also strongly influenced by the fact that the ionized hydrogen of the earth's outer atmosphere is embedded in, and interacts with, a continuous flow of similar ionized hydrogen particles moving outward from the sun. This flow, called the *solar wind,* makes up the thin outer atmosphere of the sun that was mentioned earlier in discussing atmospheric temperatures. The magnetosphere shows sharp boundaries at its outer edge because the earth's magnetism tends to repel charged solar particles and thus to prevent them from entering the upper atmosphere. In addition, the pressure of the solar wind causes the magnetosphere to be much more compressed on the sunlight side of the earth than on the dark side, where it extends many times farther out into space (Figure 5.20). The result is that the boundary of the magnetosphere is extremely elongate in the direction away from the sun.

Most of the charged particles in the magnetosphere, like those of the lower ionosphere, have relatively little energy; even though they are free to move great distances,

Figure 5.20 Shape of the magnetosphere. The widely scattered, charged hydrogen particles making up the outer magnetosphere are forced into an extremely elongate "tail" by the solar wind, and are also attracted towards the earth at the magnetic poles.

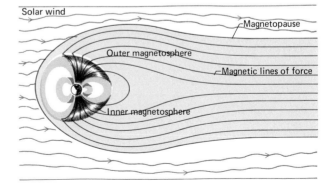

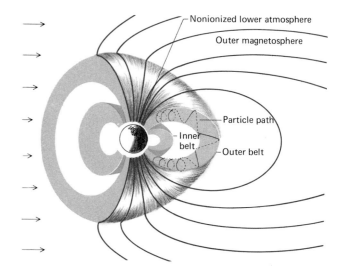

Figure 5.21   The "radiation belt" of the inner magnetosphere. This zone contains high-energy charged particles which move rapidly back and forth between the earth's magnetic poles.

they move rather slowly. One of the most dramatic discoveries of the first artificial earth satellites launched in the late 1950s was a zone of unusually energetic charged particles surrounding the earth in the lower or "inner" magnetosphere (Figure 5.21). The charged particles in this zone move very rapidly for they have energies hundreds or thousands of times greater than most charged particles in the atmosphere. This zone has come to be known, rather confusingly, as the earth's **radiation belt;** in spite of the name, it is *not* composed of electromagnetic radiation, but merely of charged gas particles which are similar to those of the rest of the magnetosphere except for their much higher energies.

Like the low-energy particles of the magnetosphere, the distribution of high-energy particles in the radiation belt is controlled by the earth's magnetic field which resembles that produced by a huge bar magnet with its poles near the earth's rotational poles. The lines of magnetic force are vertical over the magnetic poles, horizontal in the magnetic equatorial regions, and have intermediate values between. The particles of the radiation belt follow these lines of force and show a similar shape, except that they are somewhat distorted, like the larger magnetosphere of which they are a part, by pressure from the solar wind (Figure 5.21). Because of their extremely high energies, particles of the radiation belt move very rapidly along paths which generally follow the lines of magnetic force from pole to pole; they move so fast that they travel from pole to pole in less than a second. At the poles the particles move vertically toward the earth until they are reflected back upward by the converging magnetic field which acts like a mirror. The particles thus tend to "bounce" from pole to pole at extremely high speeds.

Measurements from many satellites have now clearly established the shape of the radiation belt and the distribution of high-energy particles within it, but there is still no really satisfactory explanation of the *source* of these extraordinarily energetic particles. Most of the charged particles moving in the solar wind have much lower energies, and there are no known mechanisms for trapping them in the earth's magnetic field and accelerating them to high energies. Small numbers of extremely high-energy particles, known as **cosmic rays,** *do* continually bombard the earth's atmosphere from uncertain sources, probably other stars beyond the sun. Because of their high energies, these particles tend to ionize small quantities of most atmospheric gases, producing new high-energy particles through several complex series of reactions. These high-energy products of cosmic ray ionization may account for some of the trapped particles of the radiation belt, but the quantities produced appear to be far too small to provide a source for all of the trapped particles. A satisfactory explanation of the origin of high-energy magnetospheric particles thus remains a primary goal of atmospheric research.

Just as in the lower ionosphere, charged particles of the magnetosphere are strongly affected by variations in the sun's energy output at times of solar flares. During flares, while the sun gives off increased electromagnetic radiation, the flux of charged hydrogen particles traveling outward in the solar wind also increases. The electromagnetic radiation travels at the speed of light and arrives almost immediately to cause *ionospheric* storms. The particles of the solar wind travel more slowly and arrive at the earth about a day later. This increased flow of particles in the solar wind tends to compress the boundaries of the magnetosphere, which has the effect of slightly increasing the earth's magnetic field strength on the earth's surface. As a result, magnetic compasses tend to behave erratically during such intervals, which have come to be known as **magnetic storms.**

5.10 Auroras

Closely related to the magnetosphere are the spectacular phenomena called **auroras,** bright, rapidly moving patterns

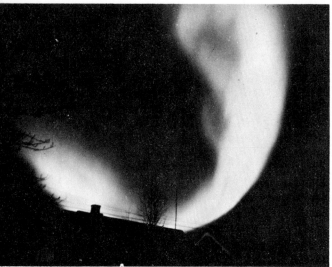
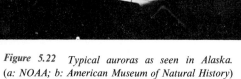

(a)   (b)

Figure 5.22   *Typical auroras as seen in Alaska.*
(a: NOAA; b: American Museum of Natural History)

of colored lights than can frequently be seen in the night sky in the earth's polar regions (Figure 5.22). These dramatic displays have been a source of wonder and speculation since man first inhabited polar lands, but initial clues to understanding their cause came only in the nineteenth century when it was observed that auroras were almost always accompanied by erratic movements in the needles of magnetic compasses. This observation showed that auroral displays were somehow related to temporary fluctuations in the earth's magnetic field. These magnetic storms now are known to be caused by large outpourings of solar wind particles during solar flares. On reaching the earth, these compress the magnetosphere and temporarily intensify the magnetic field. The strong auroral displays that accompany magnetic storms must also be caused by the solar flares but, as yet, there is no satisfactory explanation of the exact mechanisms involved. Much has been learned, however, about the local sources and distribution of auroral light through detailed programs of observing and photographing them undertaken during the past 20 years. The height of an auroral display can be computed by triangulation using simultaneous photographs taken at different stations located at known distances from each other. Thousands of such measurements show that the *base* of the lights usually occurs near the mesopause at heights of about 100 km (60 mi); from there they extend upward to heights of about 300 km (180 mi). They are thus confined to the lowermost ionosphere (see

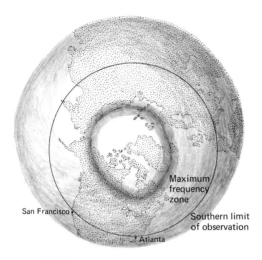

Figure 5.23 The auroral zone of the northern hemisphere. A similar zone surrounds Antarctica.

Figure 5.24 Concentration and energy of auroral and radiation belt particles. (*Modified from King and Newman,* Solar and Terrestrial Physics, *1967*)

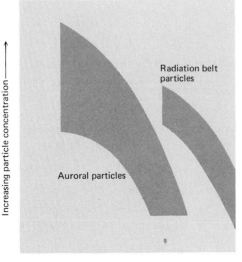

Figure 5.16). Auroras are observed most frequently in circular belts, known as **auroral zones,** that occur about 2,400 km (1,500 mi) from the magnetic poles (Figure 5.23).

The local cause of auroral light is known to be the absorption of energy by oxygen atoms and nitrogen molecules, the two dominant gases in the altitude range where auroras occur. When an atom or molecule of any gas collides with electromagnetic radiation or high-energy charged particles, the gas may give off visible light as it is raised to a higher energy level. This is the principle exploited in "neon" lighting tubes in which an electric current raises the energy level of the confined neon gas. When this occurs, different gases give off different wavelengths of light. Thus, by analyzing the wavelengths of auroral light, it is a comparatively simple matter to determine the composition of the gases that produced the light. Such analysis shows that the red and green colors that predominate in auroras come from atomic oxygen; the less common blue colors originate from molecular nitrogen. Other colors occur as combinations of these three principal wavelengths.

Because auroras are confined to polar regions where the lines of force of the earth's magnetic fields descend toward the surface (see Figure 5.20), it is clear that the energy required to excite the oxygen atoms and nitrogen molecules comes from charged particles which, like the high-energy particles in the radiation belt, are trapped and concentrated by the earth's magnetic field. As in the radiation belt, the particles producing auroras have energies far higher than those of the solar wind or the earth's normal magnetosphere, and the principal difficulty is explaining their source. When the radiation belt was first discovered, it was suggested that some of *its* high-energy particles might produce auroras by descending to low levels in polar regions during magnetic storms. Recent rocket measurements of the energy and concentration of charged particles in auroral displays show them, however, to have generally higher concentrations and lower energies than do those of the radiation belt (Figure 5.24). Thus the radiation belt and auroras appear to be distinctive, but related, products of a poorly understood interaction between the magnetosphere and solar wind which produces charged particles of exceptionally high energy.

SUMMARY OUTLINE

Atmospheric Composition

5.1 *Composition of the lower atmosphere:* below 100 km air, a mixture composed mostly of nitrogen and oxygen, predominates; both constituents are involved in cycles mediated by organisms; important minor constituents are ozone, carbon dioxide, water vapor, and aerosols.

5.2 *Composition of the upper atmosphere:* above 100 km turbulent mixing ceases and gases are stratified by weight; the lowest layer is predominantly atomic oxygen; above are layers of helium and hydrogen.

5.3 *Atmospheric density and pressure:* the compressibility of gases causes the bulk of the atmosphere's mass to be concentrated near the earth's surface by gravity; density and pressure decrease upward by a factor of 10 with each 20 km (12 mi) of altitude increase.

Atmospheric Temperatures

5.4 *Energy in the atmosphere:* 34 percent of solar radiation reaching the earth is reflected into space, 19 percent is absorbed by the atmosphere, and 47 percent is absorbed by the earth's surface; ultimately the 66 percent that is absorbed is radiated back into space from rocks, waters, and the atmosphere.

5.5 *Thermal structure of the lower atmosphere:* a warm layer near the earth's surface is caused by radiation from rock and water and subsequent absorption by atmospheric carbon dioxide and water vapor; a second warm layer at 50 km (30 mi) is caused by radiation absorption by ozone; cold layers occur between.

5.6 *Upper atmosphere temperatures:* the thin gases of the upper atmosphere have high temperatures due to the energy of motion imparted by intense solar heating.

Electrical Structure of the Atmosphere

5.7 *Charged particles in the atmosphere:* are formed when high-energy solar radiation removes electrons from an atom or molecule, producing a charged ion.

5.8 *The ionosphere:* ions occur in small quantities at all altitudes higher than about 80 km (50 mi); the maximum concentration is at 300 km (180 mi); ion concentration varies with day-night differences in solar energy and during solar flares.

5.9 *The magnetosphere:* the thin gases of the uppermost atmosphere are mostly ionized; their low concentration and electrical charge allows them to be moved by the earth's magnetic field and by the solar wind; the radiation belt is a zone of exceptionally energetic gas particle movement high in the atmosphere.

5.10 *Auroras:* are brightly colored lights formed in the upper atmosphere by high-energy charged gas particles that descend over polar regions where the earth's magnetic lines of force converge.

ADDITIONAL READINGS

Bates, D. R. (ed.): *The Planet Earth,* Pergamon Press, Oxford, England, 1964. A series of brief, well-written popular essays; many cover atmospheric phenomena.

Craig, R. A.: *The Upper Atmosphere, Meteorology and Physics,* Academic Press, Inc., New York, 1965. A standard advanced text.

*Dobson, G. M. B.: *Exploring the Atmosphere,* Oxford University Press, Oxford, England, 1968. An excellent popular introduction; one of the few treating the properties of the entire atmosphere, rather than merely weather phenomena.

Fairbridge, R. W. (ed.): *The Encyclopedia of the Atmospheric Sciences and Astrogeology,* Reinhold Publishing Corp., New York, 1967. Contains many good review articles.

Fleagle, R. G., and J. A. Businger: *An Introduction to Atmospheric Physics,* Academic Press, Inc., New York, 1963. A clearly written advanced text.

*Goody, R. M., and J. C. G. Walker: *Atmospheres,* Prentice-Hall, Inc., Englewood Cliffs, N.J., 1972. An authoritative and up-to-date introduction to the general properties of planetary atmospheres, emphasizing the earth.

* Available in paperback.

6

Weather

Weather, the earth's changing patterns of wind, temperature, and precipitation, is concentrated in the troposphere, the lowermost layer of the atmosphere that extends from the earth's surface to an average height of about 15 km (9 mi) (see Figure 5.10). Not only is the bulk of the atmosphere's mass concentrated in this zone near the earth's surface (see Figure 5.7), but also much of the sun's energy is trapped there by surface heating and the "greenhouse effect" (see Figure 5.11). As a result of this energy concentration, the gases of the troposphere are continuously stirred into turbulent motions. In discussing these weather-producing motions, it will be useful to consider first those that are predominantly *vertical* with respect to the earth's surface; these serve to transport water from the ocean into the atmosphere and thus are the ultimate cause of clouds and precipitation. We shall then consider the more complex *horizontal* motions that produce winds and, through them, transport heat and moisture from place to place on the earth's surface. These horizontal wind motions, acting on many different scales, are the direct cause of the earth's changing patterns of weather.

VERTICAL MOTIONS AND ATMOSPHERIC MOISTURE

Vertical motions of the troposphere are caused by relatively local, small-scale solar heating of the earth's surface. Because about half of the sun's energy penetrates the atmosphere and is absorbed by the land and ocean, the troposphere is largely heated from *below* by radiation from the earth's warm surface rather than from above by direct solar radiation. This heating is the primary cause of weather-producing vertical motions because *fluids heated from below tend to be unstable in the earth's gravitational*

Cumulus clouds (U.S. Department of Agriculture)

199

field. A moment's reflection will show why this is true. All fluids, and most solids as well, expand and become less dense as they are heated. When heat is supplied to a fluid from below, the heated, less dense region is overlain by cooler, heavier layers; gravity then causes the underlying lighter layer to rise and the overlying heavier layer to sink in the process known as *convection.* Because of gravitational convection, fluids surrounding the earth tend to be *stable* when their temperature *increases* upward (warmest layers on top) and *unstable* when their temperature *decreases* upward (warmest layers below).

This principle of convective instability applies to both the ocean and the atmosphere. In the ocean, however, convective motions are far less important because ocean water is heated from *above* by solar energy and thus tends to have its highest temperatures and lowest densities farthest from the earth's center of gravity. In contrast, convective motions take place whenever the earth's surface, and the adjacent air, are strongly heated by the sun.

6.1 Vertical Stability of the Troposphere

Convective motions of the troposphere occur whenever unstable thermal conditions prevail near the earth's surface. Such conditions, in turn, all relate to the degree of difference between temperatures near the surface and those of the overlying layers of air; stated another way, they depend on the *rate of temperature decrease* away from the earth's surface.

The troposphere shows a normal upward decrease in temperature from a mean of about 10°C (50°F) at the earth's surface to about −50°C (−60°F) at the temperature minimum which marks the tropopause at an average height of 15 km (9 mi). Thus the usual rate of upward temperature decrease, called the **normal rate,** is about 4°C for each kilometer increase in height (about 2.5°F/1,000 ft). The troposphere is usually very stable where this normal rate, or a lower rate, prevails. But if the earth's surface is unusually hot, as it may be on a bright summer afternoon, the cooling rate may considerably *exceed* the normal value. Under such conditions, warm air at the surface may expand and become light enough to rise spontaneously, leading to strong vertical motions; such motions are responsible for the gusty winds and scattered, billowy, "fair-weather" clouds that so commonly occur on summer afternoons in temperate latitudes and throughout the year in the warm tropics.

To understand more precisely the relation between rates of temperature decrease and vertical stability, we must consider another fundamental property of gases: when a gas expands under decreased pressure it spontaneously becomes cooler; when it contracts under increased pressure it spontaneously becomes warmer. These changes of temperature, called **adiabatic changes,** are caused *only* by the pressure change. In the atmosphere they are independent of the initial temperature of the air, and are unrelated to the temperature of surrounding materials, such as other masses of air, or the earth's surface. Because of such adiabatic temperature changes, when a particular "bubble" of air starts to rise into the less dense overlying atmosphere (recall that density and pressure *everywhere* decrease upward in the atmosphere) it spontaneously expands and becomes cooler, a process called **convective cooling.** For *dry* air in the troposphere, convective cooling has a constant value of 9.8°C temperature decrease for each kilometer rise in altitude. When this constant value, known as the **dry adiabatic rate,** is exceeded by the *actual* rate, as on the summer day we have mentioned, then a bubble of dry air set in motion by solar heating of the surface tends to *continue* to rise because, even though cooled by expansion, it remains warmer and less dense than the surrounding, stationary air (Figure 6.1). Such rising continues until the air bubble reaches a layer in which the surrounding air is as cool as the bubble itself.

Note that we have discussed only the stability of *dry* air—the presence of water vapor in the air further complicates the relationships. Energy is required to convert liquid water to gaseous water vapor; a pot boiling on a stove is a familiar example. At the earth's surface the sun shining on the oceans has a similar effect, particularly in warm tropical regions. The solar energy, like the heat from a stove, causes evaporation and the addition of water vapor to the air in contact with the ocean surface. Such evaporation is closely related to the vertical stability of the troposphere because the energy originally required to produce the water vapor *remains stored* in the molecular motions of the vapor and, if the vapor is converted back to liquid water, this energy is released from the molecules as heat. For this reason when *condensation*, the change of water from the gaseous to the liquid phase, takes place in the atmosphere heat is always released and the surrounding air is *warmed*. This heat energy given off when water condenses is called the **latent heat of condensation.**

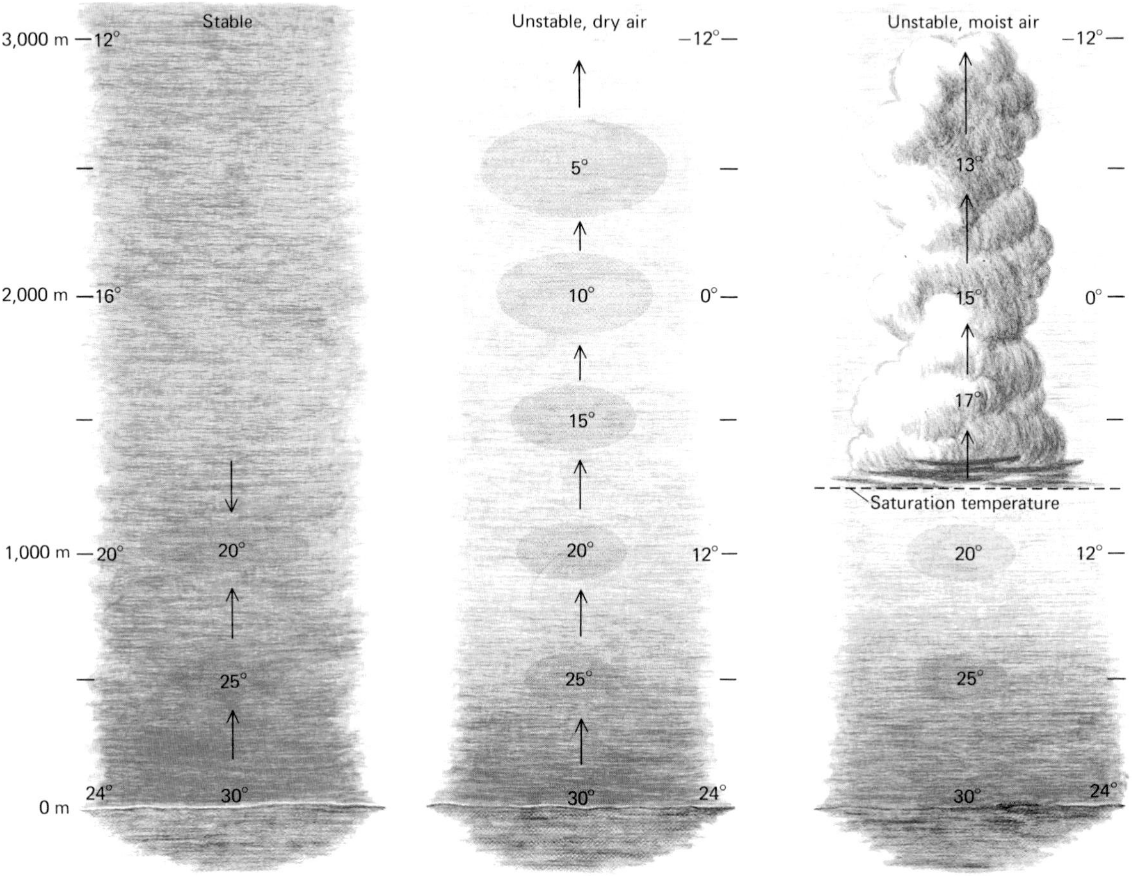
Stable
3,000 m —12°
2,000 m —16°
1,000 m —20°
0 m
20°
25°
24°
30°
Unstable, dry air
−12°—
5°
10°
0°—
15°
20°
12°—
25°
30°
24°
Unstable, moist air
−12°—
13°
15°
0°—
17°
Saturation temperature
20°
12°—
25°
30°
24°

Figure 6.1 *Upward temperature decrease and atmospheric stability. When atmospheric temperatures decrease upward more slowly than heated air bubbles formed at the surface are cooled by expansion (about 10° C/1,000 m in dry air; see text), rising bubbles are quickly surrounded by air of the same temperature and cease to rise, causing vertical stability (left). When the temperature decreases upward more rapidly than the rising bubble, it remains warmer than the surrounding air and continues to rise, causing unstable convection (center). If the unstable rising air contains moisture, it will eventually cool to the saturation temperature, causing condensation and forming a cloud (right). This condensation releases heat and causes still less rapid cooling and greater instability of the rising air.*

Condensation, in turn, takes place in the atmosphere almost exclusively as a result of cooling because *cool air can hold far less water vapor than warm air,* a principle that has familiar demonstrations in the moisture that accumulates around cold objects on a humid summer day, and in "seeing one's breath" when warm moist air is released from the lungs on cold, dry, winter days. Consequently, air that is rising and cooling adiabatically will, if it contains any water vapor, eventually cool to the temperature, known as the **saturation temperature** or **dewpoint,** below which the amount of contained water can no longer remain in the gaseous state; when this happens condensation, usually in the form of *clouds,* takes place. This condensation, in turn, releases heat *which tends to counteract the adiabatic cooling;* cooling still predominates, but the release of heat by condensation causes it to take place at a lower rate. Moist air cooled to the saturation temperature therefore begins to cool more slowly than dry air and thus tends to remain warmer than the surrounding air for longer periods and to rise to greater altitudes. These relations can be readily observed in the atmosphere: the bases of the puffy, white, fair-weather clouds, which form on our exemplary summer afternoon, mark the altitude at which the rising air bubbles reach the saturation temperature (Figure 6.1). The tops of the clouds mark the level at which the bubbles cease to rise because they have attained the same temperature as the surrounding air. In between, condensation and cloud formation take place.

The fact that rising air may be warmed by condensation at the same time that it is cooled by expansion leads to still another cooling rate, the **wet adiabatic rate,** which applies to moist rising air that has reached the saturation temperature. It is always lower than the dry adiabatic rate, but, unlike that rate, it is not constant, but varies with temperature because temperature controls the amount of water the air can hold and thus the amount of heating by condensation.

6.2 Clouds

Vertical atmospheric motions tend to occur spontaneously only when the earth's surface is strongly heated by the sun, as it is through most of the year in the tropics and during the warmer months in temperate latitudes. Such vertical motions play a crucial role in the earth's weather because they are the principal means by which water from the oceans is added to the atmosphere and, ultimately, to the land surface.

At any one time only a very small fraction, about $\frac{1}{1000}$ of 1 percent, of the earth's surface water is present in the atmosphere, either as invisible water vapor or visible liquid or solid cloud droplets; most of the rest (97 percent) is in the oceans, but about 3 percent occurs on land in rivers, lakes, ground water, and ice sheets (Figure 6.2a). All of the water present on land originates in the oceans and ultimately returns there as rivers flow into the sea. This runoff of water from the land is balanced by atmospheric motions which continuously transfer water from the oceans to the land through the processes of evaporation, condensation in clouds, and precipitation as rain or snow (Figure 6.2b). Much of our discussion of climates, landscapes, and sediments in Part 3 will center around the influence of this **hydrologic cycle** on the solid surface of the land. For the moment, we shall be concerned only with

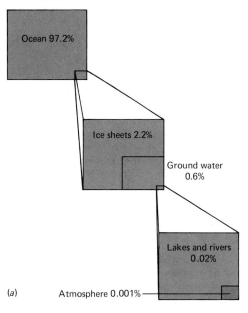

(a)

Ocean 97.2%

Ice sheets 2.2%

Ground water 0.6%

Lakes and rivers 0.02%

Atmosphere 0.001%

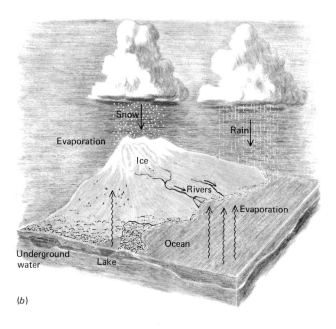

(b)

Snow

Evaporation

Rain

Ice

Rivers

Evaporation

Underground water

Lake

Ocean

Figure 6.2 Atmospheric moisture. (a) Distribution of water on the earth. Only a small fraction of 1 percent occurs in the atmosphere at any one time. (b) The hydrologic cycle, by which water is moved from the ocean to the atmosphere by evaporation and convection, then returned to the ocean by precipitation and land run-off.

the actual processes of evaporation, condensation, and precipitation that make up the fundamental steps in the cycle.

Because a specific volume of *warm* air can hold far more water vapor than the same volume of cooler air, and because evaporation also increases at higher temperatures, most water vapor in the atmosphere is found in the lowermost levels of the troposphere over the warm tropical and temperate oceans. In these regions strong solar heating of both the ocean surface and the overlying air causes evaporation to take place almost continuously. Vertical convective motions are essential for maintaining this evaporation because, if the air near the water surface were stationary or moved only horizontally, it would quickly become saturated with water vapor and evaporation would cease. Vertical convection transports the vapor-laden air away from the evaporating surface where it is replaced by downward motions of drier air. This process allows a continuous addition of water vapor to the lowermost atmosphere in regions where the earth's surface is strongly heated by the sun. Then, as this air rises from the surface and is cooled by adiabatic expansion, the saturation temperature is ultimately reached, condensation occurs, and clouds are formed.

It should be emphasized that the water vapor added to the troposphere by evaporation and local vertical motions may or may not *immediately* condense as clouds, depending on the local stability relations. Once added to the lowermost atmosphere by vertical motions, the vapor may be transported great distances by horizontal winds before it rises high enough to be cooled below the saturation temperature and form clouds. Such lifting and cooling usually results from one of three primary processes (Figure 6.3).

The first of these, *convective cooling*, was discussed earlier (see Figure 6.1, right). In this process the cooling is related only to local vertical transport of heated bubbles of air. Such movements, we have seen, are the primary means of addition of water *vapor* to the atmosphere. They also commonly, but not always, lead immediately to cloud formation. Convective clouds always have a distinctive appearance relating to their origin: because they are caused by local vertical updrafts, they are normally not continuous but are scattered with clear air in between. Furthermore, they are mostly *vertical* rather than horizontal in their development. The billowy floating white clouds that are so characteristic of the tropics, and of summer afternoons in temperate regions (Figure 6.4), are the principal clouds of convective cooling. Known as **cumulus clouds,** these usually accompany mild, fair weather. On occasion, however, the updrafts become so violent that local **thunderstorms** result. These are huge, towering, convective clouds, with strong winds, lightning, and thunder, often accompanied by rain or hail (Figure 6.5).

The second and third principal cooling processes responsible for cloud formation result not from local vertical updrafts, but from much broader uplifts of larger masses of air. Such uplifts occur when *horizontal* air movements are forced to rise by either of two processes. One, known as **synoptic cooling,** occurs when large bodies of air with differing temperatures converge horizontally (Figure 6.3b). When this happens, the cooler, more dense mass of air tends to remain near the surface and the warmer, less dense air is forced upward over it, leading to cooling and condensation. This is the process usually responsible for the extensive horizontal cloud developments, known as **stratus clouds,** that provide most of the precipitation in middle latitudes, where cold polar air frequently converges with warm tropical air (Figure 6.6).

In the final process, called **orographic cooling,** the air is forced upward by mountains or other prominences on the earth's solid surface (Figure 6.3c). The clouds that perpetually cover mountains standing in the path of oce-

Figure 6.3   The three principal processes that lift and cool air to form clouds.

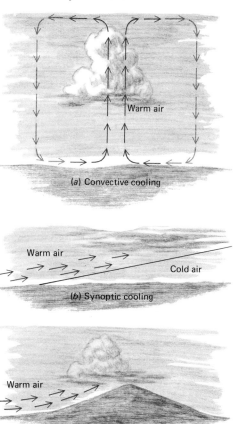

(a) Convective cooling

(b) Synoptic cooling

(c) Orographic cooling

(a)

(b)

(c)

Figure 6.4 Cumulus clouds (a) and towering cumulus thunderstorms (b). Cumulus clouds and towering thunderstorm viewed from the air (c). (a, c: Josef Muench; b: NCAR)

anic winds are a familiar example (Figure 6.7). The clouds produced by orographic cooling may be either of the vertical cumulus or horizontal stratus type, or both.

Both cumulus clouds and stratus clouds usually have their bases at altitudes of 300 to 3,000 m (1,000 to 10,000 ft) above the earth's surface (Figure 6.8). Thick stratus clouds seldom extend higher than 4,500 m (15,000 ft), although thin, wispy, stratuslike clouds made up of ice particles and called **cirrus clouds** are common between 4,500 and 12,000 m (15,000 and 40,000 ft). The only thick clouds of the upper troposphere are vertical cumulus clouds, particularly of the towering thunderstorm type, which commonly reach to the top of the troposphere (about 15,000 m). Clouds are rare above the troposphere because very little moisture escapes condensation by the cold tropopause. This cold zone acts as a moisture barrier and confines the earth's clouds and precipitation to the underlying troposphere.

There is one final kind of atmospheric condensation to be considered. This is **fog,** condensation that takes place not above the land surface as clouds, but at or very near ground level. The formation of fog is always related to local temperature differences between the earth's surface and the overlying air. When the land or ocean surface ceases to be warmed by the sun after sunset, the air immediately in contact with the surface may cool below the saturation temperature (Figure 6.9a). *Dew* is the most familiar result, but when the cooling is more extreme and the air very moist, fog can form by the same process. Such *radiation fog* usually disappears early in the morning as the lowermost air layers are again heated above the saturation temperature by the sun. More persistant *advection fogs* form when winds carry warm moist air over cooler water or land (Figure 6.9b). Such fogs are common along temperate coastlines during the spring and fall.

6.3 Precipitation

So far we have implied that condensation of water vapor, and the resultant formation of clouds, *always* occur when moisture-bearing air is cooled below the saturation temperature. Laboratory experiments and cloud observations show, however, that moist air may be cooled far below the saturation temperature without condensation if the air lacks tiny solid particles to serve as *nuclei* for condensation. Such cooling below the saturation temperature without condensation, known as **supersaturation,** is rela-

Figure 6.5 Stages in thunderstorm development.
Strong updrafts first build a cumulus cloud that
towers into the upper troposphere (left). With further growth, ice crystals and snow form in the cooled
upper regions; some of these descend in strong downdrafts, producing heavy rain or hail (center). Ultimately moisture-supplying updrafts cease, downdrafts
predominate, precipitation becomes less intense, and
the cloud dissipates (right).

tively rare because aerosols (small solid or liquid particles;
see Section 5.1) are present throughout the atmosphere.
These normally provide enough potential nuclei to cause
condensation whenever moist air is cooled to the saturation temperature. Not all aerosols, however, are equally
effective in causing condensation.

Most condensation and cloud formation takes place
around aerosol particles which, because of their composition and structure, have a special affinity for water. Such
particles are of two main sorts. The most abundant and
important are tiny *salt crystals* which originate from the
evaporation of spray from ocean waves and are widely
distributed in the lower atmosphere by the wind. Because
of its molecular structure, salt has a strong affinity for
water, as is demonstrated by clogged salt shakers during

Figure 6.6 Stratus clouds. (NOAA)

Figure 6.7 A small orographic cloud. (NOAA)

humid weather. As a result of this affinity, atmospheric condensation readily begins around salt nuclei. Salt particles are the principal *natural* condensation nuclei, but in industrial regions man-made atmospheric pollutants, particularly various oxides of sulfur and phosphorus, may be even more effective in converting water vapor to liquid droplets. The all-too-familiar result is the noxious mixture of water particles and irritating smoke known as **smog.**

Condensation around solid nuclei converts invisible gaseous water vapor into the tiny visible droplets of liquid water that make up clouds and fog. These initial liquid droplets are extremely small, so small, in fact, that they would tend to remain suspended in the air indefinitely if there were no mechanism for aggregating them into the larger, heavier particles which exceed the buoyancy of the surrounding air and thus fall to the earth's surface as rain or snow. Fortunately, such mechanisms do exist; if they did not, the troposphere would be quickly saturated by evaporation and the earth would be perpetually shrouded by clouds and fog. By steadily removing water from the air, precipitation keeps the troposphere continually undersaturated and maintains the balance of the all-important hydrologic cycle. In spite of the importance of precipitation, both for a basic understanding of the earth and for its universal influence on man's affairs, the exact processes which convert the tiny water droplets of clouds and fog to large raindrops and snowflakes are still not fully understood.

The principal difficulty in explaining precipitation is that the tiny liquid droplets of clouds and fog, once formed, grow by further condensation extremely slowly. They thus tend to remain very small. The exact reason for this is not certain, although there is some evidence that the tiny initial particles develop a slight electrical charge which causes them to repel each other and remain separated in the same way that like poles of two bar magnets repel each other. The average raindrop or snowflake has a mass a *million* times as great as these tiny liquid cloud droplets (Figure 6.10). The forces responsible for rain and snow must, therefore, work strongly in opposition to the observed tendency of cloud particles to remain as separate tiny droplets.

The most recent theories propose that large precipitation particles result from two processes which may occur either independently or together in the same cloud (Figure 6.11). In the first process, the large particles originate

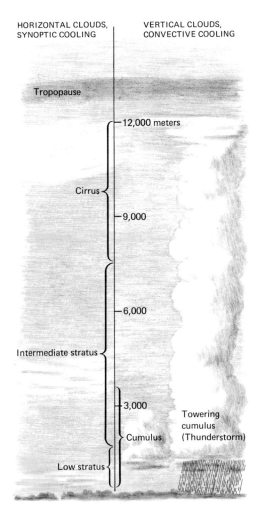

HORIZONTAL CLOUDS, SYNOPTIC COOLING

VERTICAL CLOUDS, CONVECTIVE COOLING

Tropopause

12,000 meters

Cirrus

9,000

6,000

Intermediate stratus

3,000

Towering cumulus (Thunderstorm)

Cumulus

Low stratus

Figure 6.8 Typical height relationships of clouds. Stratus clouds and typical cumulus clouds mostly form in the lower troposphere below 4,500 m (15,000 ft). Only thin cirrus clouds and towering thunderstorms normally form in the cold upper troposphere.

from condensation around unusually large, but relatively rare, salt nuclei. Because of the large size of the initial nucleus, the particle quickly exceeds the buoyancy of air as water is added to it by condensation; as a result the particle begins to fall. As it falls it begins to accumulate much smaller droplets and thus to grow larger. Soon it grows so large that it becomes unstable and breaks into several medium-sized droplets that continue to fall, grow, and break apart in a sort of chain reaction. Laboratory experiments suggest that as few as one large salt nucleus per 100,000 smaller nuclei could initiate rainfall by this process. This process is thought to play a large role in rainfall in the tropics, where large salt particles are relatively common and where clouds are seldom cooled below freezing, the necessary condition for precipitation by the second process.

When stratus clouds, or the upper portion of vertically developed cumulus clouds, are cooled below 0°C (32°F), the tiny liquid water droplets may be converted to ice. The qualifying "may be" is necessary because, just as with the vapor-liquid transition, water need not change phase instantaneously, but may remain liquid at temperatures *far below* 0°C if there are no suitable nuclei present to initiate the growth of icy crystals. Laboratory studies show that small *dust* particles, rather than the salt particles which facilitate the vapor-liquid transition, are most effective for initiating ice crystal growth. They do not, however, become fully effective until the temperature falls to −26°C (−15°F); temperatures below this level, rather than merely below freezing, are therefore usually required to initiate freezing of cloud droplets.

Ice crystal formation at such low temperatures is believed to be a primary cause of precipitation in temperate and polar regions because, as the icy crystals begin to form in a cloud of supercooled liquid droplets, they tend to attract adjacent droplets and thus grow much larger than the original liquid particles (Figure 6.11). This tendency of ice crystals to grow at the expense of liquid water droplets is a fundamental property of water related to the greater stability of the solid over the liquid phase. When such rapidly growing ice particles exceed the buoyancy of the surrounding air, they begin to fall as snow or hail. If the temperature near the surface is below freezing (or if the particles are very large), they reach the ground still frozen; otherwise they melt on their downward journey and reach the ground as rain.

Figure 6.9 Fog. (a) Two common types of fog. In radiation fogs, cooling of the land surface at night chills the overlying air below the saturation temperature. In advection fogs, winds bring warm moist air over cooler water or land. (b) Radiation fog. (c) Advection fog. (b: Photo Researchers, Inc.; c: United Press International)

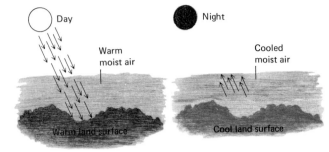

Radiation fog

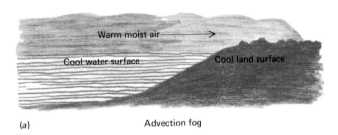

Advection fog

(a)

(b)

(c)

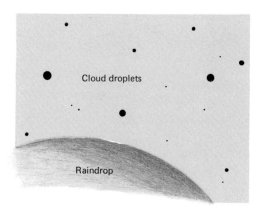

Figure 6.10 *Proportional size of average raindrops and cloud droplets. About a million cloud droplets must aggregate to form a raindrop.*

Figure 6.11 *The two principal processes of raindrop formation: condensation around unusually large nuclei (left); ice crystal growth by attraction of surrounding droplets (right). The ice crystal will melt and fall as rain unless temperatures at the earth's surface are below freezing.*

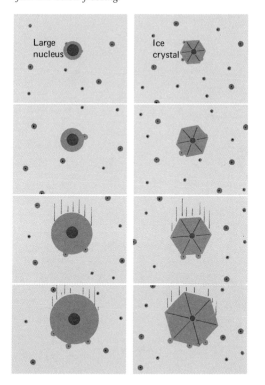

HORIZONTAL MOTIONS: WINDS AND GENERAL CIRCULATION

For the remainder of this chapter we shall consider horizontal wind motions, which are a far more obvious feature of the troposphere than are vertical movements. These horizontal motions occur on several scales. The largest are relatively constant world-encircling wind patterns that make up the *general circulation* of the troposphere. Superimposed on these large-scale wind patterns are a series of progressively smaller and less regular circulations which lead to cyclonic storms, hurricanes, and other, more local, weather phenomena.

Some idea of the differing magnitudes of horizontal and vertical atmospheric motions can be gained by the following comparison: the troposphere has a *horizontal* extent of about 10,000 km (6,000 mi) from equator to pole and through this great distance tropospheric air is continuously in horizontal motion at an average speed of 32 km/hr (20 mi/hr). In contrast, the average *vertical* extent of the troposphere is only 15 km (9 mi), and through this much shorter distance the air is continuously in vertical motion at an average speed of only 32 *meters*/hr (105 ft/hr). Horizontal winds are therefore by far the most prominent tropospheric motions. Because of these different scales of movement, it is convenient to discuss horizontal and vertical atmospheric motions separately. It should be emphasized, however, that the motions continuously interact as, for example, when moisture or clouds caused by vertical motions are simultaneously transported great distances by winds.

6.4 Horizontal Stability of the Troposphere

Horizontal atmospheric motions are caused by unequal solar heating of the earth's spherical surface. Far more energy is delivered to equatorial regions, where the sun's rays are approximately vertical, than to polar regions, where the rays meet the earth's surface at a low angle (Figure 6.12). Because of this imbalance, the energy excess of warm equatorial regions tends to flow toward the energy-deficient poles where it is reradiated into space. This transfer is accomplished primarily by horizontal wind motions, which are thus ultimately driven by the planetary

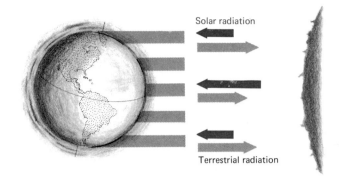

Figure 6.12   Distribution of solar energy on the earth's surface. Equatorial regions, where the sun's rays are approximately vertical, receive far more energy per unit area than do polar regions (color arrows). The imbalance is corrected by horizontal wind motions that transport energy to polar regions where the excess is reradiated to space (gray arrows).

temperature gradient. Most of the energy transferred by this process is radiated back into space from the cold polar regions, but a small fraction, about 1 to 2 percent of the total, is consumed in driving the winds which accomplish the energy transport.

Before considering the earth's general wind patterns, we shall digress briefly to consider the interrelationship of wind motion and atmospheric pressure. The same four forces that cause ocean movements—gravity, friction, pressure gradients, and the Coriolis effect (Section 4.7)—also interact to produce the motions of the fluid gases of the atmosphere. For the atmosphere, however, their relative importance differs. We have already considered the effect of the earth's gravity (Section 5.3); unlike the relatively incompressible waters of the oceans, the compressible gases of the atmosphere respond to the earth's gravity by expanding or contracting until their density at a particular distance from the earth's surface exactly balances the gravitational forces at that point. Because gases have this property of readily altering their density by expansion or contraction, gravitational forces are generally balanced throughout the atmosphere. Exceptions occur only when the atmosphere is made vertically unstable by excessive local heating through the processes discussed earlier in this chapter. Gravity thus plays a role in the *vertical* motions of the atmosphere, but may be considered to be everywhere balanced by the rapid upward decrease in atmospheric density and pressure for the analysis of *horizontal* motions.

The three remaining forces—friction, pressure gradients, and the Coriolis effect—are primarily responsible for horizontal wind motions. Of these three, friction becomes important only for motions at or very near the

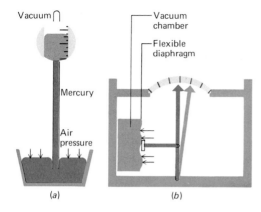

Figure 6.13 Barometers. (a) Schematic diagram of a mercury barometer. Atmospheric pressure changes cause the column of mercury to rise or fall in the vacuum tube. (b) Schematic diagram of an aneroid barometer. Atmospheric pressure changes move the flexible diaphragm, causing motions of the pointer.

Figure 6.14 "Radiosonde" balloons being released at sea and from an arctic weather station. The instrument package radios to a ground station continuous information on pressure, temperature, and humidity as the balloon ascends through the troposphere. (a: NOAA; b: National Film Board of Canada).

earth's surface. Because of friction, winds near the surface are generally far less intense and more irregular than are winds just a hundred meters or so above the surface, a fact that is easily confirmed by a visit to an open observation area atop any tall building. Most atmospheric wind motions take place above the thin zone where friction with the surface dominates and, for this reason, frictional forces are commonly omitted in analysis of wind motions, leaving only pressure gradients and the Coriolis force to be considered.

We have seen that there is everywhere in the atmosphere a strong *vertical* pressure gradient, with pressures decreasing rapidly away from the earth's surface. This vertical gradient does not ordinarily cause motions because the gradient is balanced by the earth's gravity. But if one measures *horizontal* pressures in the atmosphere at a constant distance from the earth's center [for example, at sea level, or at 6,000 m (20,000 ft) above sea level], it is found that there are also horizontal pressure differences in the atmosphere. These are, however, far smaller than are the pressure differences occurring through only a short vertical distance. At sea level, for example, horizontal pressure changes have a maximum range about equal to the normal vertical pressure drop observed in ascending to an altitude of only 1,000 m (3,300 ft). These subtle horizontal pressure differences are ultimately caused by the unequal solar heating between the earth's equatorial and polar regions and are responsible for the earth's wind patterns. It is the changing patterns of these slight horizontal pressure differences that are recorded by **barometers,** sensitive pressure-measuring instruments that play a fundamental role in analyzing and predicting patterns of wind and weather (Figure 6.13). From simultaneous barometer readings at many points on or above the earth's surface (barometers are routinely carried aloft by weather balloons; Figure 6.14) it is possible to construct maps showing the distribution of horizontal pressure at any level in the troposphere at that time. From the patterns of lines of equal pressure, called **isobars,** shown on these maps it is possible to infer wind patterns and many other aspects of the weather at that time. Indeed, wind motions are seldom measured directly in the atmosphere, but are almost always calculated from such maps based only on pressure measurements. These **weather maps,** as they are called, are the primary tool of wind and weather analysis (Figure 6.15).

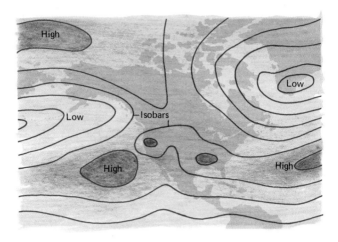

Figure 6.15 A generalized weather map, showing patterns of atmospheric pressure at the earth's surface. Similar maps are prepared daily over much of the earth and for various levels of the troposphere.

Figure 6.16 The relation of atmospheric pressure (P) to wind direction. (a) On a nonrotating earth pressure gradients alone would be active and winds would blow directly between areas of high and low pressure. (b) The earth's rotation leads to a Coriolis deflection (C) to the right (in the northern hemisphere) and causes winds to blow between the pressure areas parallel to the isobars. Such "geostrophic" winds dominate all the troposphere except near the earth's surface, where friction (F) causes winds to blow at a slight angle to the isobars (c).

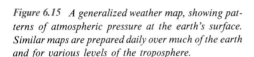

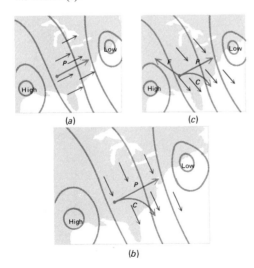

At first glance it seems that winds should always flow directly from regions of high pressure to regions of low pressure as shown on weather maps (Figure 6.16). This would be the case if the earth were not rotating but, because it is, the Coriolis effect causes winds moving from areas of high pressure to areas of low pressure to be deflected to the right in the northern hemisphere and to the left in the southern hemisphere (see Section 4.7). Above the earth's surface, where frictional effects complicate the pattern, this Coriolis deflection continues until the motion is at *right angles* to the pressure gradient. The effect of the Coriolis force is thus to cause winds to flow not directly *between* areas of high and low pressure, but rather *around* them, in directions which parallel, rather than cross, the isobars (Figure 6.16). Such winds, flowing parallel to isobars as a result of a balance between pressure gradients and the Coriolis force, are called **geostrophic winds.** Observations show that geostrophic wind motions dominate the troposphere except near the earth's surface where friction causes wind motions to cross the isobars at a low angle and thus to flow partially *between,* rather than *around,* regions of high and low pressure (Figure 6.16).

6.5   General Circulation of the Troposphere

Because the wind motions of the troposphere result primarily from the interaction of horizontal pressure differences and the Coriolis effect, it might seem that it would be relatively simple to deduce mathematically, using the principles of fluid dynamics, the broad patterns of global wind flow. In practice, however, such computations are

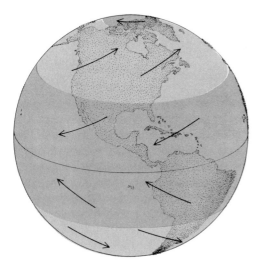

Figure 6.17 *Generalized wind directions at the earth's surface. Winds from the west predominate in middle latitudes (color); winds from the east in equatorial and polar regions.*

extraordinarily complicated. Although primarily caused by differential heating with latitude, wind-producing pressure differences are also affected by other factors, such as seasonal temperature variations and surface topography; the latter plays a complicating role because local patterns of heating differ over land and water, mountain and plain, ice pack and rain forest. The earth's wind patterns are influenced by *all* of these variables, and thus are extremely difficult to describe mathematically. Even though some progress has been made toward approximating such descriptions, most of our knowledge of wind motions still rests on direct observation, rather than on theoretical analysis.

From the early world-encircling voyages of sailing ships, it was discovered that surface winds around the earth tend to blow in consistent directions in several great belts that parallel the equator (Figure 6.17). In tropical regions the winds blow predominately from the *east;* in midlatitude belts north and south of the tropics, winds from the *west* predominate. Patterns of surface winds are less consistent in polar regions, but there is a tendency for winds from the east to dominate as in the tropics. Many theoretical models of the general circulation of the troposphere were devised from these surface wind patterns, but it is only with the greatly increased observations of winds *above* the surface, made over the past 20 years by means of weather balloons (Figure 6.14) and aircraft, that the *actual* circulation patterns have been established.

These observations show that surface winds, because of the frictional effect of the land and ocean, are not representative of winds occurring through most of the troposphere. Above the surface, the midlatitude zones of westerly winds tend to expand north and south over the equatorial and polar belts of easterly winds (Figure 6.18). Furthermore, the velocity of this dominant westerly flow increases rapidly with height, reaching its maximum near the tropopause in relatively small zones, called **jet streams,** where the velocities may reach as high as 400 km/hr (250 mi/hr). In contrast, the zones of easterly flow maintain relatively low velocities throughout. It is evident, therefore, that most of the energy of tropospheric winds is concentrated in the high-velocity midlatitude westerlies; these high-energy westerly winds also show one other significant feature. Although their flow is *dominantly* westward, when looked at in three dimensions the patterns of maximum flow are not *directly* from west to east, but follow a sinuous, continuously changing, wavelike pattern around the

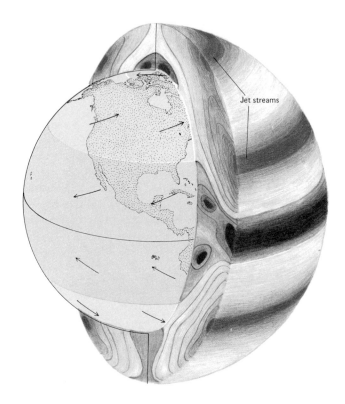

Jet streams

Figure 6.18 *Generalized wind directions through the troposphere. The belts of westerly winds* (color) *expand and intensify with altitude, reaching maximum velocities in thin, sinuous jet streams blowing near the tropopause.*

earth (Figure 6.19). This wavelike irregularity is responsible for most of the changing weather of the earth's middle latitudes.

The patterns of circulation and wind flow in the troposphere are now well documented, but there is still no satisfactory explanation of exactly how the absorption, transfer, and radiation of solar energy leads to the observed patterns. The principal difficulty is to account for the enormous north-south, or equator-to-pole, transfers of energy required for overall energy balance by means of winds that are predominantly east-west in their motions. This larger problem really has two subparts: first, why are the motions predominantly east-west but with westerlies dominant? and, second, how can these winds, once explained, transfer enough energy *poleward* to keep the earth in radiative balance?

The first problem, the reason for the east-west flow, is the more easily attacked. In a general way, the greater energy supplied to equatorial regions should cause heating, expansion, and relatively lower pressures at any horizontal level than in cooler, poleward regions. Indeed, this

Figure 6.19 Typical patterns of sinuous jet stream flow. When jet stream waves are poorly developed, east-west flow is dominant and little heat is transported from warm equatorial zones to cool polar regions (a). When the waves intensify (b) more heat is transported poleward in the sinuous loops.

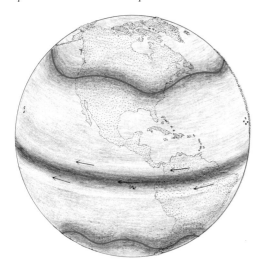

(a)

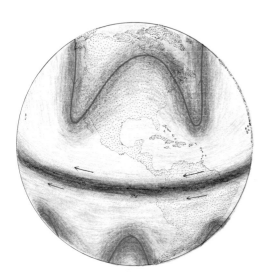

(b)

is the pattern seen on long-term averaged weather maps; the tropics have, overall, lower pressures than cooler midlatitudes. This difference would, on a nonrotating earth, lead to simple north-south flow between the belts of high and low pressure. Recall, however, that on the rotating earth the Coriolis effect deflects the flow until it is at *right angles* to the pressure gradient; in the case of the general north-south pressure gradient of the earth, we would expect the flow to be east-west as observed. There remains, however, the still more difficult question of why there is a dominance of strong westerly winds. There are good reasons why the flow cannot be *all* in one direction, either east or west. If it were all from the west, the atmosphere would tend to gradually gain momentum from friction with the earth's surface, which is rotating in the same direction at a much higher velocity. If this were the case, wind speeds would increase to velocities approaching those of the earth's rotation. Conversely, winds blowing from the east *against* the earth's direction of rotation would tend to gradually lose momentum and cease. From such considerations it is clear that both east and west flow are necessary to maintain balance, but the precise causes of the observed pattern of east-west flow are still unknown.

Still more fundamental is the second question: how can the predominantly east-west flow of the troposphere provide a large transfer of energy from the warm equator to the cool poles? Here it is clear that the large-scale sinuous waves in the midlatitude westerlies are the key to the dynamic balance of the entire troposphere. This strong westerly flow is concentrated in the transitional zone between the continuously warm tropics and the continuously cold polar regions. When, as is sometimes the case for periods ranging from several days to several weeks, the westerly flow shows weak sinuous waves and flows mostly from west to east, little heat is transferred from equatorial regions and, at the same time, midlatitude weather is relatively mild and stable (Figure 6.19a). At other times, the winds show extreme north-south deflections which result in a net northward transport of warm tropical air and a southward movement of cold polar air (Figure 6.19b); at such times midlatitude weather is very unstable, with extremes of high and low temperatures. Theoretical calculations show that these large-scale wave motions are probably the principal means of north-south transport of heat energy in the troposphere.

6.6   Motions above the Troposphere

Before considering smaller-scale motions of the troposphere, we should briefly consider how its large-scale
circulation interacts with the overlying atmosphere. Fortunately there is a considerable body of information about
winds in the lower regions of the adjacent stratosphere
(see Figure 5.10) because modern weather balloons commonly ascend through the tropopause. Data gathered with
these balloons shows that the temperature minimum of
the tropopause does not act as a sharp boundary or ''lid''
on tropospheric circulation. Instead, the dominant westerly wind patterns of the troposphere extend well into the
lower stratosphere. It is significant, however, that jet
streams, the zones of maximum wind velocities, occur very
near the tropopause, with velocities decreasing above that
level. More limited observations suggest, rather surprisingly, that higher in the stratosphere *easterly* winds dominate. In general, however, the stratosphere appears to be
more stable with less wind motion than the troposphere,
as would be expected since it is a zone of upwardly increasing temperatures where, as in the ocean, vertical,
convective motions are minimized.

In contrast with the stratosphere, the overlying
mesosphere is again a zone of upward temperature decrease, just as in the troposphere. Like the troposphere,
it is heated primarily from below but not, of course, from
the earth's surface, but by the absorption of solar energy
by ozone at heights around 50 km (30 mi). Because of
this heating from below, the mesosphere, although made
up of far thinner gases at much lower pressures than the
troposphere, can be predicted to have as complex a pattern of general circulation, but not necessarily the *same
pattern*, as does the troposphere. Limited rocket observations suggest that the pattern *is* different, for strong and
persistent *north-south* winds, which are rare at lower levels, have been reported.

One additional result of balloon observations into
the stratosphere has been information about the height
and variability of the tropopause itself, which is the zone
of minimum temperatures separating the troposphere and
stratosphere. As would be expected, the troposphere is
higher over tropical regions, where solar energy input and
vertical convective motion are greatest, than over the poles
(Figure 6.20). The tropical tropopause has an average

*Figure 6.20   Generalized height of the tropopause,
which is high over the heated air of equatorial regions
and lower over the cool poles. There is a "gap" where
the tropopause temperature minimum is poorly defined
in the region of the middle-latitude jet streams.*

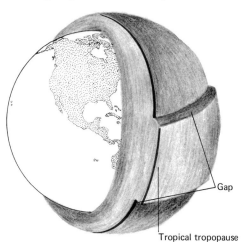

Gap

Tropical tropopause

Man and Weather

The all-encompassing influence of weather on human activity has led many nations to cooperate in establishing a worldwide system of weather stations that continuously report local weather conditions to each other. By combining such reports, most of the stations then attempt to *predict* what their local weather will be for various future intervals. The most important uses of such predictions relate to the operation of aircraft, the avoidance of storm effects, and to agriculture.

Aircraft are particularly vulnerable to weather phenomenon and, as a result, the bulk of weather reporting and predicting is aimed at facilitating aviation. Although modern jet aircraft cruise near the tropopause, far above most weather, they must take off, climb, descend, and land through the turbulent troposphere where three kinds of weather phenomenon—fog, ice, and thunderstorms—are their principal hazards. Fog (and extremely heavy rain or snow) at the earth's surface can so reduce visibility that landing, even with the most sophisticated instrument systems, becomes impossible. Likewise, ice accumulating on an aircraft in flight can increase weight and decrease stability with disastrous results. Finally, the violent convective motions associated with severe thunderstorms can damage or destroy even the largest aircraft. Because of these hazards, much weather prediction is geared toward anticipating potentially dangerous thunderstorms, ice, and fog.

A less routine but equally important function of weather prediction is to anticipate severe storms, such as tropical cyclones or tornadoes, that might pose hazards to life and property. The paths of tropical cyclones are closely monitored by both aircraft and weather satellites so that coastal areas in their path can be given warning of their arrival. Tornadoes are more difficult to anticipate, but tornado warnings are issued over wide areas whenever potentially tornado-producing atmospheric conditions are observed. Finally, weather predictions can be of great value in agricultural operations, particularly near times of planting and harvesting when unexpected wind, rain, or hail could destroy an entire crop.

Most weather prediction is based on extrapolating present weather patterns into the future. Such extrapolations are usually quite accurate for periods of several hours and reasonably accurate for periods of several days. Beyond that, weather is too changeable for any but the most general extrapolations from existing weather trends. If, however, the atmosphere shows any sort of long-term repetition of weather events, it should be possible to project weather far into the future, provided the exact nature and sequence of the repeating cycles could be determined. Much effort has been expended in recent years in computerized searches for such predictable cycles but, as yet, with only very limited success. Some workers now believe that long-term weather patterns are not cyclic at all, but are instead essentially random in their occurrence. If so, all attempts at long-range prediction will probably prove futile.

"Storm room" of the U.S. National Hurricane Center. The center gathers data on hurricane positions and issues advisories and warnings. (NOAA)

Aircraft departing in snowstorm. (Ellis Herwig; Stock, Boston)

Hurricane striking the island of Tobago in the Caribbean. (Fritz Henle)

height of about 18 km (12 mi); over the poles the average height is only 8 km (5 mi). In the intermediate midlatitudes, the tropopause is more variable and less clearly defined. In this region, where jet streams and maximum wind velocities occur, there appears to be considerable interchange between the very cold air of the lower stratosphere and the somewhat warmer air of the underlying troposphere. These interchanges undoubtly play an important role in the general circulation of the troposphere and thus in the earth's weather patterns but, as yet, they remain poorly understood.

HORIZONTAL MOTIONS: SMALLER CIRCULATIONS

The general circulation and broad wind patterns of the troposphere involve large-scale, average motions of air. These motions are important for they reflect the generalized energy transfers that cause *all* horizontal motions of the troposphere. From the point of view of man living on the earth's surface, however, these broad flow patterns are less apparent than are the relatively minor, turbulent disturbances which occur in, and are carried along with, the overall flow. Local, day-to-day changes in weather at the earth's surface are caused by these smaller disturbances. Unlike the dominant east-west flow of the general circulation, these smaller circulations tend to follow *circular paths*, and can be likened to the circular eddys and whirlpools that develop and are carried along with the flow of a large river. The exact reasons for the tendency toward smaller circular disturbances in the overall tropospheric motion are not yet fully understood but, like the eddys of a moving river, they are probably related to frictional interactions of the moving fluid with the earth's solid surface.

　　The three principal types of circular, weather-producing circulations differ from each other in size, frequency, and intensity (Figure 6.21). The largest, most frequent, and least intense, in terms of the speeds of the circular wind motions, are known as *middle-latitude cyclones* and *anticyclones*. These have average diameters of about 1,500 km (900 mi), and seldom have wind speeds greater than 65 km/hr (40 mi/hr). Smaller and less frequent, but much more intense, are *tropical cyclones*, known more familiarly as "hurricanes" or "typhoons," which have diameters of only 80 to 160 km (50 to 100

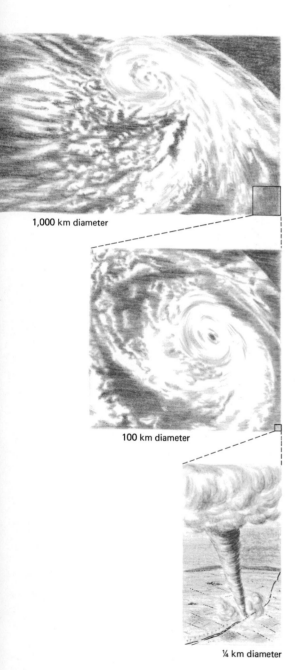

1,000 km diameter

100 km diameter

¼ km diameter

Figure 6.21 Relative sizes of middle-latitude cyclones, tropical cyclones, and tornadoes.

mi) and wind speeds as high as 240 km/hr as (150 mi/hr). Smallest, least common, and most intense of all are *tornadoes,* with diameters of less than a kilometer but with wind speeds as great as 800 km/hr (500 mi/hr).

6.7 Middle-Latitude Cyclones and Anticyclones

Most of the earth's changing weather patterns, outside of tropical regions, are caused by large circular disturbances of the lower troposphere which are continuously forming, moving, and dissipating in the strong, sinuous, westerly wind flow of the earth's middle latitudes. These disturbances are of two sorts. In some the circular winds tend to spiral *inward and upward;* these are known as **cyclones.** In others the predominant spiral motion is *outward and downward;* these are called **anticyclones.** In the northern hemisphere the Coriolis force causes the spiral winds in cyclones to blow counterclockwise, whereas in anticyclones the flow is clockwise (Figure 6.22). In the southern hemisphere these relations are reversed.

Cyclones and anticyclones are recognized by their characteristic patterns of isobars (lines of equal pressure) seen on weather maps. In an anticyclone, the outward and downward spiral flow causes pressures to *increase* towards the center as cooler, more dense air descends from above. Conversely, pressures *decrease* towards the center of a cyclone as warmer, less dense, ground-level air rises and spirals upward near the center. For this reason cyclones are often called *low-pressure systems* or merely "lows," and anticyclones are called *high-pressure systems* or "highs." In both cyclones and anticyclones the isobars assume "closed" or circular patterns in contrast to the linear pattern of noncircular flow.

Because of their characteristic patterns of air motion, the typical weather associated with anticyclones differs from that of cyclones. In anticyclones the general downward spiral of air leads to adiabatic *warming* and, ordinarily, clear, cloudless skies. Conversely, the upward spiraling air of cyclones normally leads to adiabatic *cooling,* condensation, and generally cloudy and unsettled weather. Furthermore the characteristic temperatures differ in the two disturbances. Anticyclones may be either cold or warm depending on whether they originate nearer the tropical or the polar side of the middle latitudes. In either case, however, the temperatures throughout the outward-spiraling air mass tend to be relatively constant and stable. The temperatures of cyclones are usually more

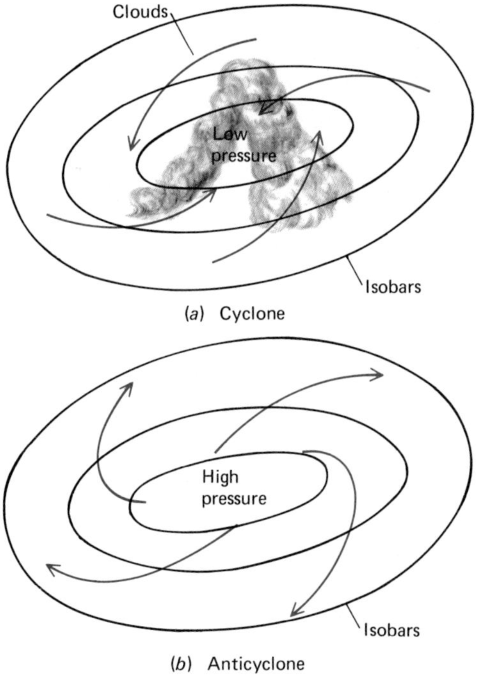

Figure 6.22 Air motions in cyclones and anticyclones. The inward and upwardly spiraling air of cyclones (a) normally produces clouds and unsettled weather; the downward and outward air spiral of anticyclones (b) usually leads to clear, stable weather.

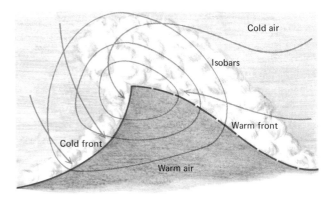

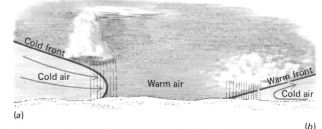

Figure 6.23 Middle-latitude cyclones. (a) Schematic
map view (above) and cross section (below). Cold
fronts are usually accompanied by a line of convective
cumulus clouds and thunderstorms; more gently sloping warm fronts normally cause large areas of stratus
clouds and precipitation. (b) Satellite photograph of
a middle-latitude cyclone. (NASA)

complex (Figure 6.23). The inward-spiraling air tends to be warm on the equatorward side of the circular flow and cold on the polar side; these warm and cool sectors normally have rather sharp boundaries, known as **fronts,** which separate the colder and warmer air.

Cyclones and anticyclones, as well as the middle-latitude westerly wind flow of which they are a part, are strongly affected by the seasons. The decreased solar energy input to the earth's winter hemisphere (see Section 13.1) leads to both an equatorward shift and intensification of the westerly flow. This, in turn, brings more frequent cyclonic disturbances and generally cooler weather to lower latitudes. The sharply contrasting seasons of the middle latitudes result from these changes in the turbulent boundary flow between the continuously warm equatorial regions and the continuously cold poles.

Most of the unsettled weather of the earth's middle latitudes results from eastward movements of cyclones with their characteristic patterns of cloudiness and sharply contrasting warm and cold air. For this reason, cyclones have been studied in far greater detail than have anticyclones with their normally clear, pleasant weather. Analyses of thousands of cyclones based on weather map patterns and, more recently, on satellite photographs of worldwide cloud patterns, have shown that their origin, movement, and ultimate dissipation normally takes place over a period of several days during which they move eastward at an average speed of about 40 km/hr (25 mi/hr). Thus they usually move hundreds or thousands of kilometers during their short lifetimes. They also tend to form in regions where there are sharp differences in patterns of surface heating. In the United States, for example, cyclones most commonly originate and move eastward from near the Rocky Mountains or along the Atlantic Coast (Figure 6.24), areas where the contrast between mountain and plain and land and water lead to sharp local differences in atmospheric heating. Cyclones also show a characteristic pattern of growth and decay that is summarized in Figure 6.25.

On a larger scale, cyclones tend to occur in groups along curving belts that spiral outward from the earth's polar regions. These large spiraling belts are, in turn, related to the large wavelike irregularities that occur in the strong middle-latitude westerly wind flow above the earth's surface. The circular air flow of both cyclones and anticyclones is most pronounced at the earth's surface and tends to die out and merge with the dominant westerly

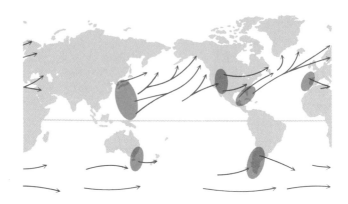

Figure 6.24 *Generalized source regions and movement paths of middle-latitude cyclones.*

Figure 6.25 *Sequence of growth and decay of a middle-latitude cyclone. The cold air zone overrides and replaces the warm zone (b–e), leaving only a whirl of cold air (f), as the cyclone dissipates.*

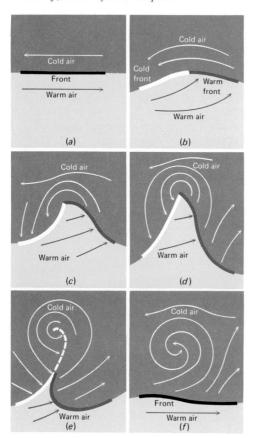

wind flow at altitudes of 3 to 6 km (1.8 to 3.6 mi). Through the remaining troposphere the moving belts of surface cyclones appear merely as moving waves in the upper air flow. When these upper air waves are well developed, surface cyclones and unsettled weather predominate in the middle latitudes. Conversely, when the waves are less well developed and the upper air flow is more directly from west to east, surface cyclones are few and middle-latitude weather is milder and more uniform. These changes tend to take place over a cycle about two weeks long.

Middle-latitude cyclones with their inward-spiraling circular air flow tend to move warm, tropical air towards the poles and cold polar air towards the tropics, and thus to play a large role in the energy transfer processes that drive *all* horizontal motions of the troposphere. As yet, however, there is no really satisfactory dynamic explanation of exactly *why* they form and *how* they interrelate with the more general circulation patterns of the troposphere. An understanding of these relationships is a primary goal of modern atmospheric research.

6.8 Tropical Cyclones

In general, weather patterns in the earth's broad equatorial zone of warm, tropical air are far more regular and stable than are those of the middle latitudes where tropical and polar air continuously converge and interchange in cyclones and anticyclones. The high temperatures and strong, constant, surface winds of the tropics lead, as we have seen, to the addition of large amounts of moisture to the lower troposphere. As a result, convective cumulus clouds and thunderstorms are common throughout the tropics. In contrast, the large-scale stratus clouds and temperature fronts of the middle latitudes, both caused by cyclonic motions, have no tropical counterparts. The tropics are not, however, entirely free of weather disturbances for they are the source region of relatively small but very intense storms of inward- and upward-spiraling air known as **tropical cyclones,** or, more familiarly, as hurricanes or typhoons.

The spiral motion of tropical cyclones superficially resembles that of middle-latitude cyclones but actually differs in scale, intensity, and origin. Tropical cyclones are normally less than 150 km (90 mi) in diameter whereas middle-latitude cyclones are 10 times as large. Conversely, in tropical cyclones winds commonly reach velocities of 250 km/hr (150 mi/hr) whereas velocities greater than 65 km/hr (40 mi/hr) are rare in middle-latitude cyclones. Fortunately, these intense tropical disturbances are also much less frequent than are middle-latitude cyclones. Any point on the earth's surface in middle latitudes is likely to experience a mild cyclonic disturbance at least once every few weeks. In contrast, any point in the tropics is likely to feel the brunt of a tropical cyclone only once every few years, both because the storms are less frequent in occurrence, and because they are much smaller than middle-latitude cyclones. In spite of their relative infrequency at any one *point* in the tropics, satellite observations of worldwide cloud patterns show there is almost always at least one tropical cyclone present somewhere in the earth's tropical regions and thus, on a larger scale, they are a constant and permanent feature of tropical weather.

Just as with middle-latitude disturbances, the exact cause of tropical cyclones is still uncertain. Detailed observations, made both at the earth's surface and in the third dimension by means of specially equipped aircraft, show that they have a characteristic structure which differs in detail from that of the larger, less intense middle-latitude cyclones (Figure 6.26). Tropical cyclones always have a relatively calm "eye" region, from 10 to 50 km (6 to 30 mi) across, in the center of the upward-spiraling air mass; in this central eye the air is *descending* instead of rising. As with all descending air masses, this air is adiabatically warmed and is free of clouds, and thus contrasts sharply with the intense clouds produced by the rising spiral of air around it. Such an eye region is not normally present in middle-latitude cyclones. Furthermore, all tropical cyclones originate and maintain their form only when moving over the oceans. They rather quickly dissipate over

(a)

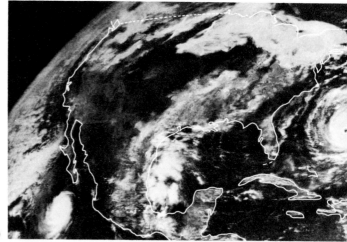

(b)

(c)

Figure 6.26 *Tropical cyclones or "hurricanes."*
(a) Structure of a typical tropical cyclone. Air
spiraling rapidly upward produces towering convective
clouds except in the central "eye," where descending
air produces a relatively calm, clear zone. (b) Sat-
ellite photograph showing two tropical cyclones ap-
proaching North America. (c) Oblique closeup view of
a tropical cyclone. (b: NOAA; c: NASA)

land, indicating that a continuous supply of evaporating moisture from below is necessary for their maintenance. These observations suggest that tropical cyclones are a kind of convective "engine" in which moist air, spiraling upward from the ocean surface, releases heat by condensation to form clouds and precipitation before it descends again in the dry central eye to complete the cycle. Just how this engine originates and maintains its concentrated energy against frictional and other dissipating forces remains poorly understood.

Coriolis forces are believed to play a role in initiating the spiral motion of tropical cyclones because even though they always form over warm tropical seas, they never originate very near the equator where the Coriolis force is negligible (Figure 6.27). Some minor, preexisting, circular flow also seems to be required. Such minor irregularities are common in the tropics but only a relatively few develop into full-scale tropical cyclones.

After they form, tropical cyclones usually move poleward and, like middle-latitude disturbances, are carried along as minor eddys in the prevailing general circulation. As long as they remain in the tropics they tend to move with the easterly wind flow, but in passing to the middle latitudes they reverse direction and move with the prevailing westerly winds (Figure 6.27). The total life cycle of a tropical cyclone lasts from a few days to a month and ends when the storm is dissipated either by passing over land, or by absorption into the larger turbulent disturbances of the middle latitudes. Because of their very high winds and very low atmospheric pressures (which tend to cause sealevel rise and flooding), tropical cyclones can be extremely destructive when they come ashore in populated coastal areas. Because they quickly dissipate over land, they seldom cause damage to inland regions, except by flooding due to heavy rains.

Figure 6.27   Generalized source regions and movement paths of tropical cyclones.

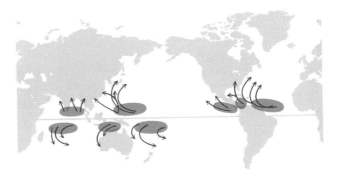

6.9 Tornadoes

Tornadoes, the final circulation that we shall consider, resemble larger cyclonic disturbances in being moving, rotating masses of air, but are far smaller and much more intense than cyclones. They are whirling, funnel-shaped vortices with a diameter of a few hundred meters and a height of only a thousand meters or so, but with maximum wind velocities ranging from 500 to 800 km/hr (about 300 to 500 mi/hr). Such winds are by far the most intense and destructive of all atmospheric disturbances but, fortunately, they are much less common than are larger cyclonic disturbances and of much shorter duration, usually lasting only a few minutes.

Tornadoes are always associated with severe thunderstorms and unusually violent convective overturn of the lower troposphere. They are especially likely to occur when cold dry polar air passes *over* warm moist tropical air. Normally such moving cold air, being denser than the warmer moist air, tends to stay near the earth's surface and pass *under* the warmer air. Under exceptional wind conditions, however, the cold dry air may begin to pass *above* the less dense warm moist air, thus creating an

Figure 6.28 Tornadoes. (a) Tornado funnel over Wichita Falls, Texas, 1958. (b) Tornado funnel over Hardtner, Kansas, 1936. (a: United Press International; b: American Red Cross)

(a)

(b)

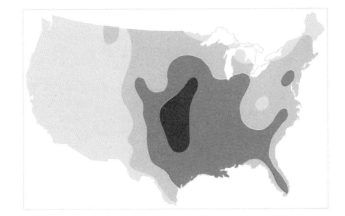

Figure 6.29 *Frequency of tornadoes in the United States. The areas of darker color receive most of the 150 or so tornadoes reported each year.*

unusually unstable situation. The result is violent convective motions, intense thunderstorms, and, sometimes, a series of swirling tornado vortices extending to the earth's surface beneath some of the thunderstorm clouds (Figure 6.28).

Tornadoes occur in middle latitudes where cold and warm air masses interact. Although they occasionally occur almost everywhere in the middle latitudes, they are common only in a few regions where conditions particularly favor violent convective interactions of warm and cold air. The central United States and central Asia are such regions and have the world's highest frequency of tornadoes (Figure 6.29). In the United States about 150 are reported each year, but because of their small size and short duration only a very few of these cause human injury or property damage.

SUMMARY OUTLINE

Vertical Motions and Atmospheric Moisture

6.1 *Vertical stability of the troposphere:* depends on the degree of heating of the earth's surface and the consequent rate of temperature decrease away from surface; this rate is complicated by adiabatic cooling and condensation warming of rising air.

6.2 *Clouds:* vertical motions transfer moisture from the earth's surface to the atmosphere; as water vapor is cooled in rising air, condensation takes place forming clouds; vertical cumulus clouds develop from convective cooling, horizontal stratus clouds from synoptic cooling; clouds are generally confined to the troposphere.

6.3 *Precipitation:* removes water vapor from the atmosphere and balances the hydrologic cycle; precipitation particles, a million times more massive than cloud droplets, form from consolidation of droplets around large salt nuclei or ice crystals.

Horizontal Motions: Winds and General Circulation

6.4 *Horizontal stability of the troposphere:* horizontal pressure differences and the Coriolis effect control wind motions except near the earth's surface, where friction also becomes important; the Coriolis effect causes winds to flow around, rather than between, areas of high and low pressure.

6.5 *General circulation of the troposphere:* the easterly winds of equatorial regions are replaced by much stronger westerly winds in middle latitudes; energy is transferred from the warm equator to the cool poles by sinuous irregularities in the westerly flow.

6.6 *Motions above the troposphere:* the stratosphere is more stable than the troposphere; limited observations suggest different wind and circulation patterns in the stratosphere and mesosphere.

Horizontal Motions: Smaller Circulations

6.7 *Middle-latitude cyclones and anticyclones:* large, circular, eddylike disturbances moving in the general wind flow cause most of the earth's day-to-day weather changes in middle latitudes; outward and downward flow in anticyclones normally produces high pressure and clear weather; inward and upward flow of cyclones produces low pressure and complex patterns of cloudy, warm air and clear, cool air separated by moving fronts.

6.8 *Tropical cyclones:* are smaller, less common, and more violent than middle-latitude disturbances; inward and upward spiraling air produces high winds and intense precipitation; they originate over tropical oceans and move poleward.

6.9 *Tornadoes:* are the smallest, least common, and most violent disturbances; they are associated with sharply contrasting cold and warm air masses and rapid convective overturn, and are most common in the central United States and central Asia.

ADDITIONAL READINGS

*Barrett, E. C.: *Viewing Weather from Space,* Longmans, Green and Co., London, 1967. An introductory survey, emphasizing the impact of satellite observations of world weather patterns.

*Battan, L. J.: *Cloud Physics and Cloud Seeding,* Doubleday & Company, Inc., Garden City, New York, 1962. A readable popular introduction.

*Battan, L. J.: *The Nature of Violent Storms,* Doubleday & Company, Inc., Garden City, New York, 1961. A popular summary emphasizing thunderstorms, tornadoes, and hurricanes.

*Hidy, G. M.: *The Winds,* D. Van Nostrand Co., Inc., Princeton, N.J., 1967. A clearly written intermediate-level introduction to atmospheric motions.

* Lehr, P. E., R. W. Burnett, and H. S. Zim: *Weather*, Golden Press, New York, 1965. A well-illustrated popular introduction.

* Miller, A.: *Meteorology*, Charles E. Merrill Publishing Co., Columbus, Ohio, 1971. A brief introductory survey.

Pettersen, S.: *Introduction to Meteorology*, McGraw-Hill Book Company, New York, 1969. Among the best intermediate texts.

* Available in paperback.

PART 3

Interactions of the Solid and Fluid Earth

Part 3 begins a consideration of the complex interactions that take place as the rocks of the solid earth come in contact with, and are profoundly changed by, the water and air of the fluid earth. The three chapters of Part 3 mostly consider these interactions as we see them today and in the immediate past. Part 4 will treat these same interactions through the much longer course of earth history.

Chapter 7 discusses the nature and causes of *climates*, long-term trends in surface weather patterns that are the primary control of interactions between the solid and fluid earth.

Chapter 8 then treats the shaping of the surface of the solid earth into differing *landscapes* by agents of the fluid earth under the influence of differing climates.

Chapter 9 concludes Part 3 with a consideration of how agents of the fluid earth move materials derived from the solid earth to new locations and deposit them, as *sediments*, to form new rocks.

7

Climates

Climates, which may be defined as long-term patterns of weather occurring at the earth's surface, control the interaction of air and water with rocks of the solid earth. Although closely related to weather phenomena, climates differ from weather in at least two fundamental respects. First, climatic patterns involve much longer time intervals than do the transient motions that produce weather. In addition, climatic studies emphasize the effects of weather on the earth's solid surface, rather than being primarily concerned with phenomena taking place within the atmosphere. In this chapter we shall first consider the patterns of climate seen on the earth today, and shall then discuss the evidence for, and possible causes of, long-term changes in these patterns. Finally, we shall introduce the differing effects of climate on the surface of the solid earth by considering the formation of *soils* through the processes that are known, collectively, as *weathering*.

PRESENT-DAY CLIMATES

7.1 Climatic Controls

In discussing weather in Chapter 6, we have already provided much of the background necessary for understanding the earth's climates as they exist today. The vertical and horizontal motions of the troposphere that cause weather also act, over longer periods of time, to produce the earth's patterns of climate. All weather phenomena, such as clouds, cyclones, and thunderstorms, are relatively minor disturbances in the more constant world-wide wind patterns that make up the general circulation of the troposphere. This circulation, acting over many years, is the primary control of climates, for it largely determines the amounts of heat and moisture that reach different parts of the earth's surface. In addition to the

Pattiwick Glacier, Greenland.
(U.S. Coast Guard)

237

general circulation, there is one additional factor that also has a strong influence on climate: the distribution of the earth's land and ocean.

Land areas affect climatic patterns because they absorb and retain much less heat from the sun than does the ocean. There are two primary reasons for this. First, the ocean, being fluid, mixes and distributes heat through a much thicker layer (the "mixed layer" of Chapter 4) than is possible in the soil and rock of the solid earth. In addition, water has the property of being able to absorb and retain far more heat than does the same volume of most other materials, including rock or soil. For these reasons, the ocean (and large lakes on land) tends to change temperature much more slowly than the land surface, and thus to have a more thermally stable climate. Because of this difference, the central parts of large continental masses, far from the influence of the ocean, tend to have climates characterized by wide extremes of temperature (Figure 7.1).

The configuration of the land also affects climate in another way. Low, flat land areas do not impede the flow of tropospheric air, but mountains and plateaus that extend far about sea level profoundly affect the general circulation because the troposphere has an average thickness of only 15 km (9 mi), while many of the earth's mountain ranges reach heights of 5 km (3 mi) above sea level, and the highest extend upwards to 10 km, (6 mi), or through two-thirds of the troposphere. These highland barriers influence climate by removing most of the moisture from the air, through orographic cooling and precipitation, along their windward sides, and by thus creating "rain-shadow" deserts on their leeward sides.

In summary, the distribution of land and water around the earth's surface profoundly affects climates by

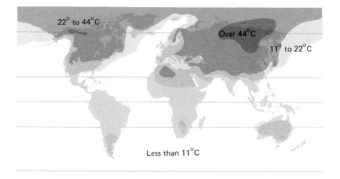

Figure 7.1 Annual range of temperatures at the earth's surface. The maximum range occurs over large continental areas located in the middle latitudes.

22° to 44°C

Over 44°C

11° to 22°C

Less than 11°C

sharply influencing patterns of both temperature and pre-cipitation. Superimposed on this "continental effect" is the larger-scale control of climate by the general atmos-pheric circulation. The distribution of climates on the earth today results from the complex interaction of these two principal controlling factors.

7.2 Climatic Classification

Many schemes have been devised to describe the climatic patterns that occur today over the earth's surface. Most are based on temperature and rainfall distribution as indi-cated by the kinds of native plants that cover the land surface. Because plants are profoundly affected by tem-perature and moisture, they provide a sensitive index to long-term climatic patterns. Most of these schemes recog-nize only a few major climatic types, each with a great many local subdivisions. The presence of hills, lakes, and other surface features can strongly influence climate over distances of only a few kilometers, and thus climatic dis-tribution can be extremely complex when viewed in detail; generalized worldwide patterns, however, can be rather simply described and explained.

Figure 7.2 shows a simplified classification of the present-day earth into four great climatic regions. Three of these, *tropical, temperate,* and *polar* climates, show a

Figure 7.2 A simplified classification of the earth's present-day climates. Cool temperate climates have generally cold winters and mild summers; warm tem-perate have mild winters and hot summers.

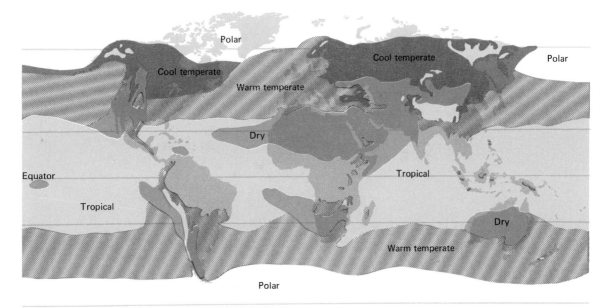

broad distribution into east-west bands that parallel the equator and are controlled primarily by the general poleward decrease in surface temperature. Tropical climates are dominated by the warm moist air surrounding the equator and polar climates are dominated by cold polar air; temperate climates occur in the relatively unstable zone of midlatitude cyclones and anticyclones that separates tropical and polar air. Seasonal changes are minor or absent in tropical climates, which have no winters, and in polar climates, which have no summers. Temperate climates, on the other hand, show sharp seasonal variations caused by north-south shifts in the westerly winds and cyclonic disturbances of the middle latitudes (see Sections 6.7 and 13.1).

Tropical, temperate, and polar climates all have at least moderate amounts of precipitation each year. In this they differ from the fourth climatic type, *dry* climates, which are somewhat more complex, since they are controlled not simply by temperature, but by a relative lack of precipitation which, in turn, results from the interaction of the general circulation with the distribution of land and ocean. An examination of Figure 7.2 shows that dry climates tend to predominate in two world-encircling belts, one on each side of the equatorial zone of tropical climates; that is, there tends to be a dry climate belt separating tropical and temperate climates over much of the area of each continent. These dry belts are usually explained as being due to an overall descent of tropical air away from the equatorial zone of maximum heating and rising. Recall that adiabatic cooling always leads to precipitation when air moves upward, and to dry cloudless skies when air is descending. Moisture-laden air raised by heating leads to precipitation in the tropics, and then the air tends to descend in adjacent regions to cause the broad belts of dry climate. This circulation can be thought of as being caused by two large convective cells, one on each side of the equator, which provide a generalized north-south movement within the more rapid easterly wind flow of the equatorial zone (Figure 7.3).

The subtropical belts of dry climate are continuous across northern Africa and into central Asia, but on all other continents, the dry climates tend to predominate on the *westward* side of the continents, whereas on the eastward side, tropical climates pass into temperate climates without an intervening belt of dry climate. Two factors probably cause this asymmetry. In the first place, the dry climate belts fall generally in the zones of easterly surface

Figure 7.3   *Origin of dry climatic belts by generalized north-south convection. Moist air at the equatorial zone of maximum heating rises and cools causing clouds and precipitation. The air descends in adjacent poleward belts causing dry climates.*

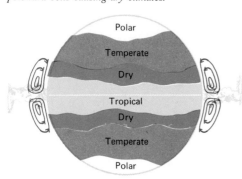

winds, which tend to transport oceanic moisture inland to cause precipitation along the eastern side of the continents. Furthermore, the major currents of the ocean's surface waters move poleward on the *western* side of the ocean basins (adjacent to the *eastern* margins of the continents) carrying with them warm tropical water as in the Gulf Stream. This poleward movement of warm tropical water tends to increase the water vapor in the air and thus favors precipitation along the eastern margins of the continents.

Note also that on three continents, North America, South America, and Asia, dry climates extend poleward far into the zone normally dominated by temperate climates. In each case these dry regions beyond the normal "dry belt" are associated with the earth's highest mountain ranges (North American Cordillera, South American Andes, and Asian Himalayas), and are largely the result of removal of moisture from midlatitude air by orographic cooling and precipitation.

CLIMATIC CHANGE

Much evidence indicates that the earth's climatic patterns are not constant but, instead, tend to change with time. Small-scale climatic changes occur so rapidly that several can be seen in the relatively short period during which man has made systematic weather records. More severe climatic changes are indicated by accounts of droughts, floods, warmings, and coolings that have taken place during the 5,000 years since the beginnings of recorded human history. Still more profound climatic variations are shown by geologic evidence which extends not just thousands, but thousands of *millions* of years into the past. In discussing these climatic changes we shall first briefly review the evidence that such changes have taken place and then look at the more complex question of their underlying cause.

7.3   Historical Evidence of Climatic Change

Man has been making systematic records of daily temperatures and precipitation for only about 200 years. In 1800 there were 5 weather stations making daily observations in the United States, 12 such stations in Europe, and none over the enormous remaining expanse of the earth's surface. Not until 1876 were the first weather ships estab-

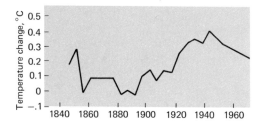

Figure 7.4 *Trends of average world annual temperatures since 1840. (After Mitchell in* Changes of Climate, *1963)*

lished to make observations at sea, and even today weather records are scarce for much of the 70 percent of the earth's surface that is covered by water.

In spite of the many gaps and uncertainties in these weather records, they do clearly show that the earth's overall climatic pattern has not been constant over the past 200 years. Instead there was a trend toward warmer weather during the late nineteenth and early twentieth centuries. Beginning about 1940, this warming trend reversed and worldwide average temperatures began dropping, a trend which continues today (Figure 7.4). These relatively short-term climatic trends have involved changes in yearly average temperatures of less than 1 °C and local changes in the amount of rainfall of only about 10 percent. Such changes are too small to have much effect on human activities, but they do indicate a sensitive balance in the dynamic forces which control climate, and suggest that still more severe cycles of climatic change could occur over longer periods of time.

Direct evidence that more severe climatic changes have taken place is provided by earlier written records which, although not generally made to record climate, nevertheless reflect dramatic climatic events. To mention only a few examples: rock paintings made about 3000 B.C. in the central Sahara, one of the driest regions on earth today, picture the hunting of water-loving hippopotamuses from canoes (Figure 7.5); Roman records made in North Africa around 200 A.D. indicate the presence of much wetter conditions with local vegetation and animals, such as lions and elephants, which are today characteristic of more moist central Africa; records from 700 A.D. indicate heavy traffic over Alpine passes that are today filled with the thick ice of mountain glaciers; in 1100 A.D. ice was common in the Nile at Cairo where today freezing temperatures are virtually unknown. These and many similar historical anecdotes reflect dramatic changes in the earth's climatic patterns over the past 5,000 years but are, unfortunately, too local and unsystematic to give a clear-cut sequence of climatic events. More useful evidence for climatic history during man's brief period of written records, and for the millions of preceding years, is provided, however, by geological evidence of climatic change.

7.4 Geological Evidence of Climatic Change

The clearest and most complete evidence of climatic change is provided by the effect of past climates on the rocks and living organisms of the earth's surface; these

Figure 7.5 *Rock painting from the central Sahara desert, made about 3000 B.C., showing men on an ox (lower left) and in a canoe (right) hunting a hippopotamus (upper left), a water-loving animal that lives only near lakes or large rivers. (H. Lhote)*

effects, when preserved in the rocks of the solid earth, provide a geological record of past climates. There are four principal kinds of such evidence: sediments, erosion features, fossils, and oxygen nuclides.

Sediments   The most abundant and useful evidence of past climates is provided by ancient sediments preserved today as sedimentary rocks. Most sedimentary rocks do not give an unequivocal indication of the climate under which they were formed, but a few types do clearly reflect climatic conditions. Sandstones made up of the former sand of desert dunes (Figure 7.6a) and *tillite*, a rock composed of a mixture of clay with angular sand and gravel particles deposited by glaciers, are examples of two of the most useful kinds of sedimentary rocks for inferring past climates.

Erosion Features   A second kind of geologic evidence of past climates is provided by erosion features preserved in ancient rocks. Moving glaciers, for example, erode valleys into a characteristic U-shaped profile, whereas the more usual valleys carved by streams typically have a V-shaped cross section (see Figure 8.12). On a smaller scale, moving

(a)

(b)

Figure 7.6 Typical geological evidence of past climates. (a) Ancient dune sands, preserved as cliffs of sandstone in southern Utah. (b) Glacial striae, grooves scoured by the flow of former glaciers. These are on granite in the mountains of eastern California. (a, b: Tad Nichols)

glaciers also leave characteristic scratched grooves on the rock over which they move because of the rasping action of pebbles frozen into the glacier ice. Such grooves, called **glacial striae,** may be preserved in ancient rocks to indicate past glacial conditions (Figure 7.6b). Other erosional climatic indicators are old shorelines and river banks which may be preserved high above the present surface of lakes or rivers. Such evidence of higher water levels suggests more humid past climates.

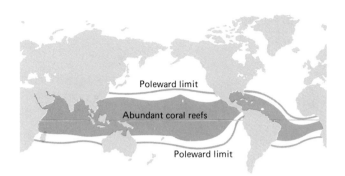

Poleward limit
Abundant coral reefs
Poleward limit

Figure 7.7   Present-day distribution of coral reefs, which are confined to warm equatorial seas. Most ancient coral reefs probably lived under similar conditions.

Fossils   Still another record of past climates is provided by **fossils,** the remains of past life preserved in sedimentary rocks. As with sediments and erosion features, not all fossils are useful for inferring climates, but remains of animals or plants that have restricted climatic requirements may provide suggestive clues. Reef corals, for example, grow today only in warm tropical seas and it may be reasonably concluded that fossil coral reefs formed under similar conditions (Figure 7.7). Among the most useful fossils for inferring climates are pollen grains, which are extremely tiny reproductive structures of flowering plants. These are sometimes preserved by the billions in muds that accumulated at the bottom of ancient lakes. Flowering plants are extremely sensitive to climatic change and, often, the sequence of pollen types found through many meters of lake sediment may be used to construct a detailed record of local climatic change (Figure 7.8).

Oxygen Nuclides   The final geological evidence of past climate is provided by a geochemical method for determining ancient temperatures that has been developed during the past 20 years. This method is based on the observation that the oxygen dissolved in the oceans today consists of two principal nuclides (see Section 1.2). Most of the oxygen occurs as the common nuclide ^{16}O, but a second, heavier nuclide, ^{18}O, is also present in small amounts—about one atom for every 500 atoms of ^{16}O. Through a complex series of chemical reactions, some of the dissolved oxygen of the oceans becomes incorporated into the calcium carbonate ($CaCO_3$) shells built by animals living in the sea. The amount of ^{16}O and ^{18}O so incorporated, however, changes slightly with the temperature of the sea water; the warmer the water, the more ^{18}O is present. The temperature at which fossil shells of ancient

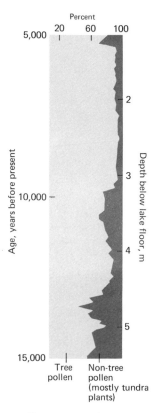

Percent

20   60   100

5,000

2

Age, years before present

Depth below lake floor, m

3

10,000

4

5

15,000

Tree pollen

Non-tree pollen (mostly tundra plants)

Figure 7.8 Pollen sequence through several meters of lake floor sediment in northern Germany. The proportion of tundra plants decreases due to a climatic warming about 10,000 years ago. (*Modified from Flint,* Glacial and Pleistocene Geology, *1957, after Muller*)

marine organisms were deposited can thus be estimated by measuring their ratios of ^{16}O to ^{18}O and comparing the results to a temperature scale established from similar measurements on present-day shells.

This method, in theory, permits a precise quantitative statement of ancient temperatures rather than the more qualitative estimates provided by sediments, erosion features, and fossil distribution; for this reason, it attracted much attention when it was first established in the early 1950s. Unfortunately, recent work has shown that there are complicating factors that make the method less than ideal. For example, the nuclide distribution in fossil shells may be altered after they are buried in the sedimentary rock, and it is usually very difficult to determine the extent of such alteration. Another difficulty is that the method assumes that the overall oxygen nuclide ratios in ancient and present-day ocean water are the same, yet there are good reasons for suspecting that this is not the case. Because of these and other complications, climatic inferences from oxygen nuclide studies have supplemented, but not replaced, the evidence provided by sediments, erosion features, and fossil distribution.

Interpreting the Geological Evidence   Each of the four kinds of geological evidence for past climates provides clues to climatic conditions only if both the *time* and *place* of their origin can be determined. The methods for establishing the relative timing of such past events will be the subject of Chapter 10; here it is important only to recognize that they usually can be related, by various methods, to a chronology that spans many hundreds of millions of years of earth history.

With regard to past climatic events, it is not so much the *time* that they originally occurred that presents difficulties, but the *place*. Because the earth has probably always had a more-or-less spherical shape that has received the bulk of its surface energy from the sun, an equator-to-pole gradient of decreasing temperatures (caused by the lesser amounts of solar energy that reach the poles; see Figure 6.12) has been present throughout earth history. Thus, at any time in the earth's past, the poles should be relatively cooler than equatorial regions, although not necessarily to the same degree as today. To make general inferences about past climates, we must therefore know the precise latitude at which a specific climatic indicator originated. For example, evidence of past glaciation at the poles would be a far less surprising and significant event than glaciation near the equator, which would suggest worldwide cold climates. The difficulty in establishing the latitudinal position of such past events arises, of course, because of the movement of lithosphere plates (Sections 3.9 and 3.10). Because of plate motions, the continents have not retained a fixed position with respect to the earth's poles of rotation throughout earth history; instead, they have repeatedly changed their positions. Furthermore, there is as yet no unequivocal means of determining the nature of these changes except during the most recent interval of earth history. For these reasons, it has so far proved difficult to integrate much of the geological evidence of past climates into a geographic summary of the earth's climatic changes.

There are, however, two extremely significant generalizations that *can* be made about the earth's climatic history. First, the presence of fossil plants in rocks representing most of the earth's long history indicates that really *extreme* climatic changes, changes that consistently exceeded the relatively limited range of temperatures tolerated by life, have not taken place. Thus climatic changes leading to average worldwide temperatures either below freezing, or above about 55°C (130°F), have probably never occurred except, perhaps, in the very earliest stages of earth history before life originated.

The second generalization is that present-day climates are probably not typical of much of earth history but, instead, are relatively cooler. Thick ice sheets today cover much of Greenland and Antarctica and there is abundant geological evidence that only 20,000 years ago, an extremely short time when compared with the thousands of millions of years of earth history, similar ice sheets covered much of northern Eurasia and North America as well (Figure 7.9). Furthermore, there is much evidence that expansion and contraction of these vast ice sheets has occurred several times over the past few million years. The geological record of sediments and erosion features left by such widespread ice is clear and unmistakable; rather surprisingly, however, only two such glacial intervals, both relatively brief, are indicated in the rock record before the last few million years—one occurred around 300 million years ago, the other about 700 million years ago (Figure 7.10). The relative scarcity of geological evidence for past ice ages is the principal reason for believing that the most recent interval of earth history, up to and including the present day, is climatically unusual. This interval of ice sheet expansion and contraction is known as the *Pleistocene Epoch*. In Part 4 we shall consider this epoch in its broader historical context. We shall

Figure 7.9 *Ice sheets of the present day (dark color) and about 20,000 years ago (light color) in the northern hemisphere. (From Bloom,* The Surface of the Earth, *1969)*

also refer to these Pleistocene glaciations throughout Part 3, for the extreme climatic fluctuations that they record have profoundly influenced the earth's present-day patterns of landscapes and sediments.

7.5 Causes of Climatic Change

Both historical records and geologic evidence make it abundantly clear that climates have shown many fluctuations in the course of earth history but, as yet, there is no generally accepted theory for the *causes* of these fluctuations. Geological evidence of changing climatic belts during the Pleistocene Epoch has shown that they reflect not local, but worldwide alternations between relatively warm and relatively cool climates. These fluctuations, in turn, can be related to changes in the position and strength of the midlatitude belt of westerly winds, which marks the unstable thermal transition between the continuously warm air of the tropics and the continuously cold air of polar regions.

The pattern of climates that we have already discussed as occurring on the present-day earth is typical of the relatively warmer, "interglacial" intervals. During the alternating intervals of maximum glaciation there was an equatorward-shift and intensification of the zone of westerlies, which led to much larger areas of cold, polar air

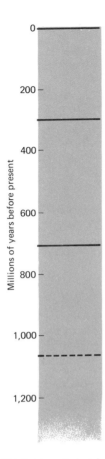

Figure 7.10 Distribution of glacial deposits (color) during the past 1,200 million years of earth history. Glaciations comparable to those of the last few million years occurred about 300 and 700 million years ago. Scattered glacial deposits are found in rocks older than 1,000 million years, but they are difficult to interrelate in time and space.

and, at the same time, brought cyclonic precipitation to subtropical dry regions that were formerly deserts. At such times the zone of hot, moist, equatorial air, and the adjacent subtropical desert belt, are much compressed (Figure 7.11). The glacial intervals end when the westerly cyclonic belts again shift poleward, bringing about a poleward expansion of both moist tropical air and the subtropical belt of dry climates. These displacements of the westerly cyclonic wind belt have been the *immediate* cause of the unusually extreme climatic fluctuations of Pleistocene time, but the more basic question still remains—what causes the circulation pattern to shift in the first place?

Changes in the general circulation pattern of the atmosphere must be caused by changes in its ultimate driving force, which is the earth's overall budget of solar energy. Such changes, in turn, can take place in three ways: the inflow of energy *into* the atmosphere may change, or the outflow of energy *from* the atmosphere may change, or, finally, the inflow and outflow of energy may remain constant with only changes in the distribution of energy *within* the atmosphere. Many theories of climatic change have been proposed with ingenious mechanisms for accomplishing one or more of these variations in the solar energy budget. The most probable and frequently proposed are summarized in Figure 7.12.

Among the most likely factors that might change the *input* of solar energy are: simple variations in the sun's energy output, slight variations in the earth's orbit around the sun, and blocking and scattering of solar energy by wind-blown volcanic dust as it passes through the atmosphere. Conversely, the *outflow* of solar energy from the atmosphere might be affected either by altering the "greenhouse effect" through changing the concentration of large, energy-absorbing atmospheric gas molecules (principally CO_2, O_3, and H_2O; see Section 5.5), or by increasing the reflectivity of the earth's surface through expanded snow and ice cover, which tends to reflect far more energy than does rock and soil. Finally, the *distribution* of energy within the atmosphere could be altered by changing the levels of the land or ocean or both. High land areas extend into cool tropospheric air and favor orographic precipitation. On the other hand, high sea levels lead to less exposed land and milder, "oceanic" climates, whereas low sea levels increase the exposed land area and cause more extreme "continental" climates.

The principal objections to each of these possible mechanisms of climatic change are summarized in Figure 7.12. Because of these objections, no single mechanism,

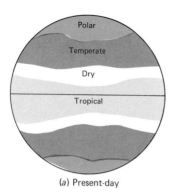

(a) Present-day

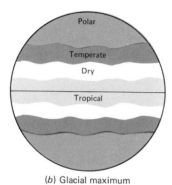

(b) Glacial maximum

Figure 7.11 *Schematic representation of the expansion of polar climates and compression of the temperate, dry, and tropical belts during intervals of maximum ice-sheet expansion.*

or combination of mechanisms, has been generally accepted as the fundamental cause of climatic change. Perhaps the most promising—and the one currently favored—is the simplest, that which calls on direct variations in the output of solar energy. Very minor variations of this sort are known to occur and these seem to correlate reasonably well with small, short-term climatic fluctuations. Still more evidence will be required, however, to firmly establish such variations as a primary cause of large-scale climatic changes.

CLIMATES AND THE LAND SURFACE

Much of the remainder of Part 3 will be concerned with differing climates as they act to control the shaping of the land surface and to influence the accumulation of sedimentary debris. These closely related processes of land sculpture and sedimentation begin when air and water, under the influence of climate, profoundly alter the rocks of the land surface through the actions known as *rock weathering;* the initial result of this rock weathering is the formation of *soil,* altered rock debris which covers the underlying crustal rock over much of the land surface.

7.6 Weathering

There are two principal processes by which air and water act to change the structure and composition of rocks at the earth's surface. The first are **mechanical weathering** processes, the most important of which is known as **frost wedging.** Most rocks exposed near the surface have many small cracks and cavities filled with rainwater. Because water expands in volume as it freezes (Section 4.2), it can exert tremendous pressures on the surrounding rock as temperatures fall below freezing. When freezing and thawing alternate, as often occurs as the land surface is warmed during the day and cooled at night in polar and cool-temperate regions, the repeated stresses of frost wedging can be a significant cause of the breakup of massive rocks into smaller fragments (Figure 7.13).

Of still greater importance than mechanical weathering are the many processes, grouped together as **chemical weathering,** whereby both the structure *and* the chemical composition of rocks are changed. Most of these processes center on the chemical actions of the earth's

Cause		Principal Objections
Variations In	Due To	
Inflow of solar energy	Changes in sun's energy output	Observational records of solar output cover too short an interval to be certain that significantly large changes occur; no adequate mechanism proposed for large variations in solar output
	Variation in the earth's orbital path around the sun	Several regular variations occur with periods of 10,000 to 100,000 years; none show convincing correlation with intervals of known climatic change
	Scattering of incoming solar energy by volcanic dust	Effect probably both too small and too transient to account for long-term climatic change
Outflow of solar energy	Changing concentrations of energy absorbing atmospheric gases ("greenhouse effect")	Requires changes in concentration of absorbing gases (principally CO_2 and water vapor) that are too large to be accounted for by any known mechanism; no independent evidence of such changes
	Changing reflectivity of earth's surface because of snow and ice	Require changing patterns of snow and ice cover and thus more likely a secondary (and perhaps reinforcing) effect of climatic change rather than a primary cause
Distribution of solar energy	Changing levels of land and sea	Changing elevation of land normally too slow to account for shorter-scale climatic fluctuation; shorter-term sea-level changes primarily due to removal of ocean water in land ice sheets, apparently secondary effect rather than cause

Figure 7.12   Some frequently suggested causes of climatic change, and the principal objections that have been raised against them.

surface waters. Absolutely pure water is chemically rather inactive, but natural surface waters are normally slightly acidic due to the presence of dissolved CO_2 from the atmosphere, some of which combines with hydrogen in the water to form a weak solution of carbonic acid. Dead and decaying plant and animal remains also contribute acids to surface waters. Like most weak acids, these surface waters are good solvents and slowly dissolve many minerals; they also enter directly into chemical reactions to form new minerals.

We saw in Chapter 3 that most of the earth's crust is composed of either silicon-rich granite or magnesium- and iron-rich basalt. The minerals of these two igneous rocks are therefore the primary materials subjected to weathering as they are exposed to acid waters at the earth's surface. Such waters ultimately convert the feldspars of granite and basalt to the fine-grained sheet silicate *clay minerals*, which contain a relatively large amount of water in their crystal structures and

(a)

(b)

(c)

(d)

(e)

(f)

Figure 7.13 The reduction of a massive limestone boulder to small fragments by mechanical frost wedging. (From Thornbury, Principles of Geomorphology, *1969; photos by Alan Pratt)*

252

are much more stable than are most igneous minerals under the temperatures and chemical environments of the earth's surface (Figure 7.14). Quartz, which is extremely resistant to chemical attack, is preserved without chemical alteration.

During alteration, much of the potassium, sodium, calcium, magnesium, and silicon of the original igneous minerals goes into solution and is removed from the place of origin by flowing water. Eventually, these elements are either deposited elsewhere in new sedimentary minerals, or added by rivers to the oceans, which contain huge quantities of land-derived dissolved materials. Of the half-dozen or so common elements in igneous rocks, only iron and aluminum are relatively insoluble and tend to persist in solid form under the attack of chemical weathering. The iron is quickly converted to insoluble, rustlike, oxide minerals, while the aluminum forms a principal constituent of clay minerals. When chemical weathering is especially intense even the silicon-oxygen tetrahedra of the clay minerals themselves may be removed by solution,

Figure 7.14 *Schematic summary of the breakdown of granite and basalt by chemical weathering. Feldspar alters to clay minerals which under very intense weathering may, in turn, alter to aluminum oxides. Pyroxene and olivine alter to iron oxides. Quartz is preserved without chemical alteration.*

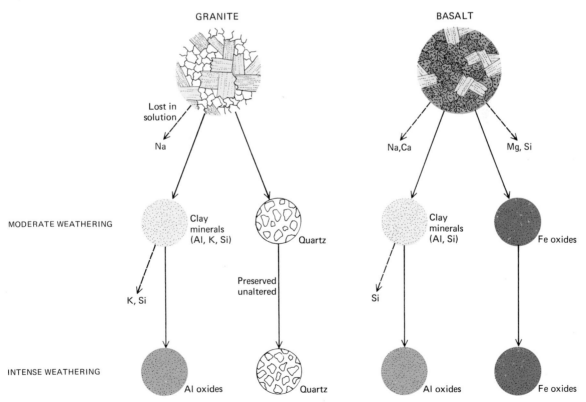

leaving only a silicon-free residue of aluminum oxide (Figure 7.14).

7.7 Soil

The result of mechanical and chemical weathering is to cover much of the land surface with a thin veneer of altered rock debris known as **soil.** Soil, as we might predict from our discussion of the products of weathering, is made up primarily of tiny particles of various clay minerals mixed, in varying proportions, with larger particles of quartz. Because quartz is the only really stable igneous mineral under the action of chemical weathering, most of the gravel and sand-sized particles of the earth's surface are composed of it.

In addition to the clay and quartz fragments produced by rock weathering, most soils also contain from 1 percent to as much as 50 percent organic carbon compounds derived from the decay of plants and animals supported by the soil. These organic constituents are extremely important components of soils for they are closely interrelated with soil texture, structure, and composition. Plants, in particular, profoundly affect soils in many ways. By utilizing dissolved elements left over from rock weathering—particularly potassium, sodium, and calcium—in their metabolic processes, they tend to retain these elements near their site of formation instead of allowing them to be dispersed by moving water. These components are returned to the soil as the plants die and decay, and are then utilized by new plants in an endless cycle. Only when this cycle is broken, for example by harvesting the plants instead of allowing them to die and decay, are these nutrient elements permanently removed from the soil. It is for this reason, of course, that supplemental nutrients in the form of fertilizer are necessary for most cultivated soils.

Plant roots and burrowing animals also tend to facilitate weathering by making the soil porous so that surface water can penetrate to unweathered, underlying rocks. In addition, the presence of organic acids from plant decay adds to the chemical activity of surface waters, as we have already noted. These processes tend, in general, to sharply increase the rate of chemical weathering on those portions of the land surface that are covered by plants. At the same time, plants facilitate the accumulation of weathered debris, as soil, by binding together the particles with their roots, and by protecting the soil surface from the erosive effects of wind and water.

*Man
and Climates*

On the time scale of a single human lifetime, climates are a relatively static and unchanging aspect of the earth. Over the long course of man's history, however, climatic changes have been a primary influence in the rise and decline of human cultures. For the future, such natural climatic changes taking place on a scale of many hundreds or thousands of years are of less immediate concern than is the prospect that industrialized man may be altering the atmosphere to produce changes over a much shorter time.

The most evident man-made climatic changes are those associated with large cities, all of which show a clear-cut rise in both average temperature and rainfall when compared with the surrounding countryside. On a larger scale there is a possibility, as yet unproved, that man's activities may be altering climates over the entire earth—for example, by the addition of large quantities of carbon dioxide to the atmosphere through the burning of coal and petroleum (see p. 188). The most serious danger in such alterations would be the possibility of large contractions or expansions of the continental ice sheets. When we consider that less than 20,000 years ago ice covered much of North America and Europe, it can be seen that another large-scale ice expansion could have a profound effect on man. Likewise, if much of the present Greenland and Antarctic ice sheets were to melt, sea level would rise 50 to 100 m (165 to 330 ft), slowly flooding the continental margins and, with them, all present-day coastal cities. Fortunately, there is no evidence that such disasters are imminent, yet the mere possibility that they could occur makes it essential that man try to understand the nature and causes of his own climatic modifications.

More positively, as climates interact with the solid earth to cause weathering they may lead to useful concentrations of essential mineral resources. In particular, the intense chemical weathering of humid, tropical regions commonly removes all rock constituents except insoluble iron or aluminum, which thus become concentrated as oxide ore minerals. Iron-rich tropical soils, called *laterites*, are mined in some areas, but are relatively less important than are other sources of concentrated iron. In contrast, almost *all* industrial aluminum comes from aluminum-rich soils, called *bauxites*, formed by intense tropical weathering.

Large, unexploited deposits of bauxite are widespread in the present-day tropics, including parts of Australia, the Caribbean area, northern South America, and central Africa. It has been estimated that these reserves will meet all demands for this important metal until well into the twenty-first century. Even then, however, there is little chance of an aluminum shortage for a potentially unlimited supply of the element occurs in the ubiquitous, but somewhat less aluminum-rich, clay minerals.

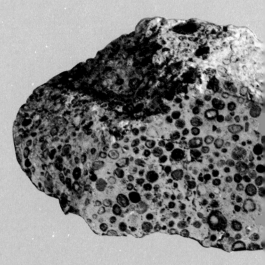

Typical bauxite ore. The specimen is about 25 cm (10 in.) across. (Reynolds Aluminum Co.)

Filling a pit from which bauxite has been mined, Jamaica. (Reynolds Aluminum Co.)

Ingots of aluminum awaiting shipment. (French Embassy Information Division)

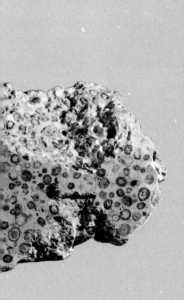

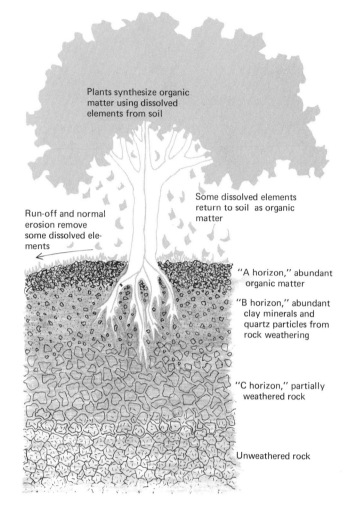

Plants synthesize organic matter using dissolved elements from soil

Some dissolved elements return to soil as organic matter

Run-off and normal erosion remove some dissolved elements

"A horizon," abundant organic matter

"B horizon," abundant clay minerals and quartz particles from rock weathering

"C horizon," partially weathered rock

Unweathered rock

Figure 7.15 A generalized soil profile.

In those many parts of the land surface where plants are present and where, consequently, the soil contains significant organic matter, soils tend to have a definite cross-sectional structure known as a **soil profile.** Typically such profiles show three horizons of changing composition and structure with depth (Figure 7.15). The uppermost and thinnest horizon contains abundant decaying organic matter from the overlying plants and a minimum of in-organic constituents, which are quickly removed by intense chemical weathering near the surface. Below is a thicker zone with less organic matter where many of the weathered constituents, such as clay materials, quartz sand, and dissolved elements, are concentrated. Below this is a still thicker horizon of only partially weathered underlying rock which generally lacks organic matter but is in the process of being converted to soil by slow chemical weathering. Superimposed on this basically simple threefold structure are a host of variations which permit soil scientists to define and classify endless minor soil types.

The kinds and intensity of weathering processes, and the soils that result from them, are closely controlled by climate. In general, chemical weathering is increased by the abundant surface water and high temperatures of tropical climates. Because of this intense weathering, tropical regions commonly have silicon-poor soils composed mostly of iron and aluminum oxides (Figure 7.16). Where iron oxides predominate the soils are red and are known as **laterite;** where aluminum oxides dominate the soils are usually yellow or gray and are known as **bauxite** (see p. 256). The intense weathering and abundant vegetation of tropical regions cause an extensive soil development; tropical soils commonly reach thicknesses exceeding 30 m (90 ft) and have an average thickness of over 3 m (9 ft) (Figure 7.16).

Chemical weathering is also the dominant process in temperate climates, but there the generally lower rainfalls and temperatures make the process less intense. In temperate soils enough silicon normally remains to make clay minerals, rather than oxides, the dominant constituents. In contrast to tropical soils, temperate soils have an average thickness of only about 1 m (3 ft) (Figure 7.16).

Figure 7.16 Interrelations of soil and climate. Thick soils develop only under temperate and tropical climates where vegetation is abundant. The high temperature and rainfall of tropical regions lead to intense chemical weathering, deep soils, and residual iron and aluminum oxides (laterite and bauxite).

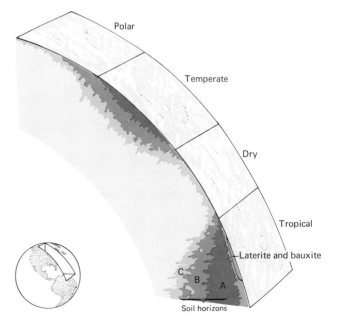

In tropical and temperate regions the presence of a permanent protecting cover of vegetation permits a relatively thick soil layer to accumulate. In polar and dry climates, on the other hand, vegetation is sparse and the products of rock weathering are exposed to continuous removal by wind and water. Under such conditions only a very thin soil cover, with abundant exposure of bare rock surfaces, normally develops. In these regions mechanical frost wedging of the exposed rock supplements chemical breakdown as an important weathering process.

SUMMARY OUTLINE

Present-day Climates

7.1 *Climatic controls:* patterns of climate are controlled by the general circulation of the troposphere as it transports heat and moisture from the equator towards the poles and interacts with the distribution of land and ocean.

7.2 *Climatic classification:* four principal climatic types, with many local variations, can be recognized today; these are tropical, temperate, polar, and dry climates.

Climatic Change

7.3 *Historical evidence of climatic change:* weather records compiled over the past 200 years, combined with scattered historical data covering the past 5,000 years, provide clear evidence that climates change.

7.4 *Geological evidence of climatic change:* sediments, erosion features, fossils, and oxygen nuclides preserved in ancient rocks provide evidence of climatic change over hundreds of millions of years of earth history; such evidence suggests that present climates are cooler than were climates through much of the earth's past.

7.5 *Causes of climatic change:* climatic change requires an overall change in the input, output, or distribution of solar energy in the earth's atmosphere; many causes for such changes have been proposed, but none has been unequivocally linked to actual climatic changes.

Climates and the Land Surface

7.6 *Weathering:* exposure to air and water at the earth's surface causes massive crustal rocks to break into small particles and change composition by both mechanical and chemical processes.

7.7 *Soil:* weathering produces an accumulation of altered rock debris, mostly clay minerals, oxide minerals, and quartz fragments, that is thickest under tropical climates, thinnest under polar and dry climates, and of intermediate thickness under temperate climates.

ADDITIONAL READINGS

Carroll, D.: *Rock Weathering,* Plenum Press, New York, 1970. An intermediate text stressing chemical weathering and soil.

*Hare, F. K.: *The Restless Atmosphere,* Harper & Row, Publishers, New York, 1961. A brief, popular introduction to climates.

Landsberg, H. E., editor-in-chief: *World Survey of Climatology,* Elsevier Publishing Co., Amsterdam, 1969 and later. A projected 15-volume reference work; volumes 1–3 will treat climatic principles and the remaining 12 volumes will describe the climates of specific regions. Chapter 5 of Volume 2 (1969) provides a good survey of climatic change.

Ollier, C. D.: *Weathering,* American Elsevier Publishing Co., Inc., New York, 1969. A readable intermediate text stressing the relationship of weathering to climates and landscapes.

Sellers, W. D.: *Physical Climatology,* University of Chicago Press, Chicago, 1965. An intermediate text emphasizing atmospheric processes that control climate.

Stringer, E. T.: *Foundations of Climatology,* W. H. Freeman and Co., San Francisco, 1972. An advanced text on climatic processes.

Trewartha, G. T.: *An Introduction to Climate,* McGraw-Hill Book Company, New York, 1968. A standard, clearly written introductory text. Part 2 provides a good survey of present-day world climates.

* Available in paperback.

8

Landscapes

Chapter 3 considered the earth's vast store of *internal* energy as it acts to move lithosphere plates and deform the edges of continents into linear mountain ranges. This chapter will be concerned with still other processes that act on crustal rocks, processes that are driven not by internal energy but by the *external* energy of the sun acting through the ocean and atmosphere. These processes tend to wear away the surface of the continents and, while so doing, to shape them into characteristic landscapes.

Weathering is the first step in the sequence of events that shapes landscapes, for weathering reduces massive rock to small particles that can be removed from their place of origin by running water, wind, and ice. Because of weathering and removal of rock debris, the land surface is constantly being worn away. Indeed, were it not for the opposing internal forces of isostasy, mountain building, and volcanism, the land surface would have long ago been reduced to a flat, featureless plain by these processes.

Weathering, we have seen, is closely controlled by climate; the same is true of the other processes that shape the land surface. The most powerful agent of land sculpture is *moving water,* and under tropical and temperate climates it is the principal agent in shaping landscapes. In dry climates, where a protective cover of vegetation is usually lacking, water still plays a dominant role, but *wind* becomes an important additional agent of land sculpture. Similarly, in polar climates, *ice* supplements moving water as a fundamental agent of land sculpture. The margins of the continents are subject to still a fourth agent of land sculpture, *ocean water,* whose action in shaping landscapes we shall consider, along with that of wind and ice, after looking more closely at land water, the principal agent of land sculpture.

263

WATER ON THE LAND

In Section 6.2 we had a brief look at the hydrologic cycle, the sequence of processes by which water is evaporated from the earth's surface, transported into the atmosphere as water vapor, and ultimately returned to the surface again as rain or snow. When considering this cycle in relation to weather, we were mainly interested in the effects of water vapor on atmospheric processes; here it will be useful to reconsider the hydrologic cycle as it relates to precipitation, which is the source of all water occurring on land.

Each year, when averaged over the entire surface of the earth, a layer of water about 1 m (3.3 ft) deep is evaporated into the atmosphere and, each year, the same meter of water is precipitated somewhere on the earth's surface as rain or snow. The amounts evaporated and precipitated differ, however, over the land and ocean. Over oceans, more water is evaporated than is precipitated; this excess falls on the land surface where, on the average, precipitation exceeds evaporation (Figure 8.1). In order to complete the cycle, this excess ocean-derived water dropped on land must ultimately return to the ocean; this occurs primarily as streams flow across the land and

Figure 8.1 The hydrologic cycle. More water is evaporated than precipitated over the ocean; the excess is transported and precipitated over land and is ultimately returned to the ocean by streams and rivers. These relatively small quantities of water moving across the land are the principal agent of land sculpture.

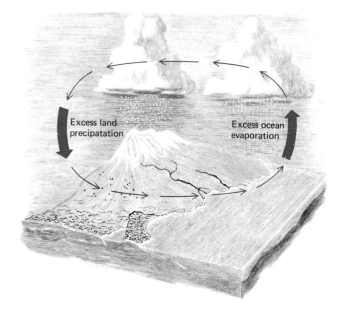

empty into the sea. The energy of this relatively small quantity of excess water, propelled by gravity as it moves over the land surface toward the sea, is the principal agent of land sculpture.

Most of the water falling on land does not pass immediately into moving streams for the return trip to the ocean. Only about one-eighth reaches streams directly as runoff from the land surface; the remaining seven-eighths is absorbed into soil and rocks to become *ground water*. Even this ground water ultimately reaches streams and returns to the ocean, but its progress is far slower and less direct than that of surface runoff.

In addition to streams and ground water, there is one additional form in which liquid water occurs on land—as lakes, relatively stationary concentrations of surface water. Neither lakes nor ground water play a significant role in shaping the land surface when compared with the far more energetic action of moving streams. Consequently, we shall only briefly consider some of their principal characteristics before looking more closely at streams.

8.1 Lakes

Viewed over millions of years of geologic history, lakes are extremely transient features of the earth's surface. They form when obstructions of rock or soil temporarily block the path of flowing water and disappear as the obstructions are worn away, and as deposition of sediments fills the depression occupied by the lake. During their relatively short existence lakes, particularly very large ones, possess many of the same phenomena as the ocean—for example, surface waves and temperatures that decrease downward due to surface warming and mixing. Most lakes differ profoundly from the ocean, however, in lacking a high concentration of dissolved materials.

Normal ocean water contains about 3.5 percent dissolved elements. A few "saline lakes" in dry regions where rains are infrequent and evaporation intense, lakes such as the Dead and Caspian "Seas" and Great Salt Lake, also have high concentrations of dissolved materials although not, ordinarily, in the same proportions as in the ocean. Most lakes have, however, only an insignificant fraction of 1 percent dissolved matter. Because of this negligible salinity, lakes differ from the ocean in several important properties. In the ocean, differences in both temperature and salinity are important in causing density gradients that lead to current movements. In lakes, only thermal differences occur and density-induced currents are less pronounced than in the ocean. Another result of the lack of dissolved materials is that lake water freezes at 0°C (32°F) rather than having the indefinite freezing properties of sea water. As a consequence, lakes in very cold regions commonly freeze solidly to depths of 100 m (330 ft) or more, whereas ice on the ocean (other than land-derived icebergs) seldom exceeds 3 m (10 ft) in thickness.

8.2   Ground Water

Under the influence of gravity, most of the water falling on the land surface sinks into the underlying soil and rock to become **ground water.** As a result, except for a relatively thin outer layer which dries between rainfalls, all cracks, voids, and pores in the rocks of the continents are permanently saturated with water. The upper surface of this zone of permanent saturation, which separates it from the periodically dried layer above, is known as the *water table.* In general, the overall depth of the water table depends on the amount of precipitation and the absorbing capacity of the rocks in a particular region. It is nearest the surface in regions of high rainfall and relatively impermeable soil and rocks, and deepest in arid regions with highly porous rocks. In any one region the depth of the water table below the surface also varies with topography, which it mirrors by rising under hills and dropping along the sides of valleys (Figure 8.2). When the water table intersects the land surface, *springs* result.

Ground water is not stationary, but moves laterally and downward very slowly under the influence of gravity until it ultimately reaches the surface at some lower elevation. Permanently flowing streams of tropical and temper-

Figure 8.2   Ground water, and its relationship to climate and stream flow.

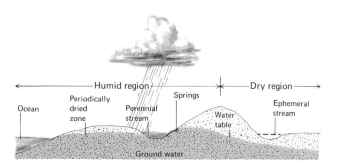

ate climates normally move in valleys whose bottoms lie *below* the local water table (Figure 8.2); most ground water ultimately reaches the surface by slow seepage into these **perennial streams.** In dry climates, on the other hand, this process is reversed. There, temporary, **ephemeral streams,** formed after the infrequent rainfalls, lose water to the ground and ultimately disappear until the next rainfall.

Except for its crucial role in weathering and soil formation, ground water plays a relatively minor part in shaping landscapes with one very important exception. This occurs in regions underlain by limestone. Limestone alone, among the common rocks of the earth's crust, is readily dissolved by dilute acids. Waters of the land surface are usually slightly acidic both because of dissolved atmospheric carbon dioxide, and because of the addition of acids from decaying plants (Section 7.6). For this reason, ground water commonly dissolves much of the underlying rock in regions where limestone is exposed at or near the earth's surface. This dissolution leads to large underground cavities, channels, and passageways which may open to the surface as *sinkholes.* Regions showing these distinctive features are said to exhibit **Karst topography,** named after the Karst region of Yugoslavia where solution features are particularly well developed (Figure 8.3). Most of the earth's large dramatic underground caverns are developed in such regions.

8.3 Streams

The third, and geologically most important, form in which liquid water occurs on land is in **streams,** relatively long, narrow concentrations moving downward across the land surface under the influence of gravity. In everyday language we normally make a size distinction between smaller "streams" and larger "rivers," but geologists, recognizing that all gradations exist between the smallest flowing trickle and the largest rivers, generally refer to *all* flowing waters on the land surface as "streams."

Streams in tropical and temperate regions derive their water from two sources. Roughly half their flow comes directly from surface runoff of rain or water from melting snow. The other half comes from slow seepage of ground water into the stream channel which normally lies below the local water table, at least in perennial, or permanently flowing, streams. It is this slow, continuous influx of ground water along stream channels that main-

Karst valley　　　　Sinkholes　　　Spring

Water table

Caverns

(a)

Figure 8.3 Karst topography. (a) Caverns, sinkholes, and related features commonly form in regions underlain by limestone, which is readily dissolved by acidic ground water. (b) Karst topography in central Kentucky. (c) Close-up of a small, deep sinkhole in northern England. (b: John S. Shelton; c: Geological Survey of Great Britain)

(b)

(c)

tains their flow between rainfalls. Conversely, stream flooding during periods of heavy rainfall is a result of the greatly increased surface runoff. In dry climates the water table is generally so deep that ephemeral streams receive none of their water from ground water seepage. Instead, they are fed only from surface runoff after the infrequent rainfalls and are dry for much of the year (Figure 8.4).

The movement of water in streams is subject to the same four basic hydrodynamic forces that are responsible for motions in the fluids of the oceans and atmosphere: gravity, pressure gradients, friction, and the Coriolis effect. In stream flow, however, only two of these forces are important. Because streams are relatively shallow and very well mixed, temperature differences, the primary cause of

(a)

(b)

(c)

Figure 8.4 Varieties of streams. (a) Perennial stream in southern England. (b) Ephemeral stream in southern Arizona in dry stage and (c) in flood. (a: Geological Survey of Great Britain; b, c: Tad Nichols)

pressure gradients in fluids surrounding the earth, are negligible and internal pressure differences are thus an insignificant force in stream movements. Likewise, because stream flow is normally confined in relatively rigid channels carved in the solid earth, streams are not free to respond to the lateral Coriolis forces that play such a large role in motions of the ocean and atmosphere. Gravity and friction, the two remaining primary forces, are therefore ultimately responsible for the characteristics of stream flow.

Gravity is the principal driving force of stream (and ground water) movements; it begins to act whenever solar energy moves water from the ocean into the atmosphere and deposits it above sea level on the land surface. The gravitational force then comes into play and moves the water downward, toward sea level, but is opposed by complex frictional forces as the water impinges on solid rock and sedimentary particles. The net result of the impelling gravitational force and the opposing frictional forces is to impart a particular speed or *velocity* to the moving water.

In seeking to analyze the dynamics of stream flow, geologists in recent years have made intensive studies of many secondary factors that influence the two primary forces of gravity and friction in stream flow. The force of gravity is primarily a function of two such secondary factors: the volume of the flowing water, which is called the *stream discharge,* and the *slope* of the surface over which it moves. The opposing frictional forces, on the other hand, are primarily related to two other factors: the amount of sediment moved by the flowing water, known as the *sediment load,* and the *shape of the channel* through which the water flows. These four secondary factors—discharge, slope, sediment load, and channel shape—are closely and complexly interrelated with each other, and with their ultimate result, the velocity of the moving water.

Analyses of records of velocity, discharge, slope, sediment load, and channel shape, from hundreds of streams have shown that the interaction of these factors is not random, but varies in a systematic way; that is, the factors are so balanced that changes in one lead to surprisingly regular and predictable changes in the others (Figure 8.5). If, for example, the discharge at some point along a stream is doubled after a rainfall, then the stream velocity will increase by a factor of 1.3, the channel shape will change by a small but predictable amount as the water rises and the channel width and depth increase, and the amount of sediment carried by the stream will increase by a factor of 8. The mere fact that such increases take place is not surprising. Streams are easily seen to become wider and deeper when discharge increases during floods. Likewise, the fact that large quantities of sediment are moved during floods is clear because most streams flow on **flood plains,** valleys partially filled with loose sand and clay that have been transported during periods of high discharge, and dropped as the stream recedes into its normal channel. It is not that such changes in shape and sediment load take place, but their geometric regularity and predictability, that has been established by careful observation and analysis.

Figure 8.5 The dynamics of stream flow at a single point along a stream. Sediment load, water velocity, and channel shape all increase regularly, but at different rates, with increasing discharge. (Modified from Leopold et al., Fluvial Processes in Geomorphology, *1964)*

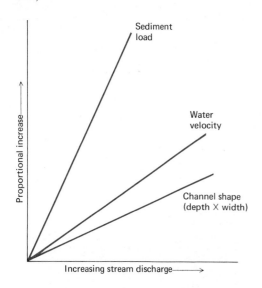

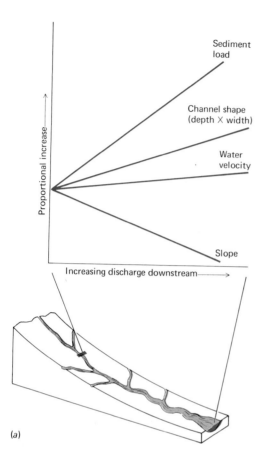

(b)

(c)

(a)

Figure 8.6 Stream flow. (a) The dynamics of stream flow with increasing distance downstream. Sediment load, channel shape, and water velocity increase, and slope decreases, with increasing discharge. (b) Stream in western Alberta in its upper course, and (c) the lower course of a stream in southern Wales. Although steep mountain streams appear to flow faster than those in relatively flat alluvial valleys, careful measurements show that this is not normally the case. (a: Modified from Leopold et al., Fluvial Processes in Geomorphology, *1964; b: G. Hunter,* Information Canada Photothèque; *c: Geological Survey of Great Britain)*

So far we have considered only changes at a single point along the course of a stream. If, instead, we consider the interrelations of discharge, velocity, channel shape, and sediment load throughout the entire length of the stream, we find equally regular, but somewhat different, relationships (Figure 8.6a). In perennial streams, discharge normally increases downstream as water is added from runoff, ground water seepage, and the inflow of tributary streams. Sediment load and channel width and depth increase regularly as discharge increases downstream. At the same time, slope, the one factor which we

Figure 8.7 Typical meanders in a small stream in Colorado. (Tad Nichols)

could not consider at a single point on the stream, decreases regularly downstream so that, in cross section, the channel has a concave-up profile that flattens in the lower part of the stream course (Figure 8.6a). One of the most surprising results of modern stream analyses was the discovery that these factors interact to cause a small but regular velocity *increase* downstream. This conclusion had not been expected since rushing mountain streams flowing on steep slopes *appear* to move faster than large rivers flowing on flat flood plains (Figure 8.6b, c). Measurements show, however, that this is not usually the case, for the downstream decrease in slope is more than compensated by the decreased friction of a larger channel.

Before moving on to consider how stream flow shapes the land, there is one rather puzzling characteristic of streams to be considered. When viewed from above, as from an airplane or on a map, streams almost never flow in straight linear channels, but instead follow a regular zig-zag series of loops known as *meanders* (Figure 8.7). These have the effect of increasing the channel length, and thus adding to the distance through which frictional forces can act; they also decrease the channel slope, with the result that the gravitational impelling force is lessened. For these reasons meanders are believed to be somehow related to the dynamic balance of frictional and gravitational forces in stream flow but, as yet, there is no satisfactory explanation of their exact cause.

LAND SCULPTURE BY LAND WATER

Shaping of the land surface takes place as rock debris produced by weathering is moved downward by gravity toward its ultimate resting place, the ocean. These move-

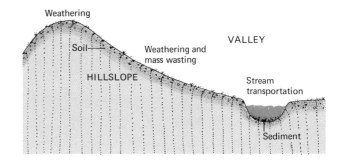

Figure 8.8 Erosion of the land surface. Weathering and mass wasting deliver rock debris to streams where it is transported as sediment.

ments of rock debris eventually involve flowing water, but they commonly begin under the influence of gravity alone, as when soil slides down a steep hillside, or when a block of rock, loosened by weathering, falls from the face of a cliff. Such direct movements of weathered debris by gravity are called **mass wasting.** In addition to mass wasting, rock debris may be moved indirectly by gravity through transportation within flowing water, air, or ice. These two processes, mass wasting and transportation, combine with weathering to cause **erosion,** the general term applied to the wearing away of the land surface (Figure 8.8). The result of erosion is the formation of *valleys,* empty spaces formerly filled with rock, bounded by *hillslopes,* curved surfaces underlain by rock that is still undergoing weathering, mass wasting, and transportation.

8.4 Mass Wasting

Soil and other rock debris formed by weathering move downward under the influence of gravity, just as water does, whenever the gravitational impelling force overcomes opposing frictional forces. On slopes steeper than about 40°, gravity exceeds frictional resistance for *all* kinds of rock debris and therefore only massive bedrock, exposed as cliffs, occurs on such steep slopes. Mass wasting also continuously takes place, however, on more gentle, soil-covered slopes. This mass wasting is of two sorts. Most commonly the frictional resistance is so great that downslope movements are very slow, from 1 to a few centimeters per year; such movements are known as **creep.** All soil-covered slopes undergo creep, which is a principal cause of tilted fences, cracked sidewalks, bent tree trunks, and other common hillslope phenomena (Figure 8.9).

Figure 8.9 Creep, slow downslope movements of soil and rock debris, and its effect on hillside objects.

273

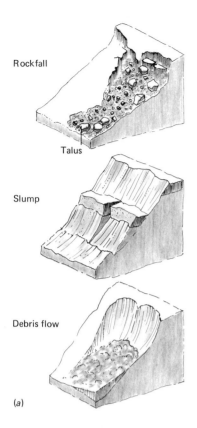

Rockfall

Talus

Slump

Debris flow

(a)

More rapid downslope movements by mass wasting are known, collectively, as **landslides.** The speed at which rock debris moves in landslides varies from several meters per *day* to several meters per *second,* but is always much more rapid than the centimeters-per-year movement of creep. Rapid landslides are especially common on slopes where frictional forces are reduced by the lubricating effects of unusual amounts of ground water. Such lubrication often occurs where soil is underlain by an impermeable layer of clay or, in polar climates, frozen ground water, both of which limit and concentrate the downward penetration of surface water. Examples of several common types of landslides are shown in Figure 8.10. Although landslides are often large and spectacular and can cause great destruction, they are less important in shaping landscapes than are the slower but more continuous movements of creep.

(c)

(b)

Figure 8.10 Landslides, rapid downslope movements of soil and rock debris. (a) Some common types; (b) slump; (c) rockfall and talus. (b: Wisconsin Conservation Department; c: Josef Muench)

274

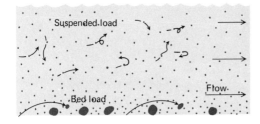

Figure 8.11 Transportation of debris by moving water.

8.5 Stream Transportation

Mass-wasting processes account for most of the *initial* movement of rock debris, but it is transportation of this debris by streams that normally carries it from its immediate area of origin, and thus makes possible a continuous wearing away of the land surface. Solid debris particles are moved by flowing water in either of two ways; they may be completely surrounded and supported by the moving water as **suspended load,** or they may be rolled, slid, or bounced along the stream bottom as **bed load** (Figure 8.11). In general, larger and heavier particles move as bed load, and smaller and lighter particles as suspended load.

In the last section we noted that the total amount of sediment moved by a stream increases very rapidly as the discharge and velocity of the stream increase (Figures 8.5, 8.6). For this reason, most sediment movement takes place at times of high water when the transporting capacity of the stream is at a maximum. Between periods of flood, the sediment is temporarily deposited in the stream channel or adjacent flood plain. Most stream valleys are at least partially filled with such sediment, called **alluvium,** that is temporarily at rest on its downslope journey. The ultimate resting place of this alluvial debris is the ocean, for sea level marks the lowest point to which streams can ordinarily flow. On entering the ocean, the stream-transported sediments are moved, sorted, and reworked by waves and currents; ultimately they come to rest in beaches and other *depositional* landscape features that will be considered later in the chapter.

The combined effect of mass wasting and stream transportation is to shape the land surface into valleys and hillslopes. Valleys result from the erosional action of the streams that flow through them; in general, two types of valley-stream relationships occur. In rugged mountainous regions where slopes are steep, the sedimentary debris supplied by mass wasting is readily transported and streams commonly flow directly on the underlying rock. When this occurs, chemical solution and mechanical abrasion cause the stream to cut actively into the rock to form steep-walled, **V-shaped valleys** (Figures 8.12a, 8.13a).

In contrast, on the more gentle slopes that characterize the lower reaches of most streams, valleys are floored by large quantities of alluvial debris that make up the relatively flat flood plain over which the stream flows.

275

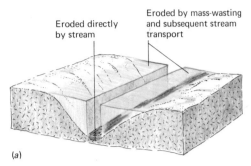

Eroded directly by stream

Eroded by mass-wasting and subsequent stream transport

(a)

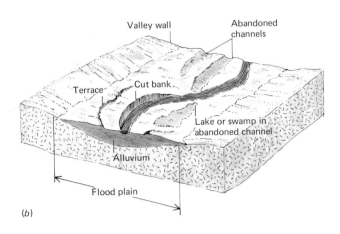

Valley wall

Abandoned channels

Terrace

Cut bank

Lake or swamp in abandoned channel

Alluvium

Flood plain

(b)

Figure 8.12 *Types of valleys.* (a) *V-shaped and* (b) *alluvial valleys. Alluvial valleys, which characterize the lower reaches of most streams, commonly show abandoned channels, cut banks, terraces, and other features caused by the shifting channel and moving alluvium.*

(a)

(b)

Figure 8.13 *V-shaped and alluvial valleys.* (a) *V-shaped valley in southern British Columbia.* (b) *Alluvial valley in southern England.* (a: National Film Board of Canada; b: Geological Survey of Great Britain)

Even during flood, such streams generally erode only these flood-plain sediments, rather than cutting downward into the underlying rock. The resulting, flat-floored, **alluvial valleys** (Figures 8.12b, 8.13b) usually show characteristic patterns of temporary lakes, terraces, and other topographic features caused by the intermittent erosion and deposition of the underlying sediment. Some of these features are summarized in Figure 8.12b.

V-shaped and alluvial valleys are easier to characterize than are the hillslopes that bound them. When

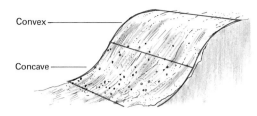

Convex

Concave

Figure 8.14 *A typical hillslope profile.*

viewed in cross section, most hillslopes have a *profile* which is convex in the higher parts and concave in the lower (Figure 8.14). The upper, convex slope is thought to be shaped primarily by weathering and creep, whereas the lower, concave portion results from fluid transportation by concentrated surface water runoff, leading to a more rapid removal of debris. Beyond this, the exact shape and form of hillslopes is a result of a complex and poorly understood interaction between the various processes of erosion, the structure and composition of the underlying bedrock, and the length of time the rock has been exposed to erosion.

8.6 Rates and Cycles of Water Sculpture

The landscapes of the earth's surface at any one time represent a balance between uplift of the land by the internal forces of crustal movement and destruction of the land by the external forces of erosion. Both of these processes act slowly, but at varying rates, over long periods of time. Therefore in order to fully understand landscapes, we must know something of the historical sequence of uplift and erosion. Today, however, we can observe and study these processes at only a brief moment in the earth's long history and interpretations of their interactions over millions of years involve many uncertainties.

The first and most basic question concerning the history of land sculpture is: how quickly does it take place? In other words, what is the rate at which the land surface is being lowered by erosion? Various schemes have been devised for estimating such rates, but none is fully satisfactory. Perhaps the most accurate is based on estimates of the amount of rock debris that is carried to the ocean each year by streams. Such estimates suggest that the United States is being lowered at an overall rate of about 8 cm (3 in.) per 1,000 years. Since the average height of the land surface is about 800 m (2,600 ft), this erosion rate would reduce the continents to sea level in about 10 million years; geologically speaking, this is a very short time. There are, however, many difficulties in such estimates. One relates to the activities of man: the clearing of much of the land surface for agriculture during the last few hundred years has greatly increased rates of erosion—but by amounts that are difficult to estimate. In addition, the opposing internal forces that cause uplift of the land must be considered. Measurements of present-day rates of uplift suggest that the continents are now rising several times faster than they are being eroded. The continents appear to stand higher today, however, than through much of earth history; this suggests that rates of uplift and erosion may have been more nearly balanced in the past.

Perhaps the most serious problem in estimating erosion rates is caused by climatic change. Climate has a profound influence on erosion; under dry and polar climates, local erosion rates have been measured that are many times greater than the 8 cm/1,000 years average for the temperate United States. As yet, however, too little is known about overall erosion patterns under such conditions to permit close comparisons with rates derived from temperate regions.

Along with these difficulties in measuring merely the *rate* at which the land is being lowered, there are, as you might expect, even greater problems in reconstructing the exact *sequence* of landscape changes which result as the land is lowered by erosion and raised by internal forces over millions of years. Because no one has seen landscapes evolve over such a long period, all attempts at such reconstruction must be based on indirect evidence and intuition. One such intuitive scheme for reconstructing landscape history dominated the study of landscapes for many years. It was developed around the turn of the century by several American geologists, the most influential and articulate of whom was W. M. Davis of Harvard University. Davis visualized landscape development in temperate climates, where flowing water is the dominant erosional agent, as commencing on relatively flat surfaces, such as former sea floors, that were raised above sea level by tectonic forces (Figure 8.15). At first, he believed, streams would cut steep, V-shaped valleys without flood plains into the surface; later the valleys would widen, and broad flood plains separated by steep hillslopes would dominate; local relief would be at a maximum in this stage. Finally the hillslopes would be reduced by erosion to a relatively flat rolling surface which he called a **peneplane.** Landscapes showing each of these characteristics exist today in temperate zones and Davis and his followers devoted much effort to classifying them according to their "stage of evolution" in his theoretical scheme of landscape development. Most geologists now feel, however, that Davis's scheme was much oversimplified. Under certain conditions the sequence he envisioned might take place, but the complications of climatic change, and the

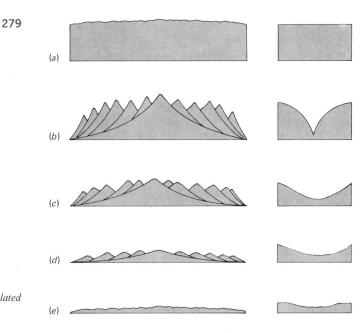

(a)

(b)

(c)

(d)

(e)

Figure 8.15 *The cycle of land sculpture as postulated by W. M. Davis.*

interaction of uplift and erosion, probably lead to the formation of valleys and hillslopes by many complex sequences of events not included in Davis's appealingly simple analysis.

LAND SCULPTURE BY WIND, ICE, AND THE OCEAN

While running water is the principal agent of land sculpture over most of the land surface, in regions of dry climate it is supplemented by wind action; similarly, land sculpture by ice is important under polar climates. Along coasts and on continental shelves, still another agent of the fluid earth, ocean water, plays a large part in shaping landscapes. In this section we shall consider these regions where wind, ice, and the ocean make large contributions toward shaping the land.

8.7 Deserts

In Section 7.2 we noted that worldwide belts of dry climate tend to develop in the zone of descending air that lies between the moist, rising air of the tropics and the turbu-

Figure 8.16 *The principal present-day desert regions.*

■ Deserts   ▨ Areas of sand accumulation

lent midlatitude belts of cyclones and anticyclones. These zones of dry climate are characterized by an annual rate of surface evaporation which equals or exceeds the annual rainfall. Under these conditions, most of the scant precipitation returns rather quickly to the atmosphere and there is no excess to form permanent streams, or support permanent vegetation. The results are **deserts,** areas with little or no moisture or vegetation which today occupy about one-fourth of the land surface (Figure 8.16).

Although we normally think of deserts as being warm, high temperatures play a role in the origin of deserts only through increasing the annual amount of evaporation. For example, regions on the warm, equatorial side of the dry climate zone may receive 75 cm (30 in.) of rainfall a year, about as much as falls in Chicago or Seattle, and still remain deserts because of their high average temperatures and consequent high rates of evaporation. In contrast, on the cooler, poleward side of the dry zones, rainfall must normally be less than 35 cm (14 in.) per year to cause complete evaporation and desert conditions.

Although the rain that falls in desert regions evaporates relatively quickly, usually in a few hours or days, it nevertheless plays a dominant role in shaping desert landscapes. Because there is little soil or vegetation to absorb the rainfall, it tends to remain on the surface where, particularly in desert mountains, it may be concentrated in normally dry channels to cause *flash floods* which subside in a few hours. Such flood waters commonly have an enormous discharge and velocity, and thus transport large volumes of sedimentary debris to the base of desert mountains where it is deposited in fan-shaped wedges called **alluvial fans** (Figures 8.17, 8.18). Because there are no permanent streams to transport the sediment beyond the base of the mountains, thick alluvial fillings, made up of superimposed alluvial fans, dominate the lower elevations of many desert regions. In contrast, relatively flat

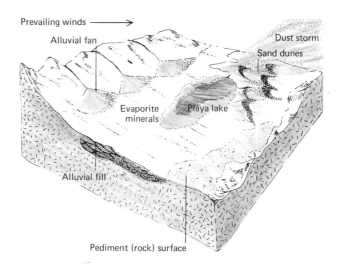

Prevailing winds

Alluvial fan

Dust storm

Sand dunes

Evaporite minerals

Playa lake

Alluvial fill

Pediment (rock) surface

Figure 8.17 Characteristic landscape features of desert regions.

surfaces of bare rock known as **pediments** surround many desert mountains. These are apparently formed by erosion, rather than alluvial deposition, during flash floods.

Since deserts generally lack vegetation, organic-rich soils do not develop even though alluvial clay, sand, and gravel are abundant. Instead of organic matter, the surface layers of desert alluvium commonly contain large quantities of calcium minerals, principally calcite and gypsum; these layers form as dissolved calcium, sulfur, and other elements picked up by ephemeral streams during their brief existence are deposited again as the water evaporates. Such evaporite minerals may accumulate to great thicknesses in desert valleys where runoff converges to form temporary **playa lakes** (Figures 8.17, 8.18).

Among the most distinctive characteristics of desert landscapes are those caused not by running water, but by the wind. Winds in desert areas are not stronger or more persistent than in most other regions, but, because of the lack of vegetation and water, exposed desert sand and clay are more readily transported by wind action. Under such circumstances, wind provides a more constant and widespread eroding agent than does moving water. Just as in flowing water, sedimentary particles may be carried by the wind as either suspended load or bed load. Because air is a far less viscous fluid than water, however, only very fine sand or clay particles can normally be carried in suspension. Strong winds blowing across desert alluvium can carry fine sand and clay several thousand meters into the air in *dust storms,* and later deposit them hundreds of kilometers away as sheets of fine dust (Fig-

(a)

(b)

Figure 8.18 *Desert landscape features in southern California. (a) Alluvial fans. (b) Evaporite minerals on a dry playa lake bed. (c) Dust storm. (d) Sand dunes. (a, c: John S. Shelton; b, d: Ramsey, Omikron)*

ures 8.17, 8.18). Larger sand-sized particles, on the other hand, cannot normally be moved in suspension by the wind, but are rolled or bounced along the surface, where they tend to accumulate as **sand dunes** on the downwind side of desert valleys. While in motion, this wind-carried desert sand actively erodes and polishes exposed rock over

(c)

(d)

which it moves. Although we normally think of sand dunes as particularly characteristic of deserts, they occupy a relatively small fraction of the earth's dry regions. Nevertheless, such sand concentrations can be impressive; the largest, in the Sahara of North Africa, occupies an area larger than Texas.

283

Figure 8.19 An ice sheet (*background*) on Baffin Island in the Canadian Arctic giving rise to two converging glaciers (*foreground*). (*Canadian Department of Energy, Mines, and Resources*)

8.8 Glaciers and Ice Sheets

Deserts form in regions where annual evaporation equals or exceeds precipitation. In tropical and temperate areas where evaporation does *not* exceed precipitation, the yearly excess of water, being liquid, does not accumulate, but moves downward to the ocean under the influence of gravity. In polar regions, on the other hand, most of the precipitation falls as solid water—snow; when the average yearly snowfall exceeds the yearly loss from evaporation and summer melting, then the snow continues to accumulate from year to year to form **glaciers** and **ice sheets,** massive ice accumulations which differ from each other in scale (Figure 8.19). Glaciers are smaller, linear ice accumulations, usually less than 300 m (1,000 ft) thick, bounded by valley walls of solid rock. Ice sheets are much larger dome-shaped accumulations that cover most of the underlying rock surface.

Man
and Landscapes

Although landscapes change rather quickly when viewed from the perspective of millions of years of earth history, during the mere few thousand years of man's recorded past they have been a relatively static feature of the earth's surface. There is, however, one dynamic feature of landscapes that is of continuous and direct human concern—the streams, lakes, and ground water that act both to shape the land and to supply man with water, his most essential resource.

Most liquid water occurs on land as slowly moving ground water which accounts for about 3,000 times the volume of water found in lakes and streams. As yet, however, less than one-third of the water used by man comes from wells tapping this large ground-water resource. The rest comes from the more easily exploited surface waters found in streams and lakes, the latter both natural and man-made.

Most of the water used by man, whether for public water supply (about 10 percent of the total used), agriculture (about 45 percent), or industry (about 45 percent), is not actually *consumed,* but is, instead, returned with varying degrees of change to streams or the ground-water reservoir where it may be reused again and again in a continuing cycle. For this reason, man's potential supply of fresh water is greater than any foreseeable need—with two important qualifications.

The first concerns the *location* of the water. Arid regions normally lack surface water in streams or lakes, and the deep-lying ground water in such areas is often prohibitively expensive to exploit by wells. Thus agriculture, or industries with large water requirements, are impractical in most arid regions and even municipal water supplies may be limited and costly. In short, water resources are not evenly distributed on the land surface and must generally be exploited where they occur.

The second important reservation concerns not the location, but the *quality* of water resources. Most human uses require relatively pure, clean waters to which wastes are then added to make the waters impure. Traditionally, these "used" waters have been returned to streams or the ground-water reservoir with little or no treatment, so that contaminants were removed either slowly, by natural processes, or more quickly by man himself before he used the waters again. Along large rivers, for example, upstream cities and industries find relatively pure water which they use and return to the river with impurities that must then be removed by consumers farther downstream. The well-known results of this practice are the polluted lakes, rivers, and streams that now plague most industrialized nations. Fortunately, there is increasing concern for both the economic and aesthetic costs of such practices and, hopefully, the pattern will be reversed in the future by requirements that waters be processed immediately *after* use so that they are clean and pure when returned to their natural sources.

Irrigation agriculture (young orange trees) based on groundwater wells, southern Arizona. (U.S. Department of Agriculture)

Regions of the United States underlain by abundant groundwater (color). The patterned area shows arid regions where groundwater resources are especially valuable. (Groundwater data from Skinner, Earth Resources, *1969, after Thomas)*

Typical view of untreated wastewaters being returned to a stream. (United Press International)

Scene in a U.S. National Park. (United Press International)

Today only two large ice sheets exist—those which cover most of Greenland and Antarctica; both have an average thickness of about 2,000 m (6,600 ft) and a maximum thickness greater than 4,000 m (13,200 ft). Although only Greenland and Antarctica are covered by ice sheets today, during the relatively recent intervals of Pleistocene glaciation similar ice sheets covered much of Canada and northern Eurasia as well (Figure 8.20). Glaciers are more widely distributed today than are ice sheets; they are particularly common in polar and cool-temperate mountains where high elevations cause both low average temperatures and abundant orographic snowfall (Figure 8.20). Because of this decreased temperature and increased precipitation, glaciers can exist in high mountains even in the tropics: Mt. Kilimanjaro in East Africa rises to an elevation of 5,800 m (19,000 ft) and has permanent glaciers even though located almost on the equator.

Although high mountains everywhere and most mountains of cool-temperate and polar regions today have extensive glaciers, it is important to emphasize that much of the earth's polar regions is *not* ice-covered. In many low-lying areas there is too little snowfall to accumulate even under extreme polar climates. For example, most of cold, northern Alaska is ice-free, while glaciers are com-

Figure 8.20 Distribution of present-day glaciers (dots) and ice sheets (Greenland and Antarctica). The maximum extent of Pleistocene ice sheets is shown in lighter color.

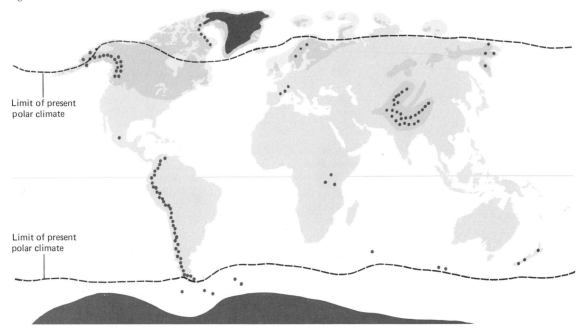

Limit of present polar climate

Limit of present polar climate

mon in milder southern Alaska where there is more precip-itation. In all, about 10 percent of the present land surface is covered by either glaciers or ice sheets. During the maximum extent of Pleistocene ice, about 30 percent of the land was covered (Figure 8.20).

As layers of snow accumulate to build glaciers and ice sheets, the structure and texture of the ice changes (Figure 8.21). Falling snow normally consists of delicate, lacy, ice crystals; when these crystals first accumulate on the surface, they make a porous, powdery mass containing only about one-fifth ice and four-fifths air. Within a few days or weeks, the combined effects of evaporation from the surface of the snowflakes, daytime melting and night-time refreezing, and compaction by the weight of overlying snow, all act to convert the powdery flakes into larger, roughly spherical ice particles with the texture of coarse sand; these particles are known as **firn.** At this stage the accumulated snow is concentrated to about half ice and half air. As the firn is buried more deeply by subsequent snowfalls, the trapped air either escapes to the surface or is concentrated in large bubbles; at the same time the adjacent grains of firn tend to fuse. The final result is a solid mass of glacier ice containing only about 10 percent air. In addition to ice, most glaciers and ice sheets contain varying amounts of sedimentary particles; wind, mass wasting, and summer meltwater commonly deposit clay, sand, and gravel on the ice surface which then become

Figure 8.21 Structural changes during the trans-formation of snow to glacier ice.

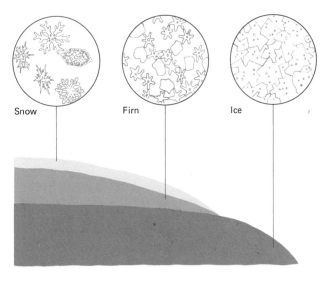

Snow   Firn   Ice

buried by the next snowfall and incorporated, in layers, into the ice.

Ice, even though a solid, responds to the downward force of gravity by flowing just as does liquid water, but far more slowly (Figure 8.22). Careful measurements show that glaciers move down their valleys at average rates of a few centimeters to a few meters per day; the steeper the valley and the thicker the ice, the more rapid the flow. Under exceptional, and poorly understood, circumstances the rate may be still faster—the maximum observed movement is several hundred meters per day. Unlike glaciers, ice caps are not confined by valleys and their flow is therefore not unidirectional but radially outward from their high, dome-shaped centers. Because of their thickness and size, flow measurements are far more difficult to make on the Greenland and Antarctic ice sheets than on glaciers, but it appears that they flow much more slowly than do most glaciers.

Figure 8.22 Flow of glacier ice in mountains of northwestern Canada. Sedimentary debris on and within the moving ice clearly shows the patterns of glacial flow. (Bradford Washburn)

The mechanisms that permit solid glacier ice to flow like a very viscous fluid have been much investigated in recent years; these studies show that at least three separate mechanisms are important in glacier flow. Some movement is caused simply by *slipping* of the entire ice mass over the underlying bedrock. Because of the small but constant outflow of the earth's internal heat at the earth's surface, the basal portion of most glaciers is not solid ice, but a slightly melted mixture of ice and water. As a result of this basal melting, glaciers are seldom "frozen to" the underlying rock but, instead, move over it on a lubricating film of water. A second factor in glacial movement is simple *fracturing* and *sliding* of large blocks of ice past each other. This fracturing takes place primarily near the surface, where the ice behaves as a brittle solid. Deep within the glacier the pressure of the overlying ice prevents brittle deformation; instead the ice deforms by still a third process—*plastic flow*. In Chapter 3 we saw that under high pressures and over long periods of time the solid rocks of the earth's crust slowly deform as if they were viscous liquids. Ice crystals under pressure behave similarly but have far less inherent strength than do the silicate crystals of rocks; for this reason ice flows even more readily, and under far lower pressures, than does rock. This internal plastic flow is probably the most important factor in glacial flow; it is known to involve several molecular mechanisms that permit slow changes of shape in both single ice crystals, and in larger ice-crystal aggregates.

As the ice of glaciers and ice sheets moves slowly downslope, it eventually either reaches the ocean or arrives at lower elevations on land where snow accumulation does *not* exceed annual evaporation and melting. In these regions the ice dissipates by melting, evaporation, or entering the ocean as icebergs (Figure 8.23). It normally requires many years for the ice which accumulates at higher elevations to flow to lower elevations and dissipate. Measurements of the ages of glacial ice, using the carbon-14 technique to be described in Chapter 10, show that in small glaciers the ice is commonly hundreds of years old; ice 50,000 years old has been found near the base of the slowly moving Antarctic ice sheet.

All glaciers and ice sheets show a dynamic balance between ice accumulation in their higher portions and loss of ice in their lower portions (Figure 8.23). When the rate of either accumulation or loss changes, as for example during climatic fluctuations, the thickness and extent of

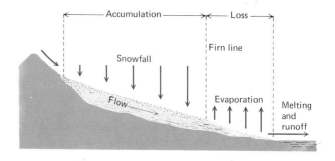

Figure 8.23 Flow and dissipation of glacier ice. As the ice flows downward it eventually reaches lowlands where loss exceeds accumulation. Fresh snow and firn are absent and the older moving ice is dissipated by evaporation, melting, and runoff.

the glacier or ice sheet also change. For this reason, fluctuations in the size of glaciers are among the most sensitive indicators of short-term climatic changes.

Ice, as a semipermanent solid, forms a dominant part of the local landscape in regions having glaciers and ice sheets. At the same time, the ice is also an agent in shaping the land, for it actively erodes and transports the rock and rock debris over which it moves. Because this

Figure 8.24 Characteristic landscape features of glaciated regions. In areas of ice accumulation, glacier flow scours and erodes the underlying rock to form U-shaped valleys (above). In areas of ice dissipation, rock debris carried by the ice is deposited to form moraines *at the dissipating edge and* drumlins *and* eskers *in cavities under the ice (below). Beyond the ice margin, glacial meltwater deposits an* outwash plain *of sedimentary debris.*

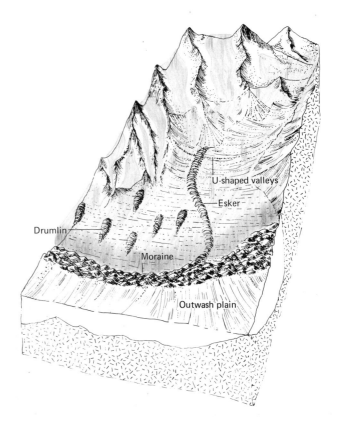

erosion and transportation occurs under great thicknesses of ice, it is difficult to study directly in present-day glaciers and ice sheets. During the several glacial expansions of Pleistocene time, however, ice covered much of Canada and northern Eurasia, regions that are presently ice-free; in these regions the erosional effect of ice is clearly evident and easily studied. Such study shows that, in general, ice acts not to *carve* valleys and hillslopes, as does running water, but, instead, to *modify* the shapes of preexisting hills and valleys. Among the most characteristic modifications seen in formerly ice-covered landscapes are steep-walled, **U-shaped valleys,** and polished, scratched bedrock surfaces (Figures 8.24, 8.25, 7.6*b*).

Most of the *erosional* effects of glaciers and ice sheets are concentrated in the areas of ice accumulation and movement; near their dissipating margins the rock debris moved by the ice is concentrated in a variety of characteristic *depositional* landscape features, some of which are summarized in Figure 8.24.

8.9 Coasts and Continental Shelves

Ocean water, the final agent of land sculpture, plays a less important role in shaping the land than do streams, wind, and ice. Because the ocean ultimately receives most of the rock debris formed by erosion of the land, the margins of the continents tend to be sites of accumulation of sediments, rather than sites of active erosion; for this reason the continental margins will be a principal theme when we consider sediments in Chapter 9. Ocean water is, however, an active agent of land sculpture along the **coast,** the long, very narrow zone where the ocean surface impinges on the land.

In Chapter 4 we saw that much energy of ocean movement is concentrated in wind-generated waves moving across the ocean surface. These waves extend downward, with decreasing motion, to depths about half as great as the distance between successive wave crests. Because wind waves on the ocean surface normally have crests less than 30 m (100 ft) apart, wave motions extend downward only about 15 m (50 ft), and thus affect the ocean floor only in shallow coastal waters. There, the energy of the wave motion is transferred to rock and sediment when the waves break and dissipate against the shore. This wave energy may be absorbed in two ways (Figures 8.26, 8.27). Where there is a local source of rock debris, the energy is expended in transporting, abrading,

(a)

(b)

(c)

Figure 8.25 Glacial landscape features. (a) A formerly glaciated U-shaped valley in southern New Zealand. (b) Dissipating margin of a glacier in the mountains of western Alberta. The ridge at left is a moraine; the sediments in the foreground form a small outwash plain. (c) A large esker in northern Canada formed under a former ice sheet. (a: Tad Nichols; b: National Film Board of Canada; c: G. Hunter, Information Canada Photothèque)

and reworking it; the result is a **beach.** Just as in alluvium-filled stream valleys, the beach sediments act to prevent local bedrock erosion by absorbing much of the energy of the moving water. When there is no local source of sediment, wave energy may be dissipated directly against solid rock; **cliffed coasts,** caused by the erosive under-cutting of waves and the fall of large blocks of rock from above, are the usual result.

Wind wave energy, concentrated in shallow coastal waters, is the most important erosional agent of ocean water, but other types of ocean movements also help to shape the land surface. Tidal motions have little direct effect, but serve to increase the vertical range over which the energy of wind waves is dissipated; for this reason, wave erosion is usually most effective in regions with large tidal fluctuations. Both waves and tides also cause local pressure differences which lead to complex patterns of

Figure 8.26 *Characteristic landscape features of coasts and continental shelves.*

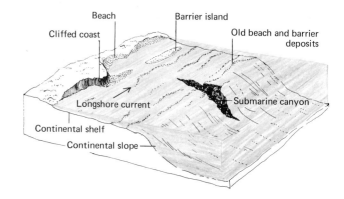

Figure 8.27 *Coastal landscape features. (a) A cliffed coast on Cape Breton Island, Nova Scotia. (b) A beach along the coast of southern Oregon. (c) A barrier island along the Texas coast. (a: National Film Board of Canada; b: Ramsey, Omikron; c: John S. Shelton)*

(a)

(b)

(c)

currents in shallow coastal waters. These currents, which usually parallel the shoreline, may erode solid rock but, more commonly, they serve only to redistribute coastal sediments and to shape them into characteristic **barrier islands** (Figures 8.26, 8.27).

Beyond the narrow coastal zone, the continental margins slope gently downward and are covered by progressively deeper ocean water. At the edges of these submerged continental margins, which are called the **continental shelves,** there is usually a much steeper zone, called the **continental slope,** leading downward to the deep ocean basins (Figure 8.26). At their outermost edges most continental shelves today are covered by about 200 m (660 ft) of ocean water. Quite significantly, however, the size of the continental shelves has changed dramatically with the climatic fluctuations of Pleistocene time because the large continental ice sheets locked up enough water to lower worldwide sea levels by at least 100 m (330 ft) and, possibly, by as much as 200 m (660 ft). During intervals of maximum glaciation much or all of the present continental shelves were therefore exposed above sea level, and the coastline lay offshore at points now deeply submerged. Conversely, it is estimated that if the present Greenland and Antarctic ice sheets were to melt, enough water would be added to the oceans to *raise* sea level about 70 m (230 ft), thus flooding much of the earth's present coastal regions. The ice sheet fluctuations of Pleistocene time have therefore led to numerous migrations of the shoreline back and forth across the continental margins; the most important result has been to deposit beachlike sediments over much of the shelf region.

Although depositional processes have dominated the continental shelves, the outermost margins of many of them show dramatic erosional canyons which are deeper and steeper than any occurring above sea level (Figure 8.26). The exact origin of these **submarine canyons** is still uncertain; some geologists believe they were formed, in their uppermost parts at least, by streams flowing on dry land during intervals of maximum glaciation when sea level was much lower. Others think it more likely that they have been carved entirely by sediment-laden bottom currents of ocean water moving down the steep continental slope.

Beyond the steep margins of the continents, the deep ocean basins are regions in which the "land" surface is little affected by erosion. Instead, the construction of new crustal rocks by volcanic action and ocean-floor spreading, and the accumulation of land-derived sediments, are the dominant processes. Nevertheless, active erosion does occur whenever volcanic rocks of the ocean build above sea level to form oceanic islands. Over many years the tops of volcanic islands may be eroded and leveled by wave action. Such wave-eroded, flat-topped volcanic mountains, called *guyots* (see Section 3.7 and Figure 3.23), are a common feature of the ocean floor. The tops of the youngest guyots stand near present sea level, but older ones have been deeply submerged by the movement of ocean floor away from midocean ridges.

SUMMARY OUTLINE

Water on the Land

8.1 *Lakes:* are temporary, stationary, accumulations of land water that show some of the properties of the ocean, but have relatively little influence on land sculpture.

8.2 *Ground water:* much of the water on land is in cracks, pores, and voids in soil and rocks; it directly influences landscape only in regions underlain by limestone, which is readily dissolved.

8.3 *Streams:* are the principal agent of land sculpture; the velocity and erosive effect of stream flow are controlled by a shifting balance between discharge, channel shape, channel slope, and sediment load.

**Land Sculpture
by Land Water**

8.4 *Mass wasting:* soil and rock debris first move downslope under the direct influence of gravity either slowly, by creep, or more rapidly, by various sorts of landslides.

8.5 *Stream transportation:* debris ultimately reaches flowing stream water which moves it from the area of origin; where slopes are steep, the debris is removed rapidly and the stream actively cuts a V-shaped valley; on more gentle slopes the debris accumulates to form an alluvial valley.

8.6 *Rates and cycles of water sculpture:* stream erosion of the land surface is offset by mountain building and internal uplift; precise cycles of erosion and uplift are difficult to reconstruct.

**Land Sculpture
by Wind, Ice,
and the Ocean**

8.7 *Deserts:* lack of covering vegetation leads to intense stream erosion and local debris accumulation during intermittent rains; more constant wind erosion accumulates sand dunes and removes finer debris in dust storms.

8.8 *Glaciers and ice sheets:* accumulate where annual snowfall exceeds evaporation; they slowly flow downward and shape the underlying land surface both by erosion and by deposition of transported debris.

8.9 *Coasts and continental shelves:* ocean water shapes the continental margins forming cliffs, beaches, and barrier islands along the coast, and influencing the submerged continental shelves as sea level changes.

ADDITIONAL READINGS

*Bloom, A. L.: *The Surface of the Earth,* Prentice-Hall, Inc., Englewood Cliffs, N.J., 1969. A brief introduction to landscapes.

Easterbrook, D. J.: *Principles of Geomorphology,* McGraw-Hill Book Company, New York, 1969. A standard intermediate text on landscapes.

*Gordon, R. B.: *Physics of the Earth,* Holt, Rinehart & Winston, Inc., New York, 1972. Chapter 5 provides a good introduction to land sculpture.

Leopold, L. B., M. G. Wolman, and J. P. Miller: *Fluvial Processes in Geomorphology,* W. H. Freeman and Co., San Francisco, 1964. An intermediate text on streams and their relation to landscapes.

*Morisawa, M.: *Streams: Their Dynamics and Morphology,* McGraw-Hill Book Company, New York, 1968. A brief introductory text.

Shelton, J. S.: *Geology Illustrated,* W. H. Freeman and Co., San Francisco, 1966. An excellent introductory text stressing landscapes; extraordinary photographs.

Thornbury, W. D.: *Principles of Geomorphology,* John Wiley & Sons, Inc., New York, 1969. A standard intermediate text on landscapes.

*Tuttle, S. D.: *Landforms and Landscapes,* William C. Brown Company, Dubuque, Iowa, 1970. A brief introductory text.

* Available in paperback.

9

Sediments

Chapters 7 and 8 have shown how air and water act to convert massive rock into small particles that are moved by streams, wind, and ice. This debris does not remain continuously in motion but ultimately accumulates, as *sediment*, either in the ocean or on low-lying parts of the land. In these areas of accumulation, older sediment commonly becomes deeply buried by younger sediment and, in the process, is converted into sedimentary rock by pressure and chemical changes.

Sedimentary rocks are extremely significant to our understanding of the earth because they provide the only evidence of *past* interactions between the solid and fluid earth. The changing atmospheric patterns responsible for ancient climates leave no direct historical record and most ancient landscapes have long since disappeared because of erosion. Ancient sediments, on the other hand, may be preserved indefinitely to reflect the differing conditions of climate and landscape under which they were deposited. Most of our understanding of the history of the earth's surface is based on such sedimentary evidence, and much of Part 4 will be devoted to historical interpretations derived from sedimentary rocks. As an introduction to Part 4, this chapter reviews the kinds of sediments being deposited on the earth *today,* and the differing environments under which they are accumulating, both on land as terrestrial sediments, and in the ocean as marine sediments.

PROPERTIES OF SEDIMENTS

Particles accumulating as sediments today have two principal sources: some originate and are transported as *solid* particles derived from weathering of the land—accumulations of such particles are called **detrital sediments;** others originate when the *dissolved* materials derived from

Beach sediments, California.
(Baker, Omikron)

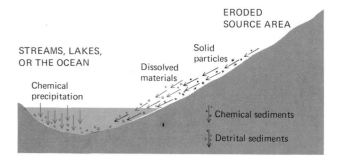

Figure 9.1   Detrital and chemical sediments.

weathering are precipitated from the waters of streams, lakes, or the ocean—accumulations of these precipitated particles are known as **chemical sediments** (Figure 9.1).

9.1   Detrital Sediments

In Section 7.6 we noted that the weathering of igneous rocks normally leads to two kinds of solid particles: *clay minerals,* extremely fine-grained sheet silicates produced by the alteration of feldspars and other igneous minerals, and *quartz particles,* which resist weathering and accumulate relatively unchanged from the original igneous rock. Because clay minerals and quartz grains are the dominant particles produced by weathering, it is not surprising that they are the principal constituents of most accumulations of detrital sediments.

Less commonly, detrital sediments have other compositions. In regions where mechanical weathering is more rapid than chemical weathering, the feldspar minerals which make up a large part of most igneous rocks may *not* be converted to fine clay minerals. Instead they may accumulate, along with quartz, as sand or pebble-sized particles. Such accumulations of both quartz *and* feldspar fragments are called *arkose.* At the opposite extreme, in regions where chemical weathering is intense, sedimentary accumulations often contain fine-grained particles of various silicon-poor iron and aluminum oxide minerals mixed with clay minerals and quartz.

Arkose and oxide-rich deposits are examples of relatively rare sediments which differ in *composition* from the quartz-clay mineral accumulations that are the dominant detrital sediments. Because most detrital sediments are made up *only* of quartz grains and clay minerals, they are not usually classified by mineral composition but, instead, by their relative proportions of clay and quartz particles

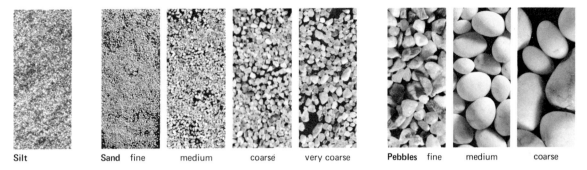

Silt   **Sand**   fine   medium   coarse   very coarse   **Pebbles**   fine   medium   coarse

Figure 9.2 Grain sizes of detrital silt, sand, and pebble particles. Still larger pebbles, to many centimeters in diameter, are not uncommon. (Fundamental Photographs)

of various sizes. Accumulations dominated by clay-sized particles are called **muds,** and those dominated by larger quartz particles are called **sands.** Muds are usually further subdivided by color, which normally ranges from very light gray to black; in general, the darker the color, the more fine-grained organic matter the mud contains. Sands, in turn, are classified by the size of their quartz fragments which may range from tiny grains of *silt* less than $\frac{1}{16}$ mm in diameter to large *pebbles* many centimeters in diameter (Figure 9.2). When such large fragments predominate, the accumulation is usually given the name **gravel** rather than sand.

Sediments composed largely of either clay or sand particles of a single size are known as **well-sorted sediments.** Mixtures of clay and sand particles of various size are called **poorly sorted sediments** (Figure 9.3). The degree of sorting of detrital sediments provides a useful index to the conditions under which they were originally deposited. Because moving fluids of differing velocities erode, transport, and deposit particles of differing sizes (Figure 9.4), well-sorted sediments normally result from slow accumulation and long exposure to varying velocities of flow. Poorly sorted sediments, on the other hand, indicate rapid accumulation and little exposure to such velocity variations.

9.2 Chemical Sediments

In contrast to the solid products of weathering that accumulate as detrital sediments, chemical sediments are derived from the more soluble elements, particularly calcium, magnesium, sodium, potassium, and silicon, that are transported to lakes and the ocean in solution. This material does not remain indefinitely in solution; instead much of it is *precipitated* as solid particles that accumulate to

Figure 9.3 A poorly sorted sandstone, made up of sand and pebble particles of varying size. (Fundamental Photographs)

Figure 9.4 Approximate velocities of water flow for erosion, transportation, and deposition of detrital particles of different sizes (100 cm/sec = about 2 mi/hr). (Modified from Turekian, Oceans, 1968, after Heezen and Hollister)

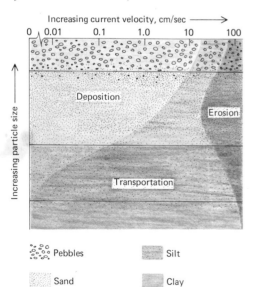

form chemical sediments. Today, much of this precipitation results from the life processes of water-dwelling plants and animals, many of which can concentrate even those dissolved elements that are present in very small quantities. Such organism-produced sediments are known as **biogenic chemical sediments.** Less commonly, the dissolved materials may become so concentrated by evaporation, for example in desert lakes or tropical lagoons, that they spontaneously precipitate without the intervention of living organisms. Accumulations of such materials are called **nonbiogenic chemical sediments.**

By far the most common chemical sediments forming today are biogenic accumulations of calcium carbonate deposited by marine animals and plants (Figure 9.5a). Many animals living in the ocean secrete hard, mineralized shells composed largely of the two calcium carbonate minerals calcite and aragonite. Corals, snails, clams, sea urchins, starfish, and crabs are a few familiar examples, but there are numerous others. In addition, many simple seaweedlike marine plants also precipitate calcium carbonate from ocean water. Over large areas of the ocean floor, areas where there is little influx of detrital sand and clay to dilute the carbonate minerals, relatively pure concentrations of calcium carbonate shells and shell fragments are accumulating today.

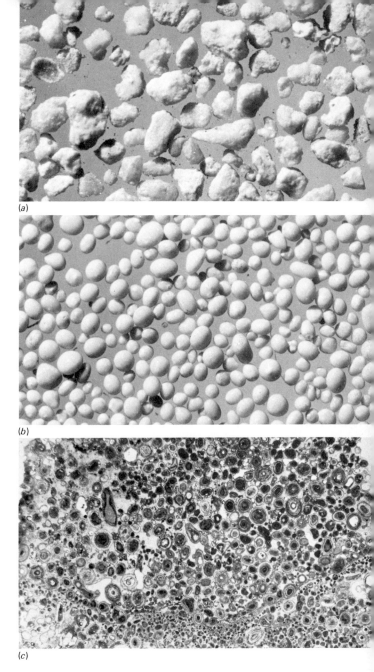

Figure 9.5 Biogenic and nonbiogenic chemical sediments. (a) Biogenic calcium carbonate sand made up largely of shell fragments. (b, c) Nonbiogenic calcium carbonate sand. The surface of the sand particles is shown in b; c shows the internal structure of concentric precipitated spheres. (a–c: Norman D. Newell, The American Museum of Natural History)

(a)

(b)

(c)

Calcium is the least soluble of the principal dissolved elements found in streams, lakes, and ocean water, and is readily precipitated not only by animals and plants, but also when the dissolved materials are concentrated by evaporation. Such nonbiogenic accumulations of calcium, usually combined in either calcium carbonate or calcium sulfate minerals, are far less abundant today than biogenic calcium carbonate, but are found in a wider range

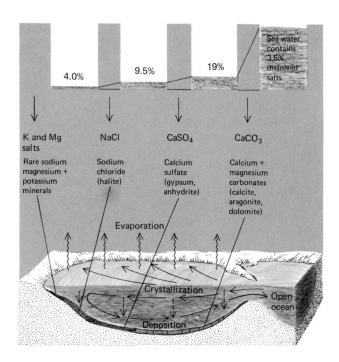

4.0%   9.5%   19%

Sea water contains 3.5% dissolved salts

K and Mg salts

Rare sodium magnesium + potassium minerals

NaCl

Sodium chloride (halite)

CaSO₄

Calcium sulfate (gypsum, anhydrite)

CaCO₃

Calcium + magnesium carbonates (calcite, aragonite, dolomite)

Evaporation

Crystallization

Open ocean

Deposition

Figure 9.6 Sequence of mineral precipitation in evaporating ocean water. Evaporation of lake water sometimes produces different minerals and sequences because of differing proportions of dissolved elements.

of environments. Desert soils and the floors of desert lakes commonly contain such calcium sediments concentrated by evaporation; they accumulate in even greater abundance in shallow ocean bays and lagoons in regions where evaporation is intense (Figure 9.5b, c).

Occasionally the evaporation of water from lakes, bays, and lagoons proceeds so far that even the most soluble dissolved elements—sodium, magnesium, and potassium—are deposited as unusual sulfate minerals and as common salt, sodium chloride (Figure 9.6). The deposition of these very soluble elements requires an almost complete evaporation of the water in which they are dissolved and leads to the accumulation of **evaporite sediments.**

Biogenic calcium carbonate deposits and nonbiogenic evaporites are the most common chemical sediments forming today, but two less common types also deserve mention. The first are deposits of *silica,* a compound of silicon and oxygen with the same composition as quartz, but lacking its stable, regular crystal structure. Quartz, as we have seen, is extremely resistant to weathering; in contrast, the silicon in more complex silicate minerals is far less stable and these minerals contribute large amounts of dissolved silicon to streams and the ocean as they are weathered. Most water-dwelling organisms that

Figure 9.7   Skeletons of radiolaria magnified 140 times. The remains of these microscopic animals accumulate in enormous quantities on some parts of the ocean floor to form biogenic silica deposits. (Eric V. Grave)

secrete shells or other hard parts construct them of calcium carbonate, but a few use this dissolved silicon instead. By far the most important silicon extractors today are *diatoms*, microscopic plants that are abundant in both the ocean and in lakes, and *radiolarians*, microscopic animals that live primarily in the ocean. Both secrete tiny, glasslike skeletons of silica that are accumulating on some parts of the ocean floor to form sedimentary silica deposits (Figure 9.7).

The final chemical sediment that we will consider also results from the activities of animals and plants, but in this case it is not the hard, mineralized skeletons that accumulate, but the soft organic materials themselves. Some organic matter from the decay of animals and plants is found in almost all present-day sediments, but such materials usually make up only a small fraction of the total volume. Less commonly, local conditions permit the accumulation of sediments with much higher amounts of organic matter. On land, deposits of *peat* may form when plant remains accumulate in acid waters which act as a preservative to prevent decay. Peat deposits often contain as much as 90 percent organic materials; the remaining 10 percent is usually detrital sand or clay. Likewise, on portions of the ocean floor organic materials produced by animals and plants in the sunlit surface waters may sink and accumulate, along with clay and sand, to make up a significant fraction of the bottom sediments. Both of these types of organic-rich chemical sediment have a significance far out of proportion to their small volume for, when buried and altered in sedimentary rocks, peat deposits become coal, and the organic matter in marine muds becomes petroleum.

9.3 Sedimentary Structures

Like detrital sand and clay, the precipitated grains of chemical sediments are usually transported by moving fluids before they finally come to rest in sedimentary accumulations. For this reason, both chemical and detrital sedimentary deposits normally show patterns of superimposed layers formed as they are sorted and deposited by fluid motions. These layers, and related features formed at the time of sediment deposition, are called **physical sedimentary structures.** An additional structural pattern is found in sediments that accumulate in aquatic environments where bottom-dwelling animals and plants are abundant, as they are on the floors of most lakes and shallow seas. Under these conditions the sediments are often reworked, burrowed, and otherwise modified by the organisms to produce **biogenic sedimentary structures.**

The most important physical sedimentary structure is the characteristic layering, called **bedding,** found in almost all sediment accumulations (Figure 9.8). Bedding results from the fact that thick sediment accumulations rarely form from a single massive deposition of particles; instead, many small depositional increments are required. A single stream flood or moving ocean current normally transports and deposits only a relatively thin sheet of particles. These sheets seldom exceed a few centimeters in thickness, and it thus requires many of them to form thick accumulations of sediment. Usually each thin increment, or *bed,* shows slight differences in particle size and arrangement which distinguish it from adjacent increments; these slight differences give sediment accumulations their characteristic bedded structure.

In the simplest case the boundaries of individual beds in a sediment accumulation are horizontal, and adjacent beds are distinguished by slight differences in color or particle size. Such **parallel bedding** is common in sedimentary accumulations but, since it can form under a variety of conditions, it gives relatively few clues to the original fluid motions that deposited the sediment (Figure 9.9). More useful in such interpretations are two equally common but less regular types of bedding—*cross bedding* and *graded bedding.*

Cross bedding is formed when the fluid motions of sediment deposition cause wavelike undulations on the

Figure 9.8 *The principal kinds of sedimentary bedding.*

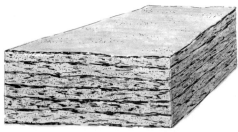

(a) Parallel bedding

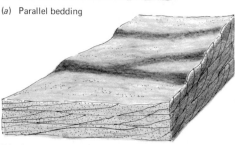

(b) Cross bedding

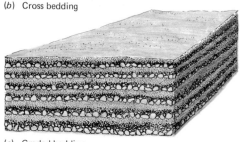

(c) Graded bedding

(a)

(b)

Figure 9.9 Parallel bedding in modern beach sand of eastern Scotland (a) and in ancient limestones of southern Utah (b). (a: Geological Survey of Great Britain; b: Tad Nichols)

(a)

(b)

Figure 9.10 Cross bedding in modern muds and silts
of eastern England (a) and in ancient sandstone of
northern Arizona (b). (a: Geological Survey of Great
Britain; b: Tad Nichols)

surface of the sediment. Such sedimentary wave forms are known as *ripples* and are common in accumulations of sand-sized sedimentary particles. When several rippled sand beds are superimposed, the individual beds, viewed in cross section, are not parallel, but have various angular relations to each other (Figure 9.10). Such bedding, called **cross bedding,** may provide important clues to the nature of the depositing fluid. For example, cross bedding formed by wind action usually shows individual beds inclined at a somewhat steeper angle than does cross bedding formed by moving water. In addition, laboratory studies of sediment rippling show that different patterns of cross bedding form under differing conditions of fluid flow. Thus it is sometimes possible to infer the approximate velocity, direction, and other features of the depositing fluid motion from patterns of cross bedding (Figure 9.11).

The second bedding type that gives important clues to the depositing fluid is **graded bedding,** in which each individual bed shows a progressive decrease in particle size upward through the bed (Figure 9.12). In most sedimentary deposits *adjacent* beds may show slight size differences but, normally, the particle size *within* an individual bed, deposited as a single event, remains constant.

Figure 9.11   Relation of bedding patterns to fluid motions and particle size. (From Allen, Physical Processes of Sedimentation, *1970)*

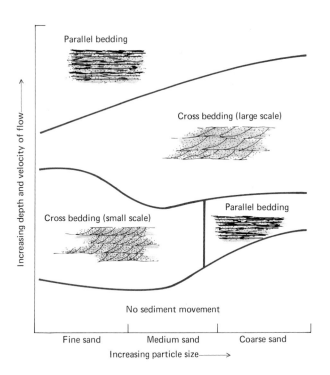

Figure 9.12   Graded bedding (upward decrease in particle size) as seen in a pebble-sized sandstone. (F. J. Pettijohn)

Figure 9.13   Origin of graded bedding from sediment slumping and turbidity current flow.

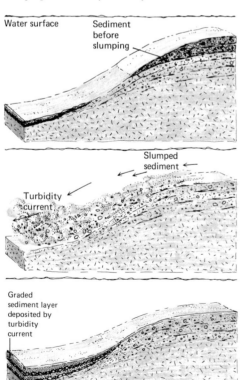

Graded beds are an exception and require a depositing mechanism which, in a single event, deposits progressively smaller particles. Field observations and laboratory experiments indicate that graded beds form when dense, sediment-laden currents are introduced into relatively still water. When this occurs, the current gradually loses velocity, depositing first the larger and heavier particles and then the finer and lighter in a gradational sequence. Such debris-laden currents, called **turbidity currents,** are common where sediment-carrying streams flow into lakes or the ocean; they also occur on the steep slopes of submerged scarps and canyons when large, unstable sediment accumulations break away and slide downward into deeper water (Figure 9.13).

Because of their high concentration of dense sediment particles, turbidity currents normally have a much higher overall density than the surrounding water; they thus move downward under the influence of gravity at high velocities and cover large areas before they are finally dissipated by friction. Their high velocities and densities also commonly cause them to erode or *scour* the previous bottom sediment before dissipating and depositing their own sediment load. This erosion usually forms distinctive channels, scratches, and gouges in the underlying sediment which, along with graded bedding, are characteristic structures of sediments deposited by turbidity currents. Because they are both formed and buried by a single sedimentary event, these erosional features are difficult to observe in present-day sediments, but are a common feature of ancient sedimentary rocks. There, the scratches, grooves, and channels are most often preserved as impressions on the base of the overlying bed, and for this reason they are known as **sole markings** (Figure 9.14).

Parallel bedding, cross bedding, graded bedding, and sole markings are the most important physical sedimentary structures; equally common are various biogenic sedimentary structures formed in the sediments shortly after deposition by living animals and plants. The two principal kinds are: tracks and trails produced by animals on the *surface* of the sediment, and burrows and borings produced by animals and plant roots *within* the sediment (Figure 9.15). Both of these types of biogenic structure can provide important clues to the original depositional environment of the sediment.

In addition to such structural traces of organic activity, animal shells and other hard parts of animals and plants are also commonly found in sedimentary accumu-

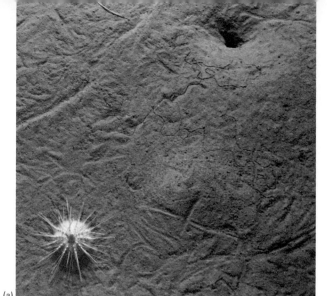

(a)

Figure 9.14 *Sole markings, caused by turbidity current scour, exposed on the base of a sandstone. (W. Hiller)*

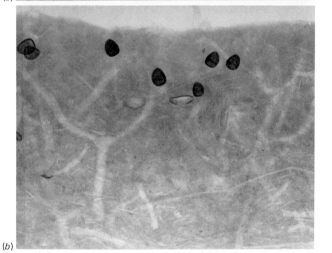

(b)

Figure 9.15 *Biogenic structures (a, b). (a) Surface tracks and trails made by bottom-dwelling animals on muds of the deep-ocean floor. (b) X-ray cross section of burrows made within lagoonal muds by burrowing animals. Some of the animals' shells can be clearly seen. (c) An accumulation of animal shells becoming buried in beach sands. (a: D. M. Owen; b, c: Donald C. Rhoads)*

(c)

lations (Figure 9.15). We have already seen that many chemical sediments, particularly calcium carbonate and silica accumulations, are formed almost entirely from the shells and hard parts of animals and plants. In addition, most detrital sediments also contain scattered animal and plant remains. Usually these biologic remains persist, as *fossils,* when the sediment is altered to sedimentary rocks and provide invaluable clues to the original depositional environment of the sediment.

TERRESTRIAL SEDIMENTARY PATTERNS

Both detrital and chemical sediments are accumulating on the earth today under a host of different environmental conditions—near polar glaciers and around tropical coral reefs, in valleys far above sea level, and in the deepest ocean basins. The principal goal in studies of these present-day sediments is to determine the characteristic associations of compositions, grain sizes, and structures that are forming under each environment so that this knowledge can be used to interpret the environments preserved in ancient sedimentary rocks.

The most important environmental distinction in both present-day and ancient sediments is between those that accumulate on land as **terrestrial sediments** and those that are deposited on the ocean floor as **marine sediments.** Our discussion of climates and landscapes has already provided a framework for understanding terrestrial sediments because the principal differences among them are those relating to deposition by permanent streams in humid tropical and temperate regions, by wind and temporary streams in desert regions, by ice in polar regions, and by combinations of these agents and ocean water around the continental margins. In this section we shall review the characteristic patterns of sediment accumulating today in each of these terrestrial environments before moving on to consider the much larger volumes of sediment accumulating on the ocean floor.

9.4 Fluvial Environments

The sedimentary environments formed by continuously flowing streams in tropical and temperate regions are known as **fluvial environments;** because of their relative accessibility, these are the most thoroughly studied and best understood of all present-day sediment accumulations. In discussing stream erosion in Chapter 8 we saw that the sand and clay carried by streams during floods are normally deposited further downstream as the flood subsides, only to be transported and deposited again during the next flood. The deposition of fluvial sediments is therefore usually only a temporary stop in a journey that leads ultimately to the ocean. Nevertheless, in certain regions, such as low-lying areas where the land surface is rapidly sinking as a result of tectonic forces, the fluvial sediment deposited by one flood may be buried *without erosion* by deposits of the next flood, and thus be removed indefinitely from the possibility of further transportation toward the ocean. Under such conditions fluvial sediments hundreds of meters thick may accumulate and, ultimately, be preserved as sedimentary rocks.

Studies of present-day fluvial sediments show that they usually consist of two strikingly different kinds of detrital particles: thick layers of well-sorted sand or gravel alternate with layers made up of poorly sorted mixtures of finer sand and mud. In addition to differences in grain size and sorting, the alternating sand and mud layers usually show a sharp contrast in sedimentary structures. The sands usually show abundant cross bedding indicating that they were deposited by swift horizontal fluid motions which strongly rippled the sediment surface; in contrast, the muds usually show a dominance of parallel bedding indicating vertical settling in relatively still water (Figures 9.16, 9.17).

Extensive studies of modern stream deposition have shown that these alternating sand and mud layers tend to form simultaneously, but in different horizontal and vertical positions. During floods the highest velocities of water movement tend to occur in the deepest parts of the stream channel; there the heaviest, coarsest particles are concentrated and moved by bouncing, sliding, and rolling

Figure 9.16   Characteristic fluvial sediments.

Channel sands

Overbank muds

(a)

(c)

along the bottom. The finer particles, on the other hand, are widely dispersed in suspension throughout the moving water which covers the entire flood plain. As a result, the coarse and fine particles are spatially separated. The coarse particles accumulate as **channel deposits**—thick masses of well-sorted sand and gravel which are rippled and cross-bedded by rapid channel flow. As the river channel slowly shifts its position, these channel deposits become covered by finer-grained **overbank deposits**—clays

(b)

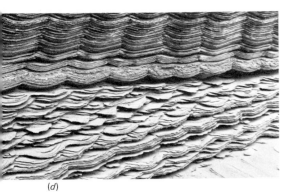

(d)

(e)

Figure 9.17 Fluvial sediments. (a) Channel sands and gravels exposed along a stream in northern Scotland. (b) Close-up view of cross-bedded channel sands exposed in a small trench along a stream in western Indiana. (c) Flooding and deposition of overbank sediments along the Missouri River in western Iowa. The normal river channel can be seen at the right. (d) Close-up view of overbank mud deposits exposed in a 1 m (3 ft) cross section along the Colorado River in western Arizona. (e) Ancient channel sands (light color) and overbank muds (dark color) preserved in sedimentary rocks exposed in southern Connecticut. (a: Geological Society of Great Britain; b: P. E. Potter; c: U.S. Department of Agriculture; d: Tad Nichols)

and fine sands that are spread in suspension over the entire floodplain and deposited there as the floodwaters recede (Figures 9.16, 9.17). Unlike the channel deposits, these overbank deposits may accumulate in a variety of subenvironments such as swamps, marshes, or temporary lakes. This horizontal and vertical alternation of channel sands and overbank muds is the characteristic pattern of fluvial sediments.

Fluvial sediments are always composed predominantly of detrital sedimentary particles. The only chemical sediments normally associated with fluvial environments are organic materials; these are rare in channel deposits, but sometimes occur in abundance in overbank muds, particularly those deposited in marshes and swamps. In such environments the muds are commonly very dark gray or black because of abundant fine-grained plant debris. In addition, larger plant fragments, such as leaves, stems, or even logs, sometimes are found in such deposits. Occasionally these larger plant fragments may accumulate in such high concentrations that they overshadow the muds to form peat, but peat occurs in only a small fraction of present-day fluvial sediments. Most contain relatively few animal and plant remains because such material quickly decays when exposed to the air between periods of flood.

9.5 Desert Environments

Much of the sediment accumulating today in desert regions is deposited by streams just as are the fluvial sediments of temperate and tropical areas. These desert sediments differ markedly from those of humid regions, however, because temporary desert streams provide little opportunity for separating the sedimentary particles into distinctive channel and overbank deposits. Most desert streams flow only during infrequent periods of heavy rainfall and, even then, they flow only over relatively short distances—usually from desert highlands down to the surrounding lowlands where they quickly disappear by evaporation and absorption into the dry, porous soil. Under such conditions the sedimentary debris moved by the streams tends to be concentrated at the base of the highlands in wedge-shaped alluvial fans (Figures 9.18, 9.19). Because there are no permanent streams to further transport alluvial fan deposits toward the ocean, they can accumulate to great thicknesses as older fan deposits are buried by younger.

Figure 9.18 Characteristic sediments of desert regions.

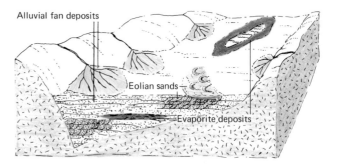

(a)

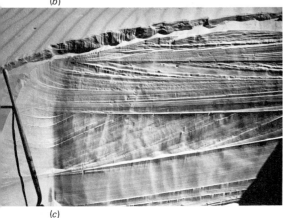

(b)

(d)

(c)

Figure 9.19   Desert sediments. (a) Cross section of poorly sorted sediments of a small alluvial fan in southern California. (b) Desert evaporite deposits, Death Valley, California. (c) Cross bedding in desert dune sands of southern California, exposed in a small trench. (d) Ancient alluvial fan deposits (above) and dune sands (below) preserved in sedimentary rocks exposed in southeastern Utah. (a, c: John S. Shelton; b, d: U.S. Geological Survey)

Alluvial fan deposits are almost always composed of a poorly sorted mixture of detrital particles ranging in size from the finest clay to extremely coarse gravel. They commonly contain abundant feldspar fragments, along with the more typical quartz and clay particles, because the relative lack of chemical weathering, and the rapid burial of the sediments, prevent much of the eroded feldspar from weathering to clay minerals. Cross bedding may be present but, in general, these poorly sorted detrital mixtures show few distinctive sedimentary structures.

In addition to alluvial fan deposits, two less common but very distinctive kinds of sediment also accumulate in desert regions. The first are chemical evaporites formed in temporary lakes that sometimes occur on the floors of desert basins (Figures 9.18, 9.19). These lakes are filled during the infrequent periods of heavy rainfall as ephemeral streams flow from desert highlands into the surrounding lowlands. Between rainfalls the waters are partially or completely lost to the atmosphere by evaporation; when this occurs the dissolved elements in the lake water are deposited as chemical evaporites. Evaporites deposited in different desert lakes show a wide range of chemical compositions because of differences in the dissolved elements transported to the lakes by local streams. Calcium and sodium evaporite minerals, such as calcite, gypsum, and halite, are the most common deposits, but some evaporites have high concentrations of unusual minerals containing such rare elements as boron and lithium.

Desert basins where evaporation is intense provide the only major setting where *non*biogenic chemical sedi-

319

ments accumulate above sea level. Biogenic chemical sediments almost never occur in desert environments, both because plants and animals are scarce, and because the lack of water, and high temperatures, lead to rapid decay of the small amounts of organic matter present.

In addition to alluvial fan and evaporite deposits, a third type of sediment may occur in desert environments. These **eolian sediments,** as they are called, originate when the sands and clays of alluvial fans are further transported and deposited by wind action (Figures 9.18, 9.19). Fan deposits normally lack a protective cover of vegetation and are therefore continually subject to varying velocities of wind flow which tend to sort them into different size groups just as does the flowing water of fluvial environments. During periods of high winds, fine silt and clay particles are transported in suspension as *dust storms* and may be carried long distances, often even beyond the desert into more humid environments, before the wind subsides.

Sand particles, in contrast, are usually too heavy for suspension transport by wind; instead, they are bounced and rolled along the desert floor to be concentrated as dunes on the windward side of desert basins. There they may be covered and preserved by later alluvial fan deposition. Wind-deposited sand accumulations show many of the characteristics of fluvial channel sands in that they are well sorted and show strong cross bedding and surface rippling caused by the rapid horizontal flow of the depositing fluid. Certain details of eolian cross bedding differ, however, from those of water-deposited sands. In particular, the angular bedding is commonly much steeper in wind deposits and this, combined with the associated alluvial fan and evaporite sediments, normally makes it possible to recognize ancient desert sands.

9.6 Glacial Environments

Sediments accumulating today in glacial environments show many of the same characteristics as desert alluvial fan deposits and the two types of deposits are often difficult to distinguish in ancient sediments. Like desert alluvium, most glacial deposits are unsorted mixtures of clay, sand, and gravel which lack distinctive sedimentary structures. Because glacial environments, like deserts, normally lack a dense cover of vegetation, wind action may locally sort the surficial glacial deposits into eolian sand dunes similar to those of desert regions. Unlike desert sediments, however, those of glacial regions rarely show evaporite deposits, which normally originate only in warm climates. As in desert environments, organic matter and animal and plant remains are rare in glacial deposits. The most characteristic feature of glacial sediments is the presence of pebbles that have been scratched and abraded by moving against solid bedrock while frozen into the flowing ice; such **striated pebbles** are the surest criterion for the recognition of ancient glacial deposits (Figure 9.20).

9.7 Coastal Environments

Only a relatively small fraction of the sediment accumulating on the earth today is found above sea level in fluvial, desert, and glacial environments. Instead, most sedimentary debris is ultimately transported to the ocean where it accumulates in various *marine* environments under the influence of ocean waves and currents. The sediments of marine environments will be the subject of the next section, but before turning to them we need to consider those environments along the edges of the continents where sedimentary accumulations are forming under the influence of *both* ocean water *and* the dominant terrestrial depositing agents—streams, wind, and ice.

All such coastal sedimentary environments occur in the narrow zone where land, sea, and air come in contact. Coasts, in general, are regions of intense and varied fluid motions: waves stir the bottom and pound the land, tides

Figure 9.20 Ancient glacial sediments exposed in southwest Africa, showing striated pebbles scratched by moving ice. (J. W. Hälbich)

rise and fall, and both act to create strong currents parallel to the shoreline. At the same time, streams flow into the ocean supplying new sediment and creating additional current motions as fresh water mixes with the heavier salt water of the sea. Because of these varied fluid motions, relatively little sediment accumulates in coastal regions. Instead, most is ultimately transported beyond the shore to come to rest in deeper water where there are fewer opportunities for fluid transport. There are, nevertheless, thick sequences of coastal sediments accumulating today in a few regions where the underlying crustal rocks are slowly sinking and thus allowing the older sediments to be buried by younger sediments before they have an opportunity to be transported farther. Studies of such accumulations show, as might be expected, that the sediments are normally well sorted because of the varying velocities of flow to which they are subjected.

In general, the larger, heavier sand and gravel particles are concentrated in linear belts parallel to the shore; such belts make up the *beaches* and *barrier islands* that are a familiar feature of so many coasts (Figures 9.21, 9.22). Both form in zones of strong wave or current action which remove the finer sedimentary particles and concentrate the coarser sand and pebbles. As in the well-sorted sands of fluvial channels and desert dunes, beach and barrier sands tend to be rippled and cross-bedded from the motions of the depositing fluid, and contain little organic matter or preserved animal or plant remains. Indeed, ancient beach and barrier deposits are often indistinguishable from channel or dune sands except by their association with other, more distinctive, sediments.

In contrast to beach and bar sands, finer silt and clay particles in coastal regions accumulate primarily in sheltered *lagoons* where fluid motions are less intense (Figures 9.21, 9.22). Lagoons develop wherever the con-

Figure 9.21   Characteristic sediments of coastal regions.

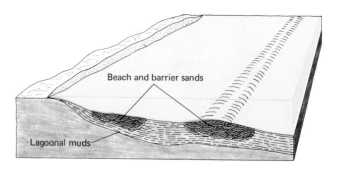

Beach and barrier sands

Lagoonal muds

(a)

(b)

Figure 9.22 *Coastal sediments. (a) A beach made up of large pebbles, New Zealand. (b) Ancient lagoonal deposits preserved in sedimentary rocks exposed in western Wales. (a: Tad Nichols; b: Geological Survey of Great Britain)*

figuration of the coastline, or the presence of energy-absorbing sand barriers, protect shallow bodies of water from strong wave and current motions. Such regions receive water and sediment both from streams on their landward side and from the ocean on their seaward side. Because of their sheltered setting, the clay and silt particles are not normally removed by fluid motions but, instead, accumulate to form thick mud or muddy sand deposits.

Lagoons usually support an abundance of plant and animal life; indeed, in very shallow lagoons, bottom dwelling plants often grow above the surface of the standing water to form *marshes*. Shells, plant fragments, and other larger remains of bottom dwelling animals and plants are therefore common in lagoon and marsh sediments. The presence of such remains helps distinguish ancient lagoonal muds from muds formed as fluvial overbank deposits, which lack abundant organic remains. The exact kinds of animal and plant remains found in lagoon sediments are determined by the proportion of fresh water and ocean water that normally occurs in the lagoon. Most ocean-dwelling life cannot tolerate fresh water and vice versa, so that lagoons dominated by ocean water contain characteristic marine organisms and those dominated by streams have characteristic fresh-water animals and plants. Lagoons which fluctuate rapidly between salt and fresh water support relatively little life because neither group can become established.

The presence of bottom-dwelling organisms in lagoons also strongly influences the sedimentary structures of lagoonal deposits. Like fluvial overbank deposits, lagoonal muds are usually deposited as horizontal layers with parallel bedding. In some lagoons that lack abundant bottom life this bedding persists but, more commonly, it is partially or wholly destroyed by the burrowing activities of bottom animals. This activity, in time, creates an abundance of biogenic sedimentary structures which are also a characteristic feature of most lagoonal sediments.

Beach and bar sands, and lagoonal muds, are the principal sediments forming today in coastal environments. In polar and temperate regions they are normally composed of land-derived quartz fragments and clay minerals transported to the shore by streams. Such deposits are also common along tropical coasts, but in tropical regions where there is no nearby source of land-derived particles beaches may be made up entirely of sand-sized fragments of biogenic calcium carbonate and lagoons may be filled with calcium carbonate mud. In these settings the sedimentary particles are not derived from rock weathering on land, but from the removal of dissolved elements from ocean water by animals and plants living near the shore.

MARINE SEDIMENTARY PATTERNS

Most sediment accumulation takes place on the ocean floor in various marine environments. In present-day marine sediments a primary distinction can be made between those deposited under relatively shallow water on the submerged continental shelves, and those accumulating beyond the continents in the deep-ocean basins. Continental shelf sediments tend to accumulate both more rapidly, and by somewhat different processes, than do those of the deep oceans, and this leads to fundamental differences in texture and composition. Shelf and deep-ocean environments resemble each other, however, in that each has some large areas where land-derived sands and muds are accumulating as detrital sediments, and other large areas dominated by chemical accumulations which, for brevity, are usually referred to as *carbonate* sediments because they are dominated by biogenic calcium carbonate. We shall consider both detrital and carbonate accumulations of the continental shelves before examining the somewhat different detrital and carbonate accumulations of the deep-ocean basins.

9.8 Detrital Shelf Environments

Most of the sand and mud delivered to the oceans by streams is deposited on the submerged shelves which surround the continents. These shelves today have an average width of about 80 km (50 mi) and slope gradually seaward to an average water depth of about 200 m (600 ft); beyond them lie the steeper slopes of the continental margins that lead to the deep-ocean basins (Figure 9.23). Because the ocean floor on the shallow, near-shore parts

Figure 9.23 Present-day continental shelves, the submerged margins of the continents where most sediment accumulation takes place. (Compiled from various sources)

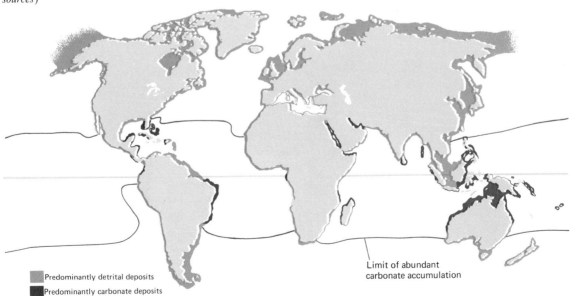

Limit of abundant
carbonate accumulation

Predominantly detrital deposits

Predominantly carbonate deposits

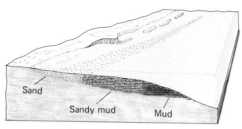

(a) Predicted shelf sediment pattern

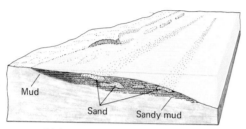

(b) Actual shelf sediment pattern

Figure 9.24 *Predicted and actual patterns of sediment distribution on the continental shelves. The relatively recent expansions and contractions of continental ice sheets have caused repeated migrations of the shoreline, and deposition of shallow-water sediment, over much of the shelves. The more regular "predicted" pattern of offshore size decrease occurs in many ancient sediments.*

of these shelves is subjected to much more intense wave and current motions than is the deeper shelf, we might predict that large sedimentary particles should be sorted and concentrated on these high-energy, near-shore bottoms as accumulations of sand and gravel. Conversely, only finer silt and mud particles should reach the outer shelf region where relatively little wave and current energy is available to transport larger particles (Figure 9.24a).

Studies of sedimentary rocks have shown that this offshore decrease in particle size *was* common in ancient marine sediments, for they normally show a concentration of sand and gravel in near-shore regions that grades seaward into finer silt and mud deposits. This pattern does *not* occur, however, on the continental shelves today. Instead, studies of modern shelf sediments (made by sampling or photographing the bottom from ships or, in shallow water, by diving with scuba gear) have shown that well-sorted *sand* is the dominant sediment over much of the present-day shelves (Figure 9.24b). Furthermore, sand is particularly abundant on the outermost shelf, where it should be rare or absent. In addition, mud accumulations are far more common today in shallow, near-shore settings than would be predicted from the distribution of wave and current energy.

We have already anticipated the reasons for this anomalous pattern—the relatively recent changes in sea level caused by the expansions and contractions of Pleistocene ice sheets. These sea-level changes led to many oscillations of the shoreline back and forth across the continental shelves (Figure 9.25). As a result, much of the sediment exposed today on even the outermost parts of the shelves was deposited in high-energy, near-shore environments at times when sea level stood much lower. Indeed, large areas of the present shelves appear to be covered with materials deposited above sea level in fluvial or, on polar shelves, glacial environments. Most present-day shelf sediments, therefore, do not reflect their present environments but, instead, are *relict* sediments of earlier environmental patterns.

Too little time has elapsed since the most recent melting of continental ice sheets, and the consequent rise of sea level, to allow sediments now being delivered to the shelves to reach an equilibrium with their environments. Presumably, if sea level were to remain at its present position long enough (an unlikely event, as we shall see in Chapter 12), fine muds now accumulating near shore would be slowly transported seaward to cover the near-

Man
and Sediments

A review of the table of principal mineral resources (p. 37) shows that sedimentary processes have been the concentrating agent for most of the materials that man removes from the earth. Indeed only copper, lead, zinc (see p. 114), and magnesium (see p. 155) have primary sources other than sediments or sedimentary rocks. The two most important sedimentary resources of industrialized societies are iron and mineral fuels and these will be discussed in Chapters 11 and 12. There remain, however, a host of other sedimentary resources—principally fertilizers, building materials, and industrial chemicals—that we shall consider briefly here.

Man's principal uses of *detrital* sediments are for building materials—sand

and gravel used for fill and making concrete, and clay used for making bricks. Such materials are consumed in enormous volumes, yet they are so universally abundant that there is no danger of their exhaustion. However, supplies conveniently close to the point of use are diminishing in many heavily populated areas and so they must be brought from increasing distances at greater costs.

Far more serious than an ultimate shortage of these materials are the widespread, landscape-marring pits and quarries that result from their removal. With careful management such blight can be avoided and the areas reclaimed for constructive uses after the materials are removed. This reclamation work is often expensive, however, and relatively few cities or regions as yet require it.

In contrast to the comparatively restricted uses of detrital sediments, *chemical* sediments, and the sedimentary rocks derived from them, have many applications. Limestone, the most abundant chemical sedimentary rock, is used chiefly as the major ingredient of cement. By far the most versatile chemical sediments, however, are evaporite deposits which provide not only gypsum, the principal component of plaster and wall board, but most mineral fertilizers and industrial chemicals as well.

As ocean or lake waters evaporate, the dissolved materials are precipitated principally as the minerals gypsum and halite. At several earlier intervals in earth history, large marine basins containing evaporite accumulations were far more extensive than they are today and, as a result, the continents' veneer of sedimentary rocks contains abundant deposits of both gypsum, for building materials, and halite, which is used principally by the chemical industry. Less common, but still adequate for any foreseeable demand, are potassium evaporite minerals and phosphorus deposits, both used mainly as fertilizers.

Still less common are deposits of native sulphur formed in only a few regions by the action of bacteria on the gypsum of certain evaporite deposits. Sulfur is a prime industrial chemical, used principally to make the sulphuric acid required for the processing of many other chemical products. For this reason, the relative scarcity of easily tapped native sulfur deposits has led to a search for alternate sources. Fortunately sulfur occurs in abundance in many other forms and is now being extracted in large quantities as a byproduct of processing both natural gas and the sulfide ore minerals of copper, lead, and zinc.

Sand and gravel being removed from under water, Maryland. (U.S. Bureau of Mines)

Limestone quarry and cement plant, southern Sweden.

Underground salt mine, Dominican Republic. (United Press International)

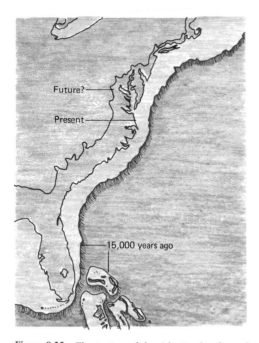

Future?

Present

15,000 years ago

Figure 9.25 *Fluctuation of the Atlantic shoreline of the United States. Fifteen thousand years ago enough ocean water was locked up in large continental ice sheets to drop sea level about 200 m (660 ft) and shift the shoreline to the edges of the present continental shelf. If the present Greenland and Antarctica ice sheets were to melt, sea level would rise and the shelves would expand to approximately the line marked "future?"*

shore sands that now dominate the shelf, leading to the expected offshore decrease in grain size. In the meantime, only a few large rivers that carry huge volumes of mud and sand are able to deposit these materials beyond the shore region by building out broad **deltas** onto the continental shelves. In contrast, the lower reaches of most streams have been "drowned" by rising sea level to create elongate, lagoonlike bodies of water known as **estuaries.** Most of the sediment carried by these streams is accumulating in these coastal estuaries, which must be filled before delta build-out, and transport of particles onto the shelf, can begin.

Because present-day continental shelves are mostly covered with sediments that were deposited under previous environmental conditions, studies of modern shelf sediments do not provide clear sedimentary patterns that can be applied to interpreting sedimentary rocks. Nevertheless, certain generalizations about shelf-deposited sands and muds are possible.

Perhaps the most characteristic feature of present-day shelf sands and muds is that they contain abundant evidence of the prolific life that is almost everywhere present on the ocean floor. Shells, bones, and other hard remains of bottom-dwelling marine animals are common in shelf sediments, as are biogenic sedimentary structures and fine-grained organic matter (Figure 9.26). About the only exceptions are those sediments, usually fine muds, that accumulate in deep, isolated basins where there is little water movement. Under such conditions there is usually too little oxygen in the bottom waters to support life and such environments are the "deserts" of the ocean floor. Fine-grained organic matter produced elsewhere, however, commonly accumulates in quantity in such environments because there is no bottom life to utilize the particles as food. The results are black, evenly bedded, organic-rich mud deposits that lack both shells and biogenic sedimentary structures (Figure 9.26).

Another characteristic feature of offshore detrital shelf sediments, in those few deltaic areas where they are being deposited today, is the frequent occurrence of small-scale alternations of parallel-bedded muddy and sandy layers, each usually from a few millimeters to a meter thick (Figure 9.26). These alternating layers are also very common in ancient shelf sediments, but their exact origin is obscure. Most probably they result from large, infrequent storms which agitate the water to unusual depths and transport near-shore sand particles seaward.

(a)

(b)

(c)

(d)

Figure 9.26 *Detrital shelf sediments. (a) Accumulation of oyster shells and (b) vertical biogenic burrowed structures preserved in ancient shelf sediments exposed in western South Dakota. (c) Ancient organic-rich shelf mud deposits, preserved as shales exposed in central New York. (d) Alternating bands of parallel-bedded mud and silt preserved in ancient shelf sediments exposed in western Wales. (a–c: Donald C. Rhoads; d: Geological Survey of Great Britain)*

Between these events, mud slowly accumulates to give the alternating muddy layers. On the more steeply sloping parts of the shelf, gravity transport of sedimentary particles by turbidity currents can also lead to alternating mud and fine sand sequences which, however, usually show graded bedding and other distinctive structures. Such **turbidite sequences,** as they are called, are less common in shelf sediments than in sediments accumulating in deeper environments beyond the shelf.

331

9.9 Carbonate Shelf Environments

Land-derived sand and mud are today the dominant sediments on polar and temperate shelves, but are replaced on many tropical shelves by extensive accumulations of calcium carbonate that have either been precipitated directly from the local ocean water or, more commonly, removed indirectly by the activities of animals and plants (see Figure 9.23). Direct precipitation is favored in tropical regions because the solubility of calcium carbonate in ocean water varies *inversely* with water temperature: the higher the temperature the less dissolved calcium carbonate it can contain, and vice versa. Only in shallow, tropical oceans does the water temperature rise high enough to cause spontaneous precipitation of calcium carbonate. Even then, some reduction in the volume of water by evaporation is probably necessary to further increase the calcium concentration and thus decrease its solubility, although the exact chemistry of the process is still poorly understood. Such conditions are met on many shallow tropical shelves today and result in sedimentary accumulations of directly precipitated calcium carbonate grains (Figures 9.27, 9.28). Most commonly these grains are rounded, sand-sized particles, called **ooliths,** which show a characteristic concentric pattern of growth (see Figure 9.5).

Under exceptional circumstances, evaporation of ocean water may proceed far enough to precipitate still more soluble dissolved elements (see Figure 9.6). Such conditions are met today only in a few tropical lagoons that receive infrequent inflows of normal ocean water. Between such influxes, intense evaporation may lead to deposition of *marine evaporite* sediments (Figure 9.27).

Figure 9.27 Characteristic sediments of carbonate shelves.

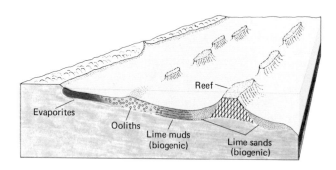

Evaporites · Ooliths · Lime muds (biogenic) · Reef · Lime sands (biogenic)

(a)

(b)

(c)

(d)

(e)

Figure 9.28 *Carbonate shelf sediments. (a) Aerial view of a part of the Bahama Bank, a broad shelf off the Florida coast dominated by carbonate deposition. (b–d) Close-up views of nonbiogenic lime mud (b), biogenic lime sand (c), and organic reef (d), all on the Bahama Bank. (e) Ancient lime mud preserved as sedimentary rock exposed in central Alabama. (a–d: Norman D. Newell, The American Museum of Natural History)*

Usually such sediments are made up mostly of the minerals halite and gypsum. Less commonly various unusual sodium, potassium, and magnesium minerals are deposited in such settings.

Directly precipitated calcium carbonate and marine evaporites are not forming today in polar or temperate regions and make up only a small fraction of the shelf sediments of tropical regions. Most present-day tropical shelves are covered instead by calcium carbonate particles of *biogenic* origin (Figures 9.27, 9.28). The exact reasons for the dominance of biogenic calcium carbonate deposits on tropical shelves are still uncertain. Probably the precipitation of calcium carbonate by animals and plants is favored in the tropics because of the lowered solubility of calcium carbonate just as is direct precipitation. Unlike directly precipitated calcium carbonate, however, shell-bearing marine animals are also common on temperate and polar shelves, yet in such regions they generally make up only a small fraction of the shelf sediments, which are predominantly detrital sands and muds. Perhaps the most likely explanation is that streams carrying land-derived sand and mud are less common in the tropics, particularly in the subtropical zones of arid climates. In these regions there is little detrital material supplied to the shelves to dilute the accumulating carbonate sediments.

When shell-bearing animals and carbonate-secreting plants die, their skeletons do not normally remain intact, but are progressively reduced to smaller and smaller particles as the surrounding organic matter decays, and by such secondary processes as wave abrasion and reworking by predators (Figure 9.29). As a result, most biogenic

Figure 9.29 *The breakdown of calcium-carbonate-depositing seaweeds and corals to produce lime sand and lime mud.* (*From Laporte,* Ancient Environments, *1968, after Folk and Robles*)

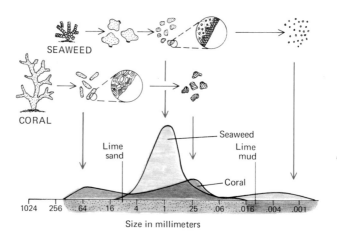

carbonate sediments consist not of whole shells, but of sand- and mud-sized particles derived from whole shells. These **lime sands** and **lime muds** are the dominant shelf carbonate sediments and generally show the same patterns of size sorting and sedimentary structures as do detrital sands and muds. The lime sands tend to be concentrated in near-shore areas of high-energy water motions and show cross bedding and other structures indicative of such motions. The lime muds, on the other hand, tend to accumulate in sheltered or deeper environments where there is less water motion. As on temperate and polar shelves, anomalous relict accumulations of lime sand, formed during recent intervals of lowered sea level, are probably common on deeper tropical shelves today, but this is not yet certain for tropical shelves have been less extensively explored than have those of temperate regions.

In addition to lime mud and lime sand, a third important type of biogenic carbonate accumulation occurs locally on most tropical shelves. These are **organic reefs** made up of closely interlocked skeletons of living and dead marine life (Figures 9.27, 9.28). Many organisms contribute to present-day reefs, but the dominant groups are corals, sponges, and certain seaweedlike plants which thrive only in relatively warm, tropical waters. Because corals are often a conspicuous part of modern reefs they are commonly called "coral reefs" but this term tends to obscure the other equally important reef-building organisms that are always present.

Unlike isolated shells and skeletons, the closely packed and cemented calcium carbonate of reefs does not readily break down into smaller particles when the reef-building organisms die and decay. As a result, massive structures composed of dead carbonate skeletons persist indefinitely to form a foundation upon which the animals and plants of the living reef are but a thin surface film. The largest such reef that is still being actively built today is the Great Barrier Reef which extends for 1,600 km (1,000 mi) along the coast of northeastern Australia; similar but smaller reefs are common on most tropical shelves.

Because of the presence of scattered organic reefs, tropical shelves usually show less regular profiles and more local relief than do the gently sloping shelves of temperate and polar regions. Basins and depressions around the massive reefs usually trap thick, *interreef* accumulations of lime sand and lime mud derived both from local, non-reef-building organisms, and by wave and current erosion of the reef itself. Such reef and interreef accumulations

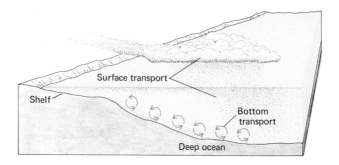

Figure 9.30 *Transport of land-derived sand and mud to the deep ocean.*

of calcium carbonate are also a common feature in ancient sedimentary rocks.

9.10 Deep-ocean Environments

Deep-ocean sediments are mostly either detrital sands and muds or biogenic carbonates just as are those of the shallower continental shelves. There are, however, significant differences in the depositional processes of these materials in the deep ocean.

Land-derived sand and clay particles reach the deep-ocean basins by either of two routes (Figure 9.30). Some are *bottom transported;* that is, they are moved seaward from the continents across the shelves by bottom currents and into the deep basins by gravity slumping down the continental margins. Such slumping results in turbidity currents that convey the sediments into the deep basins, usually through submarine canyons, and distribute them over wide areas in thin, graded beds. Most of the deep basins of the Atlantic Ocean are covered with thick sequences of such graded sand and mud layers transported across the continental shelves (Figures 9.31, 9.32).

The second process by which land-derived sands and muds reach the deep oceans is by transport not along the bottom, but by motions in either the upper layers of water or the overlying atmosphere. Clay and fine sand particles may be carried long distances by such movements. When the current motions or winds that transport them cease, they slowly sink to accumulate on the ocean floor as *surface-transported deposits.* Limited observations suggest that wind transport is the more important contributor to such deposits, for sand and clay seem to be transported across the shelves in surface waters only during very large storms.

Relatively thin, evenly bedded accumulations of surface-transported sediments are characteristic of many deep basins far from the continental margins, particularly

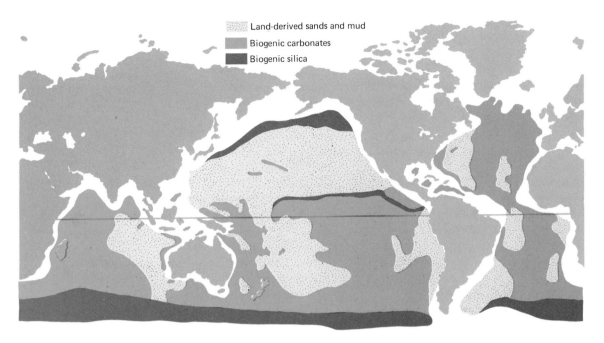

Land-derived sands and mud
Biogenic carbonates
Biogenic silica

Figure 9.31 Present-day deep-ocean sediments.
(Modified from Hill, The Sea, vol. 3, 1963)

those of the Pacific Ocean (Figures 9.31, 9.32). Unlike the Atlantic, much of the Pacific is bordered by deep trenches that extend downward thousands of meters below the surrounding ocean floor. These trenches tend to trap the bottom-transported materials that move down the continental margins and thus prevent them from spreading throughout the deep basins as in the Atlantic. As a result, thin, slowly accumulating deposits of surface transported muds and silts tend to floor most deep Pacific basins beyond the deep marginal trenches which contain thick, bottom-transported sequences. In the Atlantic, the principal surface-deposited accumulations occur in equatorial basins east of the Midatlantic ridge, where east-blowing winds transport and deposit much debris from the adjacent deserts of northern Africa.

Although the volume of land-derived sand and mud ultimately reaching the deep ocean by either bottom or surface transport is large, it nevertheless represents only a small fraction, probably about $\frac{1}{10}$, of all such particles delivered by streams to the ocean. The great bulk accumulates instead on the continental shelves.

In addition to land-derived sands and clays, the deep ocean, like the continental shelves, also has large areas dominated by biogenic deposits which, however, differ fundamentally from the biogenic sediments of the shelves. The principal difference is that deep-sea biogenic

(a)

(b)

Figure 9.32 Deep-ocean sediments. (a) Land-derived bottom muds on the floor of the deep Atlantic. Note the abundant animal tracks and burrows. (b) Biogenic calcium carbonate sand on a submarine hill in the deep Atlantic. The smaller particles are skeletons of foraminifers and the larger, light-colored particles are pteropod shells. (a, b: Richard M. Pratt, U.S. Geological Survey)

sediments are generally formed not from fragments of the relatively large skeletons of bottom-dwelling animals and plants, but from the tiny skeletons of the floating animals and plants that are collectively called *plankton.* Such organisms thrive in enormous numbers in the sunlit surface waters of the oceans. As they die and their soft tissues decay, their tiny skeletons sink to the ocean floor and accumulate there.

On those parts of the ocean floor that receive large quantities of land-derived silt and mud, such as the Atlantic basins, the steady rain of planktonic skeletons is diluted and makes up only a small fraction of the bottom materials. In regions where the land-derived influx is small, however, planktonic skeletons may accumulate as the dominant sediment. In the Atlantic such biogenic accumulations are found mostly on submarine hills and on the slopes of the Midatlantic Ridge (Figures 9.31, 9.32). Bottom currents periodically transport any accumulated clay particles from these higher areas into the surrounding basins, leaving a residue of the larger and less easily transported skeletons. In the Atlantic, the principal contributors to such deposits are two groups of floating animals called *foraminifers* and *pteropods,* and one group of tiny plants called *coccolithophores,* all of which build skeletons of calcium carbonate (Figure 9.32).

Carbonate deposits similar to those of the Atlantic also occur in some regions of the Pacific; unlike the Atlantic, however, the deep Pacific also has large areas dominated by planktonic skeletons composed not of calcium carbonate but of *silica.* Two groups of planktonic organisms secrete skeletons of silica, the plantlike *diatoms* and the animallike *radiolarians.* Diatoms are particularly common in surface waters of polar regions and their shells dominate bottom deposits in the north Pacific and in a broad belt under the ocean that surrounds Antarctica. Radiolarians, in contrast, are most abundant along the equator and accumulations of their shells floor large regions of the equatorial Pacific (Figure 9.31).

SUMMARY OUTLINE

Properties of Sediments

9.1 *Detrital sediments:* fine-grained clay minerals and quartz particles of various sizes accumulate as muds, sands, or poorly sorted mixtures of the two.

9.2 *Chemical sediments:* materials in solution in lakes and the ocean may be precipitated directly as their concentrations are increased by evaporation, or indirectly through removal by animals and plants.

9.3 *Sedimentary structures;* detrital and chemical particles moved and deposited by fluids show various kinds of layered bedding which give clues to the nature of the depositing fluid; other structures originate from the actions of sediment-dwelling organisms.

Terrestrial Sedimentary Patterns

9.4 *Fluvial environments:* are dominated by alternations of well-sorted sands deposited in the stream channel, and overbank muds deposited during floods.

9.5 *Desert environments:* show poorly sorted alluvial fan deposits, well-sorted dune sands, and lake evaporites.

9.6 *Glacial environments:* have sediments similar to those of desert alluvial fans, but often show striated pebbles abraded by moving ice.

9.7 *Coastal environments:* are dominated by well-sorted sands and gravels deposited on beaches and bars, and muds that accumulate in sheltered lagoons.

Marine Sedimentary Patterns

9.8 *Detrital shelf environments:* normally show an offshore decrease in particle size; modern shelves have anomalous deep sands caused by recent sea-level fluctuations.

9.9 *Carbonate shelf environments:* replace detrital sands and muds on many tropical shelves; most are composed of fragments of carbonate-secreting animals and seaweeds.

9.10 *Deep-ocean environments:* have accumulations of land-derived sands and muds as well as biogenic carbonate and silica derived from small, planktonic animals and plants.

ADDITIONAL READINGS

*Allen, J. R. L.: *Physical Processes of Sedimentation,* American Elsevier Publishing Co., Inc., New York, 1970. An authoritative intermediate text on detrital sediments and their environments of deposition.

Berner, R. A.: *Principles of Chemical Sedimentology,* McGraw-Hill Book Company, New York, 1971. An advanced text on chemical sediments.

Blatt, H., G. Middleton, and R. Murray: *Origin of Sedimentary Rocks,* Prentice-Hall, Inc., Englewood Cliffs, N. J., 1972. A comprehensive and up-to-date intermediate text.

Degens, E. T.: *Geochemistry of Sediments,* Prentice-Hall, Inc., Englewood Cliffs, N.J., 1965. An intermediate text on chemical sediments.

Dunbar, C. O., and J. Rodgers: *Principles of Stratigraphy,* John Wiley & Sons, Inc., New York, 1957. A standard advanced text with excellent, but now somewhat dated, chapters on sedimentary processes and environments.

*Laporte, L. F.: *Ancient Environments,* Prentice-Hall, Inc., Englewood Cliffs, N.J., 1968. A brief, readable introduction to sedimentary environments.

Pettijohn, F. J., and P. E. Potter: *Atlas and Glossary of Primary Sedimentary Structures,* Springer-Verlag New York, Inc., Berlin, 1964. Outstanding photographs of sedimentary structures.

* Available in paperback.

PART 4

The Earth through Time

Parts 1, 2, and 3 have discussed the structure, dynamics, and interactions of the solid earth, ocean, and atmosphere as we see them today or in the immediate past. Part 4 adds the dimension of time by considering the origin of the solid and fluid earth and by tracing their interaction through the billions of years that have passed since the earth began.

Chapter 10 provides an introduction to the techniques used to decipher *earth chronology*, and concludes with a discussion of the time scale of earth history that has been established by such techniques.

Chapter 11 then considers the *early history of the earth*, a span of approximately four billion years which antedates the rise of abundant higher forms of life and which is, for this reason, less clearly understood than is the much shorter interval that followed.

Chapter 12 treats the more fully documented interactions between *evolving geography, life, and environments* that have taken place during the most recent half-billion years of earth history.

10

Earth Chronology

Our understanding of events in the earth's past rests almost entirely on studies of crustal rocks exposed today at or near the earth's surface. The volatile fluids of past oceans and atmospheres are never preserved to leave a historical record whereas crustal rocks, particularly those making up the continents, have been slowly accumulating throughout most of the earth's long history. Many of these rocks have survived with little change for millions or even billions of years and may thus provide revealing insights into the earth's past. For example, plutonic igneous and metamorphic belts indicate intervals of mountain-building deformation deep within the crust, while volcanic and sedimentary rocks reflect changing landscapes at the earth's surface at the time they were formed. Sedimentary rocks are particularly useful because they may suggest past climates, and thus provide *indirect* clues about earlier states of the ocean and atmosphere. In order to decipher the long sequence of events recorded in the many kinds of crustal rocks, however, we must have some means of determining their relative ages so that a worldwide chronology of events can be established.

There are two complimentary, but basically independent, methods for establishing the age relations of ancient rocks. The first is applicable primarily to sedimentary rocks and relates to the sequential nature of sediment deposition and to progressive changes in the animal and plant remains preserved in ancient sediments. The second method applies primarily to igneous and metamorphic rocks and relates to the steady rates of decay of certain radioactive nuclides incorporated into the rocks as they were formed. We shall look briefly at both of these techniques for determining earth chronology before considering the universal time scale of earth history that they have provided.

Folded sedimentary layers,
Wyoming. (John S. Shelton)

SEDIMENTARY ROCKS AND
EARTH CHRONOLOGY

10.1 Physical Relationships

The physical process of sediment deposition provides a means of establishing relative chronology that is not available for most rocks that solidify from molten magmas. Sediments, we have seen, normally accumulate under the influence of gravity as relatively thin horizontal sheets called *beds*. Each such sedimentary bed is usually deposited on top of older beds and is, in turn, buried by younger beds. This simple relationship means that when thick sequences of sediment are preserved without deformation, the *underlying beds are always older than those overlying them*. (Note that this relationship is not necessarily true of igneous rocks which are normally melted from below and may crystallize while overlain by older rocks; see Figure 10.1.) This seemingly obvious principle—that younger sediments always overlie older—is known as the **law of superposition** and provided the first key to deciphering crustal history when it was recognized late in the seventeenth century.

Sequences of ancient sediments thousands of meters thick, representing many different ages and sedimentary environments, are found on every continent. Although about 95 percent of the total *volume* of rocks making up the continental crust is of igneous or metamorphic origin, the remaining 5 percent of sedimentary rock is spread as a veneer that covers much of the *surface area* of the continents (Figure 10.2). As a result, igneous or metamorphic rocks are directly exposed on only about 25 percent of the continental surfaces; the remaining 75 percent is covered by a layer, averaging about 2 km in thickness, of sedimentary rocks that lie on top of a much greater thickness of igneous and metamorphic rocks (see Figures 3.2, 3.3). Once the law of superposition was understood, it became possible to infer sequences of ancient sedimentary events by merely observing the progressive upward changes in sediment types preserved in this continental veneer.

A major difficulty in such studies stems from lack of adequate *exposures* of the rocks. On most parts of the land surface the sedimentary rock veneer is, itself, overlain

Figure 10.1 The law of superposition. (a) In sequences composed entirely of sedimentary rocks, younger beds always overlie older. The sequence shown thus becomes progressively younger from A to E. (b) Igneous rocks, on the other hand, are not necessarily older than overlying rocks. In the sequence shown the igneous unit (G) was emplaced into the sedimentary layers (A–F) and thus is younger than sediments which overlie it.

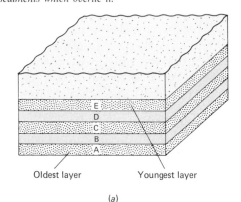

Oldest layer   Youngest layer

(a)

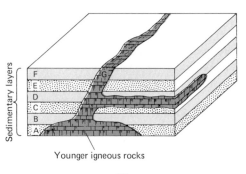

Sedimentary layers

Younger igneous rocks

(b)

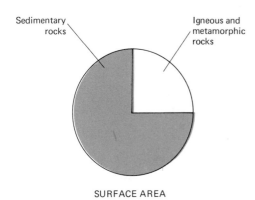

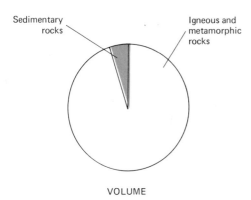

Figure 10.2 Surface area versus volume of sedimentary rocks of the continental crust. Although they make up only 5 percent of the volume of crustal rocks, sedimentary rocks cover 75 percent of the continental surfaces.

by a much thinner veneer of soil and vegetation which largely hides the underlying rock. In such regions the sedimentary sequence is exposed only where streams cut into "bedrock" along hillsides, or where deep, man-made excavations, such as road cuts, quarries or mines, extend into the underlying rock (see Figure 1.34). Even where such local exposures of bedrock are abundant, they commonly do not reveal more than a surficial fraction of the underlying sedimentary rock which has an average thickness of about 2,000 m. Only in mountainous regions of high relief is the complete sedimentary sequence normally exposed for study. As a result of these limitations, it is always necessary to **correlate,** that is, attempt to interrelate, the sedimentary rocks observed at scattered exposures.

The simplest kinds of sedimentary rock correlations are those based on the physical nature of the rocks themselves. Where exposures are continuous or very closely spaced, as sometimes occurs in desert regions, it may be possible to follow single beds or groups of beds over long distances. More commonly, physical correlations are made by recognizing similar rock types, or sequences of rock types, in discontinuous exposures (Figure 10.3).

The difficulty with physical correlations is that they are useful only for rocks originally deposited over relatively small areas and in the same sedimentary environment or sequence of environments. Physical techniques are useless between continents, or over distances of thousands of kilometers on a single continent, where sedimentary environments and resulting sediment accumulations must have varied at any time in the past just as they do today. Physical correlations based only on similarities in rock type, in short, can never provide a worldwide chronology of sedimentary events, even though they sometimes permit local events to be worked out in great detail. What is needed to establish a worldwide chronology is an independent means of determining the relative ages of sedimentary rocks wherever, and in whatever environment, they were deposited. A close approach to such a chronology is provided by the fossil remains preserved in many sedimentary rocks.

10.2 Fossil Contents

In Chapter 9 we saw that animal and plant remains commonly become buried in sediment accumulations where they may be preserved indefinitely as *fossils*. Such fossil

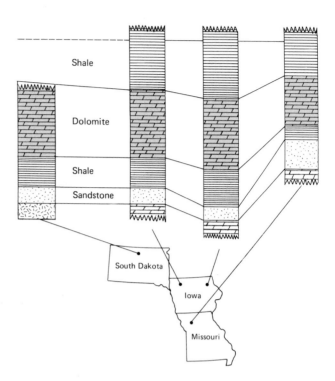

Figure 10.3 *Physical correlations of sedimentary rocks. Similarities in rock sequence often permit interrelation of the continent's sedimentary veneer over wide areas. The illustration shows a distinctive sequence of sandstone (oldest), shale, dolomite, and shale (youngest) that can be recognized over much of the north-central United States.*

remains have long provided the most useful and widely applicable means of establishing the relative ages of sedimentary rocks. Late in the eighteenth century, about a hundred years after the concept of superposition was first appreciated, an English surveyor named William Smith noticed that each unit of sedimentary rock that he encountered in excavations for a canal west of London was characterized by a distinctive assemblage of fossil shells. These characteristic fossils allowed each unit to be easily recognized wherever it occurred. Using this knowledge, and the law of superposition, Smith over many years worked out the complete sequence of rocks and fossils exposed in England and Wales and, in 1815, he published a geological map illustrating his findings that is a landmark in our understanding of earth chronology (Figure 10.4).

Smith's map clearly demonstrated what later came to be known as the "law of faunal succession," and paved the way for a worldwide geologic time scale, for it was soon discovered that sequences of fossil assemblages very much like those in England occurred in the sedimentary veneer of the European continent and even as far away as North America. This discovery showed that widely sepa-

Figure 10.4 *William Smith's landmark geological map, published in 1815. (a) Title sheet. (b) Rock units of the London region. (c, d) Some characteristic fossil shells of the London Clay (c) and Chalk (d). (Fossils from Smith,* Strata Identified by Organized Fossils, *1816)*

London Clay

Brickearth

Chalk

(b)

(c)

(d)

347

rated sedimentary rocks could be correlated simply by comparing their fossil remains with those found in other regions where the fossil sequence had been carefully established. As a result of Smith's discovery a relative time scale of earth history, based on the fossil contents of sedimentary rocks, was established during the first half of the nineteenth century and has been employed with only minor modifications ever since.

Smith and his followers, who established the sedimentary time scale of earth history, based their work on progressive changes in the kinds of fossil animals and plants, but had little understanding of *why* such changes occurred. We now recognize that the underlying cause of faunal succession is **organic evolution,** the expansions, contractions, and modifications of the living world that have taken place since life originated early in earth history (see Sections 11.8, 12.2, and 12.4). Because of this continuous change of living organisms, the fossil record of past life provides a dynamic framework for judging relative time, a framework that is not available from only the physical properties of the sedimentary particles themselves. So successful has been the "fossil dating" technique that it is still the principal means of unraveling crustal history in mines and wells that penetrate the sedimentary veneer, and in surface rock exposures in relatively unexplored areas of the continents. For much of the earth's continental surface, however, the general age relations of the sedimentary veneer have long been established by such studies.

In spite of its enormous value, a chronology of earth history based only on fossils also has serious limitations. In the first place, not all sedimentary rocks contain fossils. In general they are most common in marine sediments, particularly those that accumulate on the shallow, submerged continents rather than in deep-ocean basins. Bottom-dwelling life, particularly shell-bearing animals, is abundant today almost everywhere on the shallow ocean floor and the same was also true for much of the geologic past. In contrast, most terrestrial sediments contain very few fossils because even hard skeletal remains rather quickly decay, weather, and disappear when exposed to the atmosphere. Only in relatively rare terrestrial sediments deposited beneath the standing waters of lakes or swamps are animal and plant remains common.

Because fossils are most abundant in marine sedimentary rocks, it follows that it is usually far easier to establish the age relations of such rocks than it is for those deposited in terrestrial environments. If sea level had

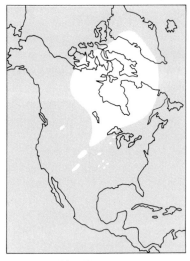  

500 million years ago   380 million years ago   100 million years ago

Figure 10.5 The distribution of land and sea on the North American continent at three intervals of the geologic past. Note the occurrence of shallow seas over much of the continental surface. Fossil-bearing marine sediments that accumulated in such seas permit long-range correlation of the continent's sedimentary veneer.

always stood as low in relation to the continents as it does today, there would be few ancient marine sediments preserved above sea level and correlations of the continents' sedimentary veneer would be correspondingly difficult. Fortunately, however, shallow seas covered much of the surface of the continents for long intervals of the geologic past (Figure 10.5). For this reason, thin layers of fossil-rich marine sedimentary rock cover a large fraction of the continental surfaces even in regions that today lie far above sea level. The age relations of this sedimentary veneer can thus be far more easily established than would be the case if terrestrial sediments made up a larger proportion of it. In those regions where thick sequences of fossil-poor terrestrial sediments *do* occur, it is often difficult or impossible to determine their relative ages.

Still other difficulties arise in applying fossil dating. Fossils, of course, are never found in igneous rocks and only rarely can they be recognized in slightly metamorphosed sedimentary rocks. More commonly, igneous or metamorphic rocks may be interbedded with, or otherwise related to, fossil-bearing sedimentary rocks in such a way that their relative ages are apparent (Figure 10.6). Usually, however, it is difficult or impossible to relate the bulk of the crust's igneous and metamorphic rocks to the ages established for the sedimentary veneer.

A still greater problem arises from the fact that abundant shell-bearing life arose relatively late in earth history. For this reason, even marine sedimentary rocks

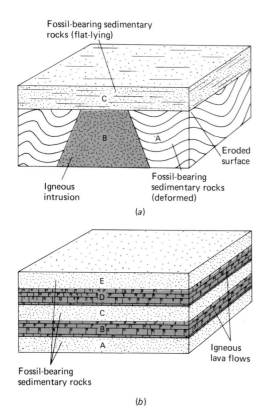

Fossil-bearing sedimentary
rocks (flat-lying)

C

B   A

Eroded
surface

Igneous
intrusion

Fossil-bearing
sedimentary rocks
(deformed)

(a)

E

D

C

B

A

Igneous
lava flows

Fossil-bearing
sedimentary rocks

(b)

Figure 10.6 *Typical relationships of igneous and fossil-bearing sedimentary rocks. (a) The age of the igneous intrusion* (B) *is "bracketed" by the ages of sedimentary unit* A, *which is clearly older than the intrusion, and sedimentary unit* C, *which is clearly younger. (b) The lava flows, having formed at the earth's surface, are clearly younger than underlying sedimentary rocks and older than those which overlie them.*

representing much of the earth's past cannot be dated by the fossil technique because they contain few fossils.

A final difficulty with the fossil time scale is that it provides only relative, not absolute ages. Fossil-bearing rocks can be readily placed in a sequence based on the continuous change of the living world, yet fossils provide no means of knowing exactly how long these changes required *in years*, or some other absolute time unit. A particular evolutionary sequence might have taken place over thousands, millions, or even hundreds of millions of years, but the fossils themselves provide no means for determining this.

Because of these difficulties in the fossil time scale, there is a clear need for an independent means of determining the age relations of rocks. The most useful would be a technique that would establish *absolute* ages of rocks and also be applicable to the great volume of crustal igneous and metamorphic rocks that cannot be dated by fossils. We shall see in Section 10.4 that a chronologic tool filling both of these requirements is provided by the slow decay of certain radioactive nuclides incorporated into rock minerals when they are formed.

10.3 Sedimentary Rocks and Past Environments

So far we have emphasized only how the physical relationships and fossil contents of the continents' sedimentary veneer permit a determination of the relative ages of ancient sediments. As yet, however, we have said almost nothing about the *kinds of events* in the earth's past that may be inferred from the sedimentary rocks once their ages are understood.

As might be predicted from our discussion of present-day sediments in Chapter 9, sedimentary rocks provide much information about ancient environments on the earth's surface at the time and place of their deposition. Thus they may answer such questions as: was the area submerged by the sea, or did terrestrial sediments accumulate above sea level? If below sea level, was the region receiving only carbonate sediments, as do many tropical shelves today, or were sands and muds dominant as on present-day polar and temperate shelves? If the sediments were deposited above sea level, do they reflect fluvial, desert, or glacial environments?

In Chapter 9 we saw that the composition and structure of the sediments themselves usually provide criteria for answering such environmental questions (see, for example, Figures 9.17, 9.19, 9.20). When these answers are pieced together for sedimentary rocks of many different ages distributed over much of the present-day surface of the continents, they provide an invaluable record of past changes in the geography of the earth's surface. Much of Chapters 11 and 12 will be devoted to reviewing the record of changing environments reflected in the continents' sedimentary veneer. Before considering the *results* of such studies, however, it will be useful to look a bit more closely at some of their difficulties and limitations.

The principal problem in reconstructing earth history from sedimentary rocks relates to areas where no sediment accumulates. Much of the surface area of the continents today is not covered by sites of sediment deposition but, instead, is being actively eroded; such areas can obviously leave no future sedimentary record. Similar conditions existed through much of the earth's past and thus the continents' veneer of sediments never reveals the *complete* sequence of surface events, but only those times and places where sediment accumulation was taking place.

As a further complication, even areas that receive sediment at one time may, because of mountain-building deformations, lowered sea level, or other causes, be subject to later erosion which destroys their record of earlier events. For these reasons, sedimentary rocks seldom reflect a continuous sequence of local sediment accumulation through long intervals of earth history; instead the sedimentary veneer at any one place always shows discontinuities representing long time intervals when no sediment was accumulating in that area. Such discontinuities are called **unconformities.** Even though unconformities reflect intervals of no local sediment accumulation, it is often possible to make inferences about the events they record from the nature of the contact between the rocks above and below the surface of unconformity. Some of the circumstances where this is possible are illustrated in Figure 10.7.

In spite of the problems of nondeposition, erosion, and unconformities, we shall see in Chapters 11 and 12 that the continents' sedimentary veneer, when viewed on a worldwide scale, preserves a remarkably complete record of events taking place through much of the earth's long history. Undoubtedly, the reason the record is so good is that through much of the earth's past the continents have stood much lower in relation to sea level than they do

(a)

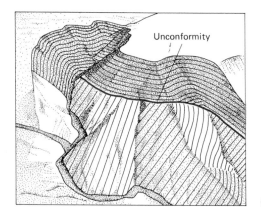

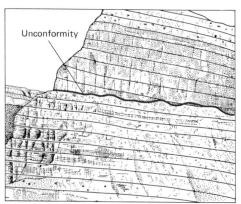

(b)

Figure 10.7   Typical unconformities. (a) *The steep tilt and irregular topography of the rocks beneath the surface of unconformity suggest that an interval of mountain-building deformation and erosion occurred between the time of deposition of the sedimentary units separated by the unconformity. The rocks are exposed in southeastern California.* (b) *The parallel relationship of the beds below and above the surface of unconformity suggests relatively minor erosion of the underlying unit during the time interval represented by the unconformity. The rocks are exposed in northern Arizona.* (a, b: John S. Shelton)

today. During long intervals shallow seas repeatedly covered the continents, thus both minimizing erosion and allowing the accumulation of the thin but widespread layers of marine sediment that make up much of the continents' cover of sedimentary rock.

NUCLEAR CLOCKS AND EARTH CHRONOLOGY

The time scale of earth history based on fossil remains has been established for well over a hundred years; for most of that time, fossils provided the *only* worldwide scheme for working out earth chronology. Within about the last 25 years, however, fossil dating techniques have been supplemented by a powerful new chronologic tool based on the decay of certain radioactive atomic nuclei.

10.4 Radioactive Decay

In discussing the structure of matter in Chapter 1, we noted that the small nucleus at the center of every atom is composed of two principal kinds of particles, uncharged *neutrons,* and positively charged *protons,* the latter being the same in number as the negative electrons which orbit around the nucleus (it may be useful here to review Section 1.2). All atoms of a particular element have the same number of protons in the nucleus, but may have different numbers of uncharged neutrons giving rise to different nuclides of the element.

Nuclides are distinguished by two important numbers. The first is the *atomic number,* which is simply the number of protons in the nucleus (or orbital electrons, since the numbers of each are always the same); the atomic number is constant for all nuclides of a particular element. The second is the *mass number,* which is the total of both protons and neutrons in the nucleus and differs for different nuclides of the same element. Calcium atoms, for example, always have 20 nuclear protons whose positive charges are balanced by 20 orbital electrons; thus the atomic number of calcium atoms is always 20. Naturally occurring calcium is, however, a mixture of six different calcium nuclides which have either 20, 22, 23, 24, 26, or 28 neutrons in the nucleus. These neutrons, added to the constant number of 20 nuclear protons, give these six nuclides mass numbers of 40, 42, 43, 44, 46, and 48 respectively. For brevity, nuclides are normally referred to by the name of the element and their mass numbers, for example, "calcium-40" or "carbon-14."

Many different nuclides of almost every element can be prepared artificially in nuclear reactors or particle accelerators by adding protons and neutrons to, or subtracting them from, atomic nuclei. Most such artificial nuclides are, however, very unstable or "radioactive," which merely means that they rapidly decay to more stable nuclides of other elements by one of the three processes summarized in Figure 10.8. During decay the original radioactive nuclide is called the *parent* nuclide and the nuclide or nuclides produced by the decay are called *daughter* nuclides. Each different radioactive nuclide decays into daughter nuclides at a constant rate which is unaffected by such physical conditions as temperature or pressure. Most artificial nuclides decay very rapidly, usually in a few hours or days, to more stable nuclides, but some may persist for many years. The average lifetime of a radioactive parent nuclide is termed its **half-life** (Figure 10.9), which is the time required for half the atoms in any mass of the nuclide to decay. Because some few atoms of even the most rapidly decaying nuclides persist indefinitely, the time required for *complete* decay is infinite.

In contrast to the many radioactive nuclides produced in reactors and accelerators, most natural nuclides

Figure 10.8   The three principal processes of radioactive decay.

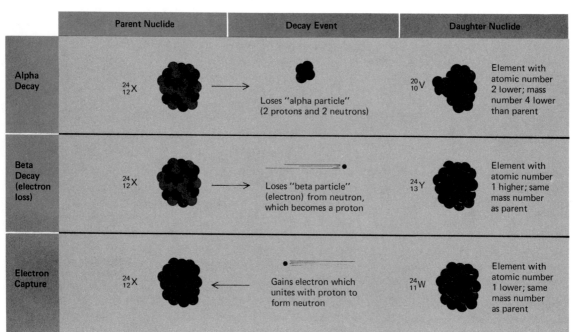

	Parent Nuclide	Decay Event	Daughter Nuclide
Alpha Decay	$^{24}_{12}X$	Loses "alpha particle" (2 protons and 2 neutrons)	$^{20}_{10}V$ Element with atomic number 2 lower; mass number 4 lower than parent
Beta Decay (electron loss)	$^{24}_{12}X$	Loses "beta particle" (electron) from neutron, which becomes a proton	$^{24}_{13}Y$ Element with atomic number 1 higher; same mass number as parent
Electron Capture	$^{24}_{12}X$	Gains electron which unites with proton to form neutron	$^{24}_{11}W$ Element with atomic number 1 lower; same mass number as parent

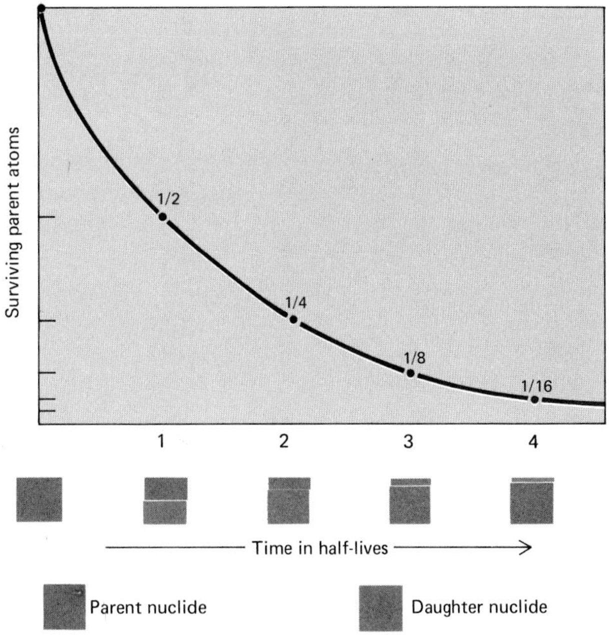

Figure 10.9 *Half-life of a radioactive nuclide. The half-life is the time required for half of any initial mass of a parent nuclide to decay to a daughter nuclide. With each half-life interval, the remaining mass is reduced by 50 percent. The initial mass never decays completely because some small proportion of the initial atoms persists indefinitely.*

occurring in rocks, the ocean, and the atmosphere are stable, that is, they have no tendency to decay to other nuclides. Most natural elements are mixtures of several, usually from two to eight, such stable nuclides of that element. If all naturally occurring nuclides were stable, however, they would be of no value for earth chronology. Fortunately, there are several *natural radioactive nuclides* that make possible the use of radioactive decay as a tool of earth chronology.

10.5 Natural Radioactive Materials

All radioactive nuclides found in rocks, the ocean, and the atmosphere come from one of two sources. The first group, called **primary nuclides,** have such extremely long half-lives that they have persisted since the earth first formed (Figure 10.10). About 20 such nuclides have been detected, but only four are widespread and abundant enough to be generally useful as chronologic tools; these are: potassium-40, which decays to argon-40; rubidium-87, which decays to strontium-87; uranium-235, which decays, through a series of intermediate radioactive nuclides, to lead-207; and, finally, uranium-238, which decays, also through an intermediate series of nuclides, to lead-206 (Figure 10.10).

Radioactive Parent Nuclide	Type of Decay (Figure 10.8)	Stable Daughter Nuclide	Half-life (years)
Potassium-40 $\binom{40}{19}K$	Electron capture	Argon-40 $\binom{40}{18}Ar$	1.3 billion
Rubidium-87 $\binom{87}{37}Rb$	Beta	Strontium-87 $\binom{87}{38}Sr$	47 billion
Uranium-235 $\binom{235}{92}U$	7 alpha and 4 beta	Lead-207 $\binom{207}{82}Pb$	0.7 billion
Uranium-238 $\binom{238}{92}U$	8 alpha and 6 beta	Lead-206 $\binom{206}{82}Pb$	4.5 billion

Figure 10.10 *The principal primary nuclides used as geologic clocks (see also Figure 1.7).*

In addition to these long-lived radioactive nuclides left over from the earth's formation, there is a second group of much shorter-lived radioactive nuclides that are continually being produced in the earth's upper atmosphere by cosmic rays. These "rays" are really extremely high-energy nuclear particles moving in space from unknown sources (see Sections 5.9 and 13.1). When such particles enter the earth's atmosphere they collide with atmospheric gas particles and produce nuclear reactions similar to those of man-made particle accelerators. Some of these reactions form short-lived radioactive nuclides; at least eight such **cosmic-ray induced nuclides** have been identified but only one, carbon-14, has so far proved to be a widely useful chronologic tool.

Carbon-14 is produced by cosmic rays acting on atoms of nitrogen, the most abundant atmospheric gas (Figure 10.11). Atmospheric nitrogen is composed of only stable nuclides, principally nitrogen-14. Through a complex series of nuclear reactions, high-energy cosmic rays may subtract a proton and add a neutron to some of these nitrogen atoms to produce radioactive carbon-14, which then decays back to stable nitrogen-14 with a half-life of only about 5,600 years.

Once produced, the carbon-14 atoms quickly combine with atmospheric oxygen to become carbon dioxide. Most atmospheric carbon dioxide is made up of stable carbon-12 or carbon-13 nuclides; the relatively small quantity of cosmic-ray produced "carbon-14 dioxide" quickly mixes with this more abundant and stable atmosphere carbon dioxide and, in this manner, enters into the earth's general carbon cycle. There it may be dissolved

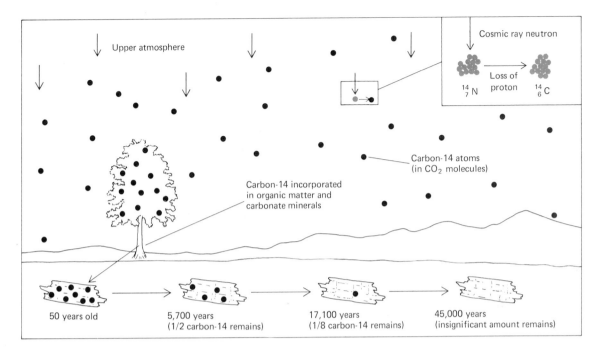

Figure 10.11   The carbon-14 cycle. Radioactive carbon-14 is produced from atmospheric nitrogen by cosmic rays (inset). It then enters into CO_2 molecules and becomes incorporated into carbon-bearing sediments and organic remains. The amount of remaining carbon-14 is used to date such materials.

in the oceans, precipitated as carbonate minerals, or utilized in the structures of animals and plants. The small amounts of radioactive carbon-14 which ultimately enter animals, plants, and minerals in this fashion provide an extremely useful tool for dating the very latest interval of earth history (Figure 10.11).

10.6   Radiometric Ages

All natural nuclides of the same element, whether radioactive or not, have the same general chemical behavior and thus tend to occur mixed together wherever the element is found. When the mixture contains a radioactive nuclide, however, small quantities of a different element—the daughter element produced by decay of the radioactive nuclide—are also present in the mixture. In theory, all that is necessary to use radioactive nuclides as geologic clocks is to measure the amounts of both the radioactive mother nuclide (normally either potassium-40, rubidium-87, uranium-235, uranium-238, or carbon-14) and stable daughter nuclide (normally either argon-40, strontium-87, lead-207, lead-206, or nitrogen-14) that are present today in the rock or mineral to be dated (Figure 10.12). Since the decay rates of the radioactive nuclides are known with

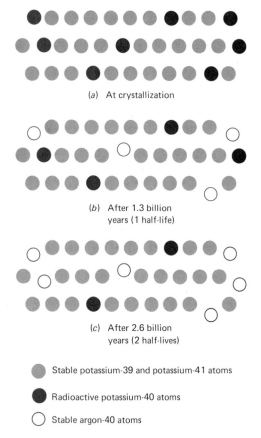

(a) At crystallization

(b) After 1.3 billion
years (1 half-life)

(c) After 2.6 billion
years (2 half-lives)

● Stable potassium-39 and potassium-41 atoms

● Radioactive potassium-40 atoms

○ Stable argon-40 atoms

Figure 10.12 *The progressive decay of potassium-40 atoms incorporated into a mica mineral at the time of crystallization. The ratio of remaining potassium-40 atoms to atoms of the daughter nuclide argon-40 is used to determine the age of the mineral (see also Figure 1.20).*

considerable precision from laboratory measurements, a simple calculation (the ratio of mother to daughter nuclides multiplied by the decay rate) gives the *length of time* since the radioactive mother nuclide was first isolated in the mineral, and thus provides a **radiometric age** for the mineral. Precise techniques have been developed for measuring very small quantities of parent and daughter nuclides, but there still remain several complications in obtaining ages by this means.

In order for radiometric ages to be accurate, the radioactive mother nuclide and the stable daughter nuclide must have been a *closed system* since their formation; that is, there must have been no later additions or subtractions of either parent or daughter. Unfortunately, these conditions are seldom met. Loss of some of the daughter nuclide is the most usual problem for, being a different element with different chemical properties, the daughter nuclide is not in equilibrium with its surroundings once formed (Figure 10.13a). Daughter nuclide escape is particularly common when the nuclide is a gas, such as argon-40, or when the mineral to be dated has been secondarily heated by tectonic activity or deep burial. Although ingenious techniques have been developed to estimate the amounts of daughter elements lost from various minerals, this remains one of the principal difficulties in radiometric dating.

Because of daughter element loss, the most accurate radiometric ages are always considered to be those where calculations on two or more mother-daughter pairs give the same result. In such cases it is unlikely that the different daughter elements, each having different chemical properties, would be lost at the same rate and thus provide the same erroneous ages. Fortunately, many rocks and minerals contain more than one of the principal radioactive nuclides so that cross checking is frequently possible. Uranium-bearing minerals are particularly useful in this regard, since they always contain both uranium-235 and uranium-238, each of which provides an independent age.

A second major difficulty in radiometric dating is that the parent nuclide must have been free of contamination by the daughter nuclide when it was originally incorporated into the mineral to be dated. That is, *all* the daughter nuclide subsequently measured must result from radioactive decay rather than original contamination (Figure 10.13b). Seldom is this condition fulfilled, but here again ingenious methods have been developed for esti-

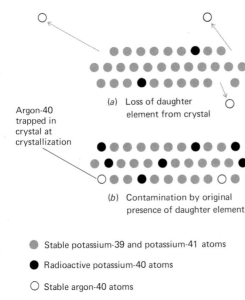

Argon-40 trapped in crystal at crystallization

(a) Loss of daughter element from crystal

(b) Contamination by original presence of daughter element

● Stable potassium-39 and potassium-41 atoms

● Radioactive potassium-40 atoms

○ Stable argon-40 atoms

Figure 10.13 *Two of the principal complicating factors in radiometric age determinations (see text).*

mating the amount of daughter element originally present, so that this difficulty can frequently be overcome.

Various techniques for getting around the daughter element loss and contamination problems have been developed for the long-lived primary nuclides, but these problems are insoluble for carbon-14, the principal short-lived nuclide that is useful in radiometric dating. Carbon-14, we have seen, decays to nitrogen-14, which is the principal constituent of the atmosphere. The daughter element is thus present in large, contaminating quantities both when the carbon-14 is produced in the atmosphere and, commonly, when it decays as well. For this reason, carbon-14 dating is not based on measurements of both mother and daughter nuclides; instead it requires the assumption that carbon-14 has been produced at a constant rate in the atmosphere and is quickly mixed, in constant proportions, with nonradioactive carbon to enter the earth's overall carbon cycle. If these assumptions are correct, then the carbon incorporated into carbonate minerals and living organisms always has a constant proportion of carbon-14, which steadily decreases by decay after the materials are formed. It is therefore necessary only to measure the present ratio of carbon-14 to nonradioactive carbon in such materials to determine their ages by this method (Figure 10.11).

So far we have said very little about the specific kinds of rocks, minerals, or other materials that contain the five principal radioactive nuclides in measurable amounts and can therefore be dated by radiometric techniques. Both the potassium-40 and rubidium-87 methods have been applied most successfully to mica minerals found in a variety of igneous and metamorphic rocks. In addition, some other igneous and metamorphic minerals, for example certain types of feldspar and amphibole, are commonly dated by these techniques. Normally, the individual mineral grains are separated from the rest of the rock to minimize contamination, but it is sometimes possible to obtain accurate dates for very fine-grained igneous and metamorphic rocks by simply analyzing crushed samples of the whole rock, rather than individual mineral grains.

The uranium dating technique was the first to be established but for many years it could be applied only to certain rare minerals that contain relatively large quantities of the element. More recently, the development of precise analytical techniques has permitted accurate measurements of very small traces of uranium that occur in the silicate mineral *zircon*. Zircon occurs as a minor accessory mineral in many igneous rocks, and thus permits a wide application of the uranium dating technique.

Radiometric dating by means of potassium, rubidium, and uranium nuclides is applicable mostly to minerals found in igneous and metamorphic rocks. In contrast, the carbon-14 technique is principally useful for sedimentary materials; carbonate sediments and fossil shells, bones, and wood fragments are the materials most commonly dated by this technique.

THE TIME SCALE OF EARTH HISTORY

So far we have been considering *how* sedimentary rocks and radioactive nuclides are used to decipher earth history. In this section and the two chapters that follow, we shall turn our attention to exactly *what* these chronologic tools have revealed about the earth's past.

10.7   The Sedimentary Time Scale

Soon after the recognition of "the law of faunal succession" by William Smith and his followers in the early nineteenth century, a system was established for subdividing the continents' veneer of fossil-bearing sedimentary rocks. This system still provides the standard time scale used in all discussions of earth history and, for this reason, it is necessary to understand it thoroughly.

The largest subdivisions of the sedimentary time scale are known as **eras;** three eras are recognized, each bounded by relatively profound, sudden, and worldwide changes in the living organisms preserved as fossils. These are: the Paleozoic ("Ancient Life") Era, Mesozoic ("Middle Life") Era, and Cenozoic ("Recent Life") Era. The three eras are further subdivided into **periods,** each bounded by somewhat less profound changes in the living world (Figure 10.14). Eleven periods are now recognized: the Paleozoic Era includes six of them [Cambrian (oldest), Ordovician, Silurian, Devonian, Carboniferous, and Permian (youngest)]; the Mesozoic Era has three periods [Triassic (oldest), Jurassic, and Cretaceous (youngest)]; and the Cenozoic Era includes only two (Paleogene and Neogene). The eleven geologic periods are, in turn, further subdivided

Eras - time scale

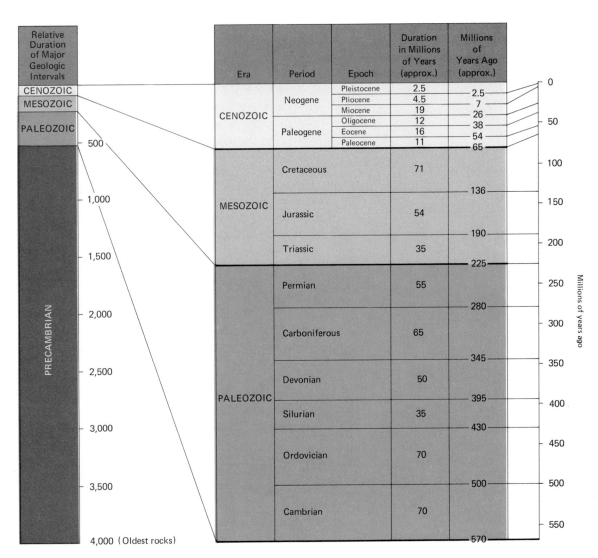

Relative Duration of Major Geologic Intervals	Era	Period	Epoch	Duration in Millions of Years (approx.)	Millions of Years Ago (approx.)
CENOZOIC	CENOZOIC	Neogene	Pleistocene	2.5	2.5
			Pliocene	4.5	7
MESOZOIC			Miocene	19	26
		Paleogene	Oligocene	12	38
PALEOZOIC			Eocene	16	54
			Paleocene	11	65
500	MESOZOIC	Cretaceous		71	136
1,000		Jurassic		54	190
1,500		Triassic		35	225
	PALEOZOIC	Permian		55	280
2,000		Carboniferous		65	345
2,500		Devonian		50	395
3,000		Silurian		35	430
		Ordovician		70	500
3,500		Cambrian		70	570
PRECAMBRIAN					
4,000 (Oldest rocks)					

Millions of years ago

Figure 10.14   The time scale of earth history.

by their fossil contents into still smaller units, called **epochs.** Most of the many names of these smaller divisions
of geologic time need not concern us. The most important
are the six epochs of the Cenozoic Era, the youngest of
which, the Pleistocene Epoch, includes the recent intervals
of ice-sheet expansion and contraction that were discussed
in Chapters 7 and 8 (Figure 10.14).

The oldest sedimentary rocks bearing abundant fossils are those of the Cambrian Period which begins the
Paleozoic Era. Wherever they occur around the world,
fossil-bearing Cambrian rocks contain the same distinctive
association of shell-bearing marine animals dominated by
the long extinct trilobites, which were distant relatives of
modern crabs and shrimps. These earliest Cambrian sediments always overlie older rocks that *lack* abundant fossils.

Commonly, early Cambrian sediments rest upon a
"basement complex" of older igneous and metamorphic
rocks which, we have seen, make up the bulk of the continental crust. In many regions, however, the earliest Cambrian fossils occur above thick sequences of sedimentary
rocks that are identical to the overlying Cambrian rocks
except that they contain no fossils. This abrupt appearance of trilobites, and other relatively advanced kinds of
animal life, in the sedimentary record is perhaps the most
significant milestone in the earth's long history, for it
serves to divide the earth's past into two great divisions.
Before this event lies "Precambrian time," whose rocks
contain few fossils and therefore provide no worldwide
time scale based on changing life. After the great early
Cambrian expansion of animal life comes "Phanerozoic
("exposed life") time," and the worldwide applicability of
the sedimentary time scale that we have been considering.

Because Precambrian igneous, metamorphic, and
sedimentary rocks make up a large percentage of all the
rocks of the continental crust, it has long been assumed
that they represent a very long span of earth history.
Before the widespread application of radiometric dating,
however, no worldwide chronology was available for this
great volume of Precambrian rocks. As with younger, fossil-bearing rocks before the discovery of the "law of faunal
succession," local sequences of Precambrian rocks, and
the events they record, had been worked out by purely
physical means, but there was still no reliable way to
interrelate these events over long distances. This situation
has been profoundly changed by radiometric dating.

Man
and Earth
Chronology

Much of our understanding of earth chronology is a direct result of the search for mineral resources. Because most such resources are hidden by overlying rock and soil, their presence must usually be inferred from the known age relations of rocks exposed on the surface or in nearby mines and wells. For this reason many geologists concerned with deciphering earth history are employed by the mining, petroleum, and other extractive industries. Much of the information so obtained is eventually released to the scientific community and thus adds to the accumulated understanding of earth history.

Still more fundamentally, knowledge of the earth's long and complex history serves to place in perspective the mere instant of time that man has begun to modify the planet. The earth has existed for at least 4,500 million years and abundant animal life about 600 million years, but man for only a half-million years or so. Within this 500,000 years, recorded human history occupies only the last 1 percent, about 5,000 years. Moreover, the time since the industrial revolution and the increasing consumption of nonrenewable resources has been only about 200 years, a mere $\frac{1}{2,500}$ of man's existence on earth. In sharp contrast to man's relatively brief existence are the hundreds of millions of years during which the resources required by industrialized societies accumulated in the earth's crust. As examples we may consider the two most essential industrial resources—iron and mineral fuels.

The bulk of the iron now used by man accumulated as sedimentary deposits over a period of about 1,500 million years, from about 1,700 to 3,200 million years ago. They thus date from the long Precambrian interval of earth history. On the other hand, most deposits of coal and petroleum, the principal mineral fuels, have formed since the Cambrian expansion of life about 600 million years ago. Nevertheless, they are still very ancient by human standards. The principal coal deposits were formed during the Carboniferous Period, from about 280 to 345 million years ago. Much petroleum is equally old, or older, but oil and gas are also produced from rocks as young as one million years.

Such considerations make it abundantly clear why most mineral resources are indeed *nonrenewable*. Crustal materials accumulated over many millions of years can never be replaced on the scale of human history where hundreds or even tens of years can be critical. This does not mean that all resources are severely limited for, as we have seen in earlier chapters, there are large potential reserves of most of the materials used by man. It *does* indicate, however, that man should approach with humility his role as modifier of an irreplaceably ancient heritage.

Geologists studying the age relations of surface rocks in a remote area of Alaska. (Jon Brenneis, Omikron)

Geologist examining a fossil-bearing rock specimen. (Jon Brenneis, Omikron)

Earth history (gray) and human history (color).

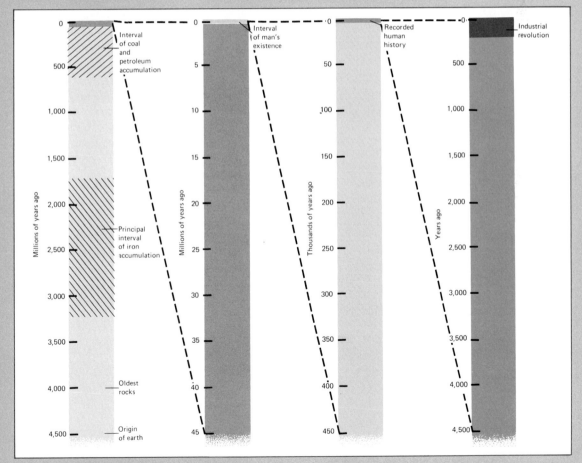

Interval
of coal
and
petroleum
accumulation

Interval
of man's
existence

Recorded
human
history

Industrial
revolution

Millions of years ago

0
500
1,000
1,500
2,000
2,500
3,000
3,500
4,000
4,500

Principal
interval
of iron
accumulation

Oldest
rocks

Origin
of earth

Millions of years ago

0
5
10
15
20
25
30
35
40
45

Thousands of years ago

0
50
100
150
200
250
300
350
400
450

Years ago

0
500
1,000
1,500
2,000
2,500
3,000
3,500
4,000
4,500

10.8 Absolute Radiometric Dating

The application of radiometric dating techniques has, over the past two decades, made three principal contributions to our understanding of earth chronology. First, these techniques have led to an absolute calibration, in years, of the relative sedimentary time scale. In addition, they have revolutionized our understanding of the enormous volume of Precambrian rocks that cannot be dated by fossils. Finally, they have greatly clarified the events of the past 50,000 years, a span of time too short to include significant changes in living organisms and thus also beyond the application of fossil dating techniques. We shall briefly consider each of these contributions before turning to the more complex question of the overall age of the earth.

Perhaps the most significant achievement of radiometric dating has been to provide absolute dates for the relative sedimentary time scale. Since the early nineteenth century, attempts have been made to calibrate the sedimentary time scale. Most of these involved estimates of the rates at which sediments are accumulating today, and extrapolation of these rates into the past to account for the total thicknesses of sedimentary rocks of different ages. Such estimates all suggested that many millions of years were required to account for the crust's sedimentary veneer, but only with the advent of radiometric dating could the precise number of years be determined.

Radiometric dating techniques are usually applicable only to the minerals of igneous or metamorphic rocks and, for this reason, radiometric dating of the sedimentary time scale requires rather unusual associations of sedimentary and igneous rocks (see Figure 10.6). Such associations are far from common but, fortunately, enough are known to give a reasonably complete calibration (Figure 10.14). These show that the early Cambrian expansion of animal life took place about 600 million years ago. The oldest *Precambrian* rocks so far dated are about 4,000 million years old, and thus the Phanerozoic time scale, based on fossils, represents only the latest 15 percent of the time since the formation of the oldest known crustal rocks. Within this 600 million years of Phanerozoic time, the Paleozoic Era accounts for about the first 400 million years, the Mesozoic Era the next 130 million years, and the Cenozoic Era only the last 70 million years. The average length of the 11 Phanerozoic periods is about 55 million years, but radiometric dating has shown that the changes in the living world which bound the periods were not regularly spaced. Instead, the periods range in age from a minimum of 35 million years (Silurian Period) to a maximum of 71 million years (Cretaceous Period) (Figure 10.14). Radiometric dating has also shown that the great ice sheet expansions and contractions of Pleistocene time, events that have had such a profound effect on the earth's present climates, landscapes, and sediments, have been concentrated in about the last 4 million years, less than 1 percent of the total of Phanerozoic time.

In addition to furnishing an absolute calibration for the sedimentary time scale, radiometric dating has provided the *only* chronologic tool applicable to the great mass of Precambrian rocks which represent over 3,400 million years of earth history. As a result, a general worldwide chronology and sequence of events are beginning to be established for this long and significant interval. Finally, the carbon-14 dating technique has permitted a refined understanding of events over the latest 50,000-year span of Pleistocene time.

10.9 The Age of the Earth

The following chapters summarize many of the conclusions about earth history that have grown from both the sedimentary and radiometric dating techniques. Chapter 11 reviews our understanding of Precambrian earth history, a history which depends almost entirely on radiometric dating. Chapter 12 then considers Phanerozoic earth history and is concerned largely with conclusions based on the sedimentary time scale, supplemented by radiometric dating. Before turning to the long sequence of events recorded in both Precambrian and Phanerozoic rocks, however, we need to consider one additional chronologic milestone—the overall age of the earth.

The oldest crustal rocks so far dated, granites from southwestern Greenland, have radiometric ages of about 4,000 million (or 4 billion) years. In other regions metamorphosed sedimentary rocks surround similar ancient granites that were injected into, and thus are younger than, the deformed sediments that surround them. It is therefore clear that crustal erosion, sediment deposition, and the formation of sedimentary rocks were all taking place very early in earth history. Apparently these processes have completely recycled, and thus obliterated, any original rocks formed as the crust first consolidated. Direct dating of crustal rocks can therefore only indicate that the earth is older than 4,000 million years; to answer the question "how much older?" we must turn to less direct evidence of two sorts.

The first comes from radiometric dating of meteorites and rocks recovered from the surface of the moon. All the solid material of the solar system, including the earth, moon, and the small particles that fall on the earth as meteorites, are believed to have had a common origin from gases and dust given off by the sun (see Sections 11.1 and 13.6). Neither the rocks of the moon, nor, presumably, meteorites have been subjected to the intense, continuous erosion and weathering suffered by rocks of the earth's crust. For this reason, some of them might be expected to give ages indicating the time that they, and by extrapolation the earth, first consolidated from solar matter. Most significantly, almost all meteorites have radiometric ages between 4,500 and 5,000 million years, suggesting that the earth has about the same overall age. In addition, the oldest rocks so far recovered from the surface of the moon have ages of about 4,500 million years, which suggests that the earth is at least that old.

There is also a second kind of indirect evidence for the overall age of the earth; this is based on the present-day abundance of the various nuclides of lead that occur in minerals of the earth's crust. Natural lead is a mixture of four stable nuclides: lead-204, -206, -207, and -208. Three of these (206, 207, and 208) are produced by the radioactive decay of uranium and other less common radioactive elements. The fourth, lead-204, is not produced by radioactive decay; *all* of it present on the earth today originated when the earth was formed, whereas only a *part* of present-day lead-206, -207 and -208 originated at that time—the rest has been slowly added through the course of earth history by radioactive decay of other elements (Figure 10.15). Now if there was some means of determining the *original* relative abundance of the four nuclides at the time the earth was formed, then the earth's age could be estimated by first measuring their *present* relative abundance and then calculating the time required for the additional lead-206, -207 and -208 to have been added by radioactive decay (the decay rates producing each are constant and are precisely known from laboratory measurements). Although there is no direct way to estimate the original lead abundances, certain meteorites which contain

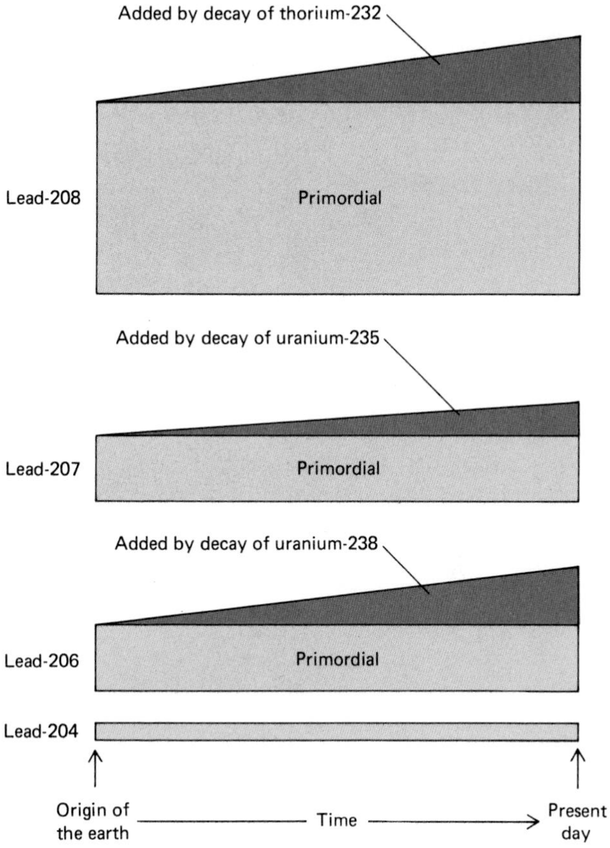

Figure 10.15 Determination of the age of the earth from its initial and present-day abundances of lead nuclides. The initial abundance is estimated from meteorites and the present abundance from the average of many rock measurements. The difference represents additions from the decay of radioactive uranium and thorium through earth history.

no uranium or other radioactive elements that decay into lead are thought to provide a reasonable approximation of the earth's original lead. Using this information, about 4,500 million years of radioactive decay would be necessary to produce lead having the average nuclide abundances found on earth today. This estimate further confirms the suggestion that the earth originated at least 4,500 million years ago.

SUMMARY OUTLINE

Sedimentary Rocks and Earth Chronology

10.1 *Physical relationships:* the sequential accumulation of sediments, younger on top of older, provides a means of establishing local sequences of events in sedimentary rocks.

10.2 *Fossil contents:* Progressive evolutionary changes in ancient life provide a worldwide chronology of fossil-bearing sedimentary rocks.

ADDITIONAL READINGS

*Berry, W. B. N.: *Growth of a Prehistoric Time Scale,* W. H. Freeman and Co., San Francisco, 1968. A brief, readable history of the development of the sedimentary time scale in the eighteenth and early nineteenth centuries.

Donovan, D. T.: *Stratigraphy,* Thomas Murby and Co., London, 1966. A concise intermediate text on the techniques of earth chronology.

*Eicher, D. L.: *Geologic Time,* Prentice-Hall, Inc., Englewood Cliffs, N.J., 1968. An introduction to all aspects of earth chronology.

*Faul, H.: *Ages of Rocks, Planets, and Stars,* McGraw-Hill Book Company, New York, 1966. A good introduction to radiometric dating.

*Harbaugh, J. W.: *Stratigraphy and Geologic Time,* William C. Brown Company, Dubuque, Iowa, 1968. A brief introduction to earth chronology.

Krumbein, W. C., and L. L. Sloss: *Stratigraphy and Sedimentation,* W. H. Freeman and Co., San Francisco, 1963. A standard advanced text on sedimentary chronology.

Weller, J. M.: *Stratigraphic Principles and Practice,* Harper and Brothers, Publishers, New York, 1960. A standard advanced text on sedimentary chronology.

*York, D., and R. M. Farquhar: *The Earth's Age and Geochronology,* Pergamon Press, Oxford, 1972. A brief intermediate-level survey emphasizing radiometric techniques.

* Available in paperback.

11

Early History of the Earth

Precambrian stromatolites.
(Paul Hoffman, Geological
Survey of Canada)

The last chapter mostly considered the principles and methods used in understanding the earth's past; this chapter, and the one that follows, summarize the results obtained by using these tools. The two chapters treat earth history in a generally chronological sequence; this one considers the events between the earth's origin, some 4,500 million years ago, and the time when animal life first became widespread and abundant at the beginning of Cambrian time about 600 million years ago. These events, although they occupy eight-ninths of earth history, are far less clearly recorded in surviving crustal rocks than are the events of the last 600 million years, which will be the subject of Chapter 12.

PREGEOLOGICAL EARTH HISTORY

Although the earth is believed to be at least 4,500 million years old, the oldest earth rocks so far discovered have ages of only 4,000 million years. Some of these early rocks are of both igneous and sedimentary origin, and most are similar to rocks being formed on the earth today. In particular, the presence of clastic sedimentary particles—mud, sand, and pebbles—among these oldest rocks indicates that erosion of still older continental rocks was already taking place at that time. Such erosion, in turn, indicates that an ocean was supplying moisture to an atmospheric hydrologic cycle that transported and deposited water on the land surface. In short, when these early rocks were formed both the solid and fluid earth must have had many of the same characteristics that we find today. This means that the initial separation of the earth into concentric spheres of core, mantle, crust, water, and air took place still earlier, during the 500 million years or so of earth history that antedated the oldest surviving rocks (Figure

373

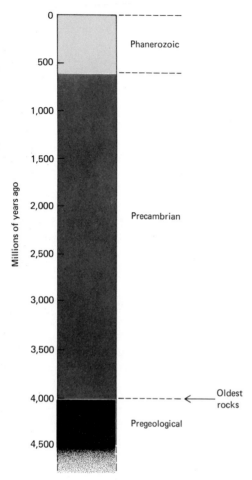

Figure 11.1 *The "pregeological" interval of earth history (color). During this interval, which antedates the oldest rocks, the initial differentiation of the solid earth, ocean, and atmosphere apparently took place.*

11.1). These "pregeological" events can only be understood by indirect inference, but it will be useful to review some speculations about them before turning to the more clearly documented history recorded in Precambrian rocks.

11.1 Differentiation of the Solid Earth

The present-day solid earth shows a clear differentiation by density: the heaviest abundant elements, nickel and iron, are concentrated in the partially liquid core while the lighter abundant elements, mostly silicon and aluminum combined with oxygen, surround the core as the massive mantle and thin surficial crust (see Figure 2.8). This density layering is apparently an early and fundamental feature of earth structure, for the oldest known rocks are mostly granites and gneisses similar to the younger rocks that today dominate the continental crust. Thus a differentiated crust and, most probably, an iron core and silicate mantle as well, were already present when the oldest known rocks formed 4,000 million years ago. The internal density layering must therefore have originated still earlier—probably either along with, or relatively soon after, the earth's original accretion from gas or dust particles surrounding the sun (see Section 13.6).

There are two general theories for the origin of the core-mantle-crust separation and both are closely interrelated with the initial accumulation of the earth from solar matter. If the earth accumulated from a *homogeneous* mixture of solar particles, then it must have initially been rather uniform throughout and separation of the core, mantle, and crust must have occurred sometime *after* initial accretion. On the other hand, it is also possible that the initial consolidation of the earth took place from *inhomogeneous* solar materials, in which case the iron core might have accreted first, to be subsequently surrounded by a mantle of lighter silicate materials. In this second model the core, mantle, and crust thus formed *as the earth grew to its present size, rather than afterward. As yet there is no compelling evidence favoring one of these models over the other, and we shall briefly review both.

Homogeneous Accretion In this hypothesis the primitive solid earth is considered to have had a rather uniform density and composition throughout; that is, the large amounts of iron and nickel now making up the core were uniformly dispersed through a larger mass of lighter silicate minerals (Figure 11.2). This initial "proto-earth" is visualized as having been rather cool throughout, so that all of the elements were initially present as solids rather than as gases or liquids. In order to separate the heavier iron and nickel and concentrate them in the core, the interior of this initially cool earth must have somehow become hot enough to melt the iron and nickel, thus allowing them to sink through the lighter, surrounding silicate minerals. These silicates melt at higher temperatures than do nickel and iron but would, nevertheless, soften and slowly flow as the earth's interior became hotter; this flow would permit the much heavier molten metals to sink and displace the lighter silicates at the earth's center.

Such a partial melting of the early earth could have been caused by heat generated through the decay of radioactive nuclides, which were present in far greater abundance early in earth history than they are today (see Section 10.5). Calculations based on reasonable estimates of the amount of radioactive material present in the initial earth suggest that soon after it formed enough heat might have been generated to begin to melt nickel and iron at depths of about 650 km (400 mi; this melting would begin first at this relatively shallow depth because melting temperatures increase rapidly as pressures become greater toward the earth's center). These molten metals would first accumulate as a layer in the outer part of the earth (Figure 11.2), but being heavier than the soft, underlying silicate material, would tend to slowly sink as great "drops" to accumulate at the earth's center. This sinking, in turn, would create additional frictional heat that might melt the metals occurring still deeper in the earth, and thus continue the process until all of the free iron and nickel became concentrated in the central core.

Figure 11.2 Differentiation of the solid earth by homogeneous accretion. In this hypothesis the earth was initially a uniform mixture of silicates (gray) and nickel-iron (color). Subsequent radioactive heating then led to a concentration of nickel-iron in the core and silicate minerals in the mantle and crust (see text).

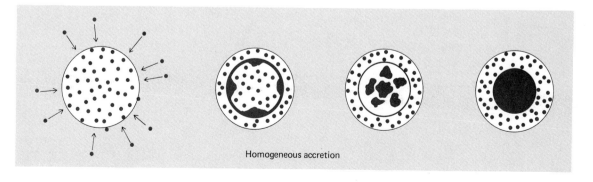

Homogeneous accretion

As the molten metals sank to form the core, the lighter silicates would have risen into the mantle which, as a consequence, would have been turbulent and unstable. Under such conditions the lightest and most easily melted elements such as sodium, calcium, potassium, and aluminum would tend to move upward through the turbulent mantle silicates. As a result, they would be concentrated near the earth's surface, where they are found today in the rocks of the crust and upper mantle.

Inhomogeneous Accretion In this second hypothesis, materials of the core, mantle, and crust are visualized as having condensed sequentially from hot solar gases, rather than forming from a uniform mass of solid solar particles (Figure 11.3). Calculations show that even though iron and nickel melt at *lower* temperatures than do most silicate minerals, they condense from the gaseous state at slightly *higher* temperatures. Thus, in a cooling cloud of hot solar gases, these metals might condense first and accumulate to form the earth's core and subsequently be surrounded by later-condensing silicate minerals to form the mantle. Finally, the last material to condense and accumulate as the temperature dropped would be the light, volatile elements that are concentrated in the crust and upper mantle.

Such condensation and accumulation might have taken place very rapidly as the original solar gases cooled—perhaps in only a few hundred or a few thousand years. This hypothesis therefore does not require the relatively long interval of radioactive heating and internal reorganization demanded by the homogeneous accretion model. In either case, however, it is clear that the internal differentiation of the solid earth took place very early in earth history.

Figure 11.3 *Differentiation of the solid earth by inhomogeneous accretion. In this hypothesis the nickel-iron of the core (color) accumulated first and was surrounded by a slightly later accumulation of the silicate minerals of the mantle and crust (see text).*

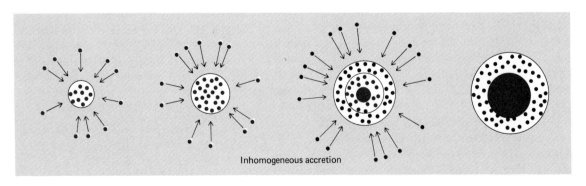

Inhomogeneous accretion

11.2 Origin of the Earth's Fluid Cover

If the solid earth condensed directly from cooling solar gases, then its ocean and atmosphere might have had a similar origin and be merely a residue of the most volatile elements that persisted as liquids or gases after the solidification of the crust. If, on the other hand, the earth accreted from a homogeneous mixture of solid particles, then its ocean and atmosphere must have a secondary source—from volatile elements subsequently released from its solid materials by internal heating. Regardless of the origin of the core, mantle, and crust, there is strong evidence that its covering fluids *did* have such a secondary origin from within the solid earth.

This evidence comes from the relative abundance of rare gaseous elements, particularly neon, argon, krypton, and xenon (see Figure 1.1). These extremely stable gases never form natural compounds with other elements (see Section 1.2) and are, in addition, too heavy to escape the earth's gravitational attraction and be lost in space. As a result, any present in the earth's earliest atmosphere should *still* be present, for there is no known way that they could be removed, either physically or chemically. They are, however, about a million times less abundant today in the earth's atmosphere than in the sun and other stars. It is therefore most probable that the earth's gaseous atmosphere is not merely a residue of solar gases, but accumulated instead from heating and "outgassing" of volatile materials originally present in materials of the solid earth.

There is abundant evidence that local heating and melting of the earth's solid crust and upper mantle do take place, for these processes are responsible for volcanoes, earthquakes, and plate motions (see Sections 2.8, 3.9). Volcanoes, in particular, release large quantities of volatile elements at the earth's surface. Such volcanic outgassing, over the long course of earth history, is the most probable source of its fluid ocean and atmosphere (Figure 11.4). The exact composition and volume of the earth's initial atmosphere are uncertain, but it may have been similar to the gases given off by present-day volcanoes that have their source deep within the upper mantle. Water vapor and carbon dioxide are the principal constituents of these modern volcanic gases; hydrogen, nitrogen, ammonia (NH_3), methane (CH_4), chlorine, and many other gases

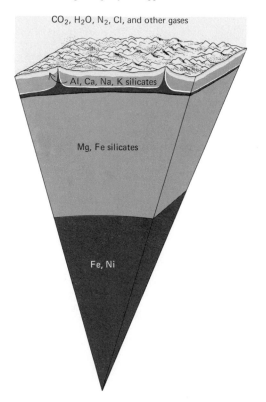

Figure 11.4 Origin of the ocean and atmosphere by volcanic "outgassing" of the upper mantle.

CO_2, H_2O, N_2, Cl, and other gases

Al, Ca, Na, K silicates

Mg, Fe silicates

Fe, Ni

occur in smaller quantities. The water vapor would have condensed to form the initial ocean as the underlying crustal rocks cooled below 100°C, leaving a primitive gaseous atmosphere composed largely of carbon dioxide.

The ocean has probably changed relatively little in composition through geologic time for it is still 96.5 percent water. Its relative volume throughout earth history is less certain. Early volcanism may have released large quantities of water to produce a large initial ocean. Alternatively, it may have grown more slowly through geologic time.

In contrast to the relatively constant composition of the ocean, today's atmosphere of nitrogen (78 percent) and oxygen (21 percent) is quite unlike the postulated earliest atmosphere. This early atmosphere was probably dominated by carbon dioxide as are both present-day volcanic gases and the present atmospheres of Venus and Mars, the neighboring planets that most closely resemble the earth in size and structure. The changes in atmospheric composition that have led to our present nitrogen-oxygen atmosphere are of two principal sorts (Figure 11.5).

Figure 11.5 *Removal of carbon dioxide from the atmosphere. The numbers show the present-day distribution of carbon in billions of metric tons. Much of the* CO_2 *released through geologic time by volcanoes has dissolved in the ocean and then precipitated as non-biogenic calcium carbonate sediments. Still more* CO_2 *is removed from the atmosphere by the metabolic activities of animals and plants, whose carbon-rich remains also accumulate in soils and sediments. Through such physical and biological processes most of the earth's surficial carbon has come to be concentrated in sedimentary rocks, rather than occurring as atmospheric carbon dioxide.*

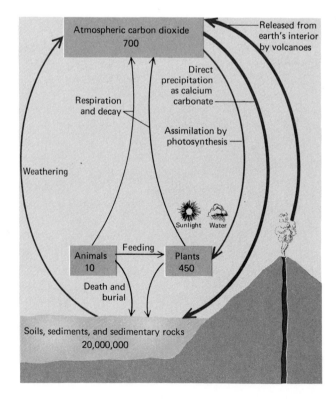

The first are physical processes of change, particularly additions to the atmosphere (and ocean) by volcanic activity throughout the long course of earth history, and subtractions from the atmosphere by reactions between atmospheric gases, crustal rocks, and ocean water to form new solid materials. The exact nature and amounts of these physical additions and subtractions are still highly speculative, but some clues to their limits are provided by ancient crustal rocks.

The second cause of atmospheric change is the origin and expansion of life. Both nitrogen and oxygen, the principal gases of the present atmosphere, as well as the much smaller quantities of present-day atmospheric carbon dioxide, are all involved in continuous short-term cycles in which they are first removed from, and then later returned to, the atmosphere by the metabolic activities of animals and plants (see Figures 5.3, 5.4, 5.6). Because of these cycles, the history of the changing composition of the atmosphere is closely tied to, and in large measure controlled by, the history of life.

11.3   The Beginnings of Life

Animals and plants not only affect the atmosphere but are also interrelated with many other aspects of the physical world: for example, weathering, erosion, and landscapes are largely determined by the kinds of plants present on the land, while many kinds of chemical sediments are produced as shell-bearing organisms remove dissolved materials from the earth's waters. The origin and subsequent development of life on earth are therefore closely interwoven with the history of its rocks, ocean, and atmosphere.

Although abundant fossilized remains of *animal* life first occur rather late in earth history, it is extremely significant that the remains of primitive *plants* are found in Precambrian rocks of all ages, including some that are almost as old as the oldest known rocks. We shall consider the Precambrian record of these primitive plants in the next section; the important point here is that plants similar to some that still survive have been present since the earliest recorded earth history. For this reason the *origin* of life on earth, like the differentiation of the solid earth, ocean, and atmosphere, must have taken place in the dim "pregeological" interval lying between the formation of the earth and the origin of the oldest surviving rocks. Even though there is no direct evidence for life's origin, the

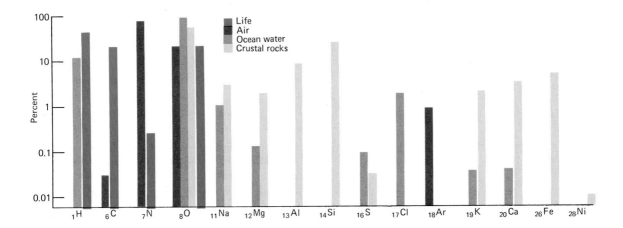

Figure 11.6 The relative abundances of the dominant elements of air, ocean water, crustal rocks, and life. The principal elements of life (color) are also abundant in the ocean and atmosphere.

subject has attracted much speculation and experiment.

All organisms are composed largely of compounds of the four light elements hydrogen, carbon, nitrogen and oxygen. These elements were undoubtedly present in large quantities in the earth's initial ocean and atmosphere just as they are today (Figure 11.6). Biologists now believe that life arose as simple compounds containing these elements were combined into more complex compounds by strong energy sources present on the early earth. This belief has been strengthened by numerous laboratory experiments conducted over the past twenty years in attempts to simulate the conditions under which life might have arisen.

These experiments employ various mixtures of probable early atmospheric gases, such as carbon dioxide, water vapor, methane, nitrogen, ammonia, and hydrogen, and various possible energy sources, such as strong solar radiation, lightning, thunder, and meteorite shock waves, any of which might have caused the relatively simple atmospheric compounds to combine into the large, complex molecules that make up life. Although none of these experiments has produced anything even approaching the complexity of the simplest organism, they *have* succeeded in showing that a variety of the complex chemical building blocks that make up life could have been present in the early ocean and atmosphere. Such building blocks form readily when any of the possible energy sources is applied to almost any gaseous mixture containing the elements hydrogen, carbon, nitrogen, and oxygen (Figure 11.7).

There is, however, one strict limitation on the origin of these biological building blocks. Their production always requires gaseous mixtures in which any oxygen present

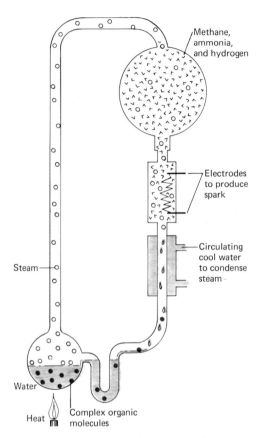

Figure 11.7 *An apparatus for producing the organic building blocks of life from a simulated "primitive atmosphere" of methane* (CH_4), *ammonia* (NH_3), *and hydrogen* (H_2). *When exposed to a high-energy electrical spark simulating natural lightning, complex organic molecules (color) are formed.*

is combined with carbon, nitrogen, or hydrogen rather than occurring as the free oxygen gas which makes up about one-fifth of the present atmosphere. In the presence of free oxygen the molecular building blocks do not form. This limitation causes few difficulties because the abundant free oxygen of the present atmosphere has apparently been produced by the metabolic activities of green plants through the long course of earth history. Before oxygen-producing plants arose, there is every reason to believe that the earth's atmosphere, like present-day volcanic gases, lacked free oxygen.

The exact processes whereby accumulations of non-living biologic building blocks were transformed into simple organisms are still uncertain. There are, however, reasonable hypotheses suggesting how such accumulations could give rise to self-duplicating organisms, the first perhaps being similar to certain very simple present-day bacteria. These would have been the precursors of the primitive fossil plants found in Precambrian rocks.

THE PRECAMBRIAN EARTH

Rocks still present in the earth's crust provide *direct* evidence of earth history beginning around 4,000 million years ago, or about 500 million years after the earth's origin. In this section we shall review the nature and historical implications of the ancient rocks formed during the long *Precambrian* interval of earth history—the 3,400 million years that elapsed between the origin of the oldest surviving rocks, and the appearance of abundant animal fossils at the beginning of Cambrian time about 600 million years ago (Figure 11.8).

11.4 Distribution of Precambrian Rocks

Precambrian rocks make up the bulk of the present-day continents.[1] Even though much of the surface area of the continents is covered by a veneer of sedimentary rocks, most of their volume is occupied by thick igneous and metamorphic sequences that lie beneath this sedimentary cover (see Sections 3.2 and 10.1). These sequences of

---

[1] No such early rocks are found underlying the ocean basins because rocks of the ocean floor are continuously destroyed and regenerated by motions of lithosphere plates (see Section 3.9). The oldest rocks underlying the present ocean basins appear to have ages of only 200 million years.

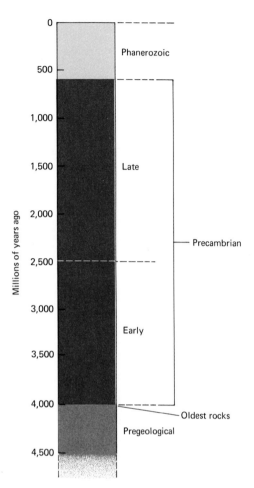

Figure 11.8 The Precambrian interval of earth history (color). The bulk of the earth's past is represented by Precambrian rocks which span the time interval from about 4,000 to 600 million years ago.

continental "basement rocks" are predominantly of Precambrian age whereas most of the overlying sedimentary veneer has been deposited since Cambrian time. If *all* the great volume of continental Precambrian rocks were covered by younger sediments we would, however, know relatively little of Precambrian earth history for the rocks could be studied only in occasional mines or drill holes that penetrate the overlying sediments. Fortunately, each of the present continents has relatively large areas where Precambrian rocks are exposed at the surface without an overlying sedimentary veneer. These are the *continental shield areas* discussed in Section 3.2—large regions where younger rocks have been removed by erosion to expose the Precambrian basement (Figure 11.9). Most of our knowledge of the earth's Precambrian history comes from studies of rocks exposed in these regions.

The Precambrian rocks of certain shield areas, particularly those of South Africa and southern Canada, have been intensively studied for many years because they contain large concentrations of iron, gold, and other valuable minerals (see p. 393). These studies led to the recognition of local rock sequences that record Precambrian volcanic activity, mountain building and varying sedimentary environments. Such events were, however, spread over an enormously long span of earth history and, until the advent of radiometric dating, there was no way to relate them into a chronologic framework that might reveal worldwide patterns of Precambrian history. Within the past two decades, hundreds of radiometric ages have been determined for Precambrian rocks from every continent. These are now beginning to provide an understanding of the complex growth and change of the continental masses through the long 3,400 million years of Precambrian time.

11.5 Precambrian Continental Nuclei and Mobile Belts

Radiometric dating in several of the principal shield areas has shown that rocks formed early in Precambrian time, between about 2,500 and 4,000 million years ago, are generally similar wherever they occur (Figure 11.8). These distinctive early Precambrian rocks typically consist of basaltic volcanics and associated clastic sedimentary rocks, particularly shales and sandy shales, both of which give evidence of having been deposited under ocean water

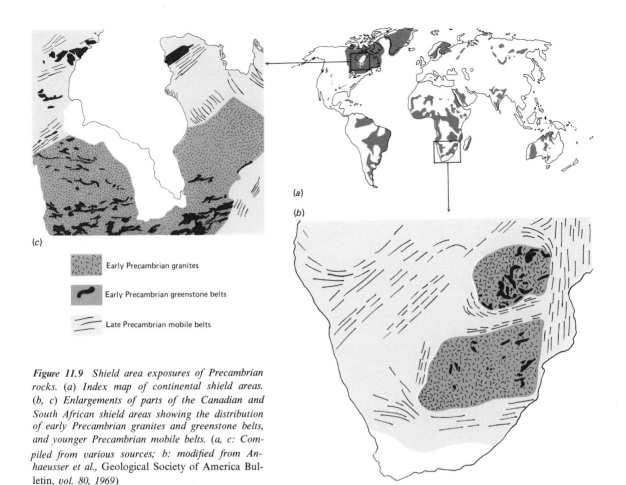

Early Precambrian granites

Early Precambrian greenstone belts

Late Precambrian mobile belts

Figure 11.9 Shield area exposures of Precambrian rocks. (a) Index map of continental shield areas. (b, c) Enlargements of parts of the Canadian and South African shield areas showing the distribution of early Precambrian granites and greenstone belts, and younger Precambrian mobile belts. (a, c: Compiled from various sources; b: modified from Anhaeusser et al., Geological Society of America Bulletin, *vol. 80, 1969)*

in long, linear troughs. Rather surprisingly, these earliest rocks are usually only slightly metamorphosed, indicating that they have never been deeply buried nor otherwise subjected to high temperatures and pressures. This is in sharp contrast to many *younger* Precambrian rocks which *have* been strongly metamorphosed. Because these early Precambrian rocks often have a characteristic greenish color, the elongate bands in which they occur have come to be known as **greenstone belts** (Figures 11.9, 11.10).

The volcanic and sedimentary rocks of the greenstone belts must have been deposited on top of some preexisting solid crust; this earlier crust is, however, never preserved. The nature of the greenstone belt rocks themselves suggests that the underlying crust was relatively thin and it has probably been destroyed by later intrusions of large volumes of granite under and around the belts.

Figure 11.10 Aerial photograph of a 4.5-km-wide (2.8 mi) portion of a greenstone belt (darker rocks) and related granites (lighter rocks) in the shield area of northern Canada. The black areas are lakes. (Geological Survey of Canada)

Because of these intrusions, the greenstone belts today occur only as isolated bands in much larger areas of granite (Figures 11.9, 11.11).

In contrast to the thin crust on which the greenstone belt rocks appear to have been deposited, seismic evidence indicates that the granites that now surround them make up the thickest and most stable parts of the continental crust. Furthermore, the structural relations and radiometric ages of these great granitic masses indicate that they, too, are very old and were emplaced *beneath* the earth's surface only shortly after the volcanic and sedimentary rocks of the greenstone belts were deposited *on* the surface. These great masses of early granite, unlike most younger granitic rocks, show little evidence of having been formed in linear belts by mountain-building deformation and melting of preexisting rocks (see Section 3.5). For this reason, they are now suspected to have formed directly from the upward migration of light, volatile elements in the last stages of crustal differentiation (Figure 11.11). If so, the separation of much of the granitic crust from the underlying mantle belongs not to the "pregeological" interval of earth history, but to the documented history recorded by Precambrian rocks.

The Precambrian greenstone belts and the much larger volume of early granite that surrounds them are

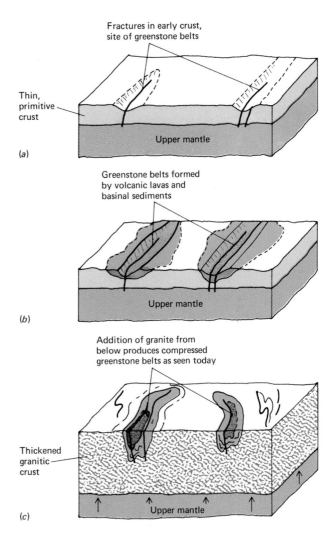

Figure 11.11 A hypothesis for the origin of the Precambrian greenstone belts and surrounding granites. (After Anhaeusser et al., Geological Society of America Bulletin, *vol. 80, 1969)*

both older than about 2,500 million years and appear to have been relatively little altered since they first formed. Together, the greenstone belts and early granites make up the oldest and most stable parts of the Precambrian shield areas and can be thought of as *original continental nuclei* which, during the 2,500 million years since they formed, have been modified and changed principally along their margins by movements of lithosphere plates similar to those that are still going on today (Figure 11.9). Each greenstone-granite continental nucleus is surrounded by younger Precambrian rocks. These are mostly highly deformed intrusive igneous and metamorphic rocks, but locally relatively undeformed sedimentary rocks also occur. All of these rocks tend to occur in linear bands, called mobile belts, which suggest that they originated as the early continental margins were subjected to mountain-building deformations by plate motions (Figure 11.9). These younger Precambrian rocks thus appear to be similar in origin to the still younger rocks formed at moving plate margins since the beginning of Cambrian time.

The details of this later Precambrian phase of earth history are still poorly understood, partly because the rocks have been intensively studied in only a few areas, and also because later deformation, and the extensive cover of younger sedimentary rocks, obscure much of their history. We still know relatively little, therefore, about the size, shape, and distribution of continents, ocean basins, and mountain ranges during Precambrian time. There are, however, several instructive inferences about Precambrian environments that can be made from a closer look at Precambrian sedimentary rocks.

11.6 Precambrian Environments

The intrusive igneous and metamorphic rocks that make up the bulk of the Precambrian rock record reveal mostly events taking place *beneath* the earth's surface. In contrast, the smaller volume of relatively unaltered Precambrian sedimentary rocks provides clues to environmental conditions at the earth's surface during the long Precambrian interval of earth history. Consequently, these rocks assume an importance that is out of proportion to their small volume.

As with sediments being deposited today, Precambrian sedimentary rocks are of both detrital and chemical origin. In general, the detrital sediments, preserved as shales and sandstones, show few features that differ from comparable modern sediments. The earliest Precambrian sediments—the shales and sandstones preserved in the greenstone belts—are detrital. These generally show graded bedding and other structures indicating they were deposited in deep basins by turbidity currents (Figure 11.12). It is these early detrital sediments, some of which are among the oldest rocks known, that show that erosion, transportation, and sediment deposition were already taking place very early in earth history. The lack of either shallow marine shelf or terrestrial deposits among these earliest sediments suggests that such settings may have been uncommon before the separation of the thick granitic masses that now surround the greenstone belts. Younger Precambrian detrital sediments show a much wider range

(a)

(b)

Figure 11.12 Early Precambrian sediments. (a) Graded sandstones (light bands) and shales (dark bands) exposed in northeastern Minnesota. (b) Close-up view of sand and shale layers. (a, b: Richard W. Ojakangas)

of depositional environments; many reflect marine shelf and nonmarine environments as well as deposition in deep basins.

One other important environmental inference can be made from Precambrian detrital sediments. Unsorted mixtures of pebbles, sand, and mud that appear to represent deposition by ice occur scattered through much of the Precambrian sedimentary record, but are most abundant near the end of Precambrian time, about 600 to 700 million years ago. These late Precambrian ice deposits, which occur on all of the present-day continents, show that an interval of extensive glaciation preceded the expansion of animal life in early Cambrian time (see Figure 7.10).

Unlike most Precambrian detrital sediments, which resemble muds and sands accumulating today, certain Precambrian chemical deposits are unlike *any* present-day sediments. These distinctive chemical sediments, although still puzzling in many ways, provide important clues to differences between the environments of Precambrian time and those that we see around us on the earth today.

Foremost among these unusual Precambrian chemical sediments are deposits known as **banded iron formations** (see p. 394). These are extensive sequences, usually hundreds of meters thick and covering hundreds of square kilometers, made up of thin layers or "bands" of dark chert (massive, noncrystalline quartz) alternating with layers of reddish iron oxide minerals. Both the chert and iron oxide layers show evidence of having been directly precipitated from ocean water over large areas of the Precambrian sea floor. The puzzling thing about these deposits is that iron can be transported in solution, and then deposited as sedimentary iron oxide, only under chemical conditions that are not present on the earth today.

In the presence of abundant free oxygen, as occurs in the atmosphere and waters of the modern earth, iron forms insoluble, rustlike oxides which cannot be transported in solution by streams to accumulate in the oceans. For this reason, the present-day oceans contain only small traces of dissolved iron. In the absence of free oxygen, however, iron is readily soluble and could be easily weathered from iron-rich rocks and transported in solution to the oceans by streams. Thus the presence of large quantities of dissolved iron in the Precambrian oceans suggests an oxygen-poor atmosphere.

The scanty fossil record of Precambrian life may also indicate that the atmosphere was low in oxygen through much of Precambrian time and thus the presence of large quantities of dissolved iron in the Precambrian oceans seems understandable. The more complex problem is exactly how this dissolved iron was converted to iron oxides which, being insoluble, precipitated to make the banded iron formations. Such precipitation seems to require a local source of oxygen in the Precambrian oceans; this oxygen may have been supplied by simple, seaweedlike plants growing on the Precambrian sea floor.

Banded iron formations are found only in Precambrian rocks older than about 1,000 million years. In the interval from 4,000 to 1,800 million years ago they are the most abundant Precambrian chemical sediment. Precambrian terrains also contain limestones, organic-rich shales, and other chemical sediments that are similar to those forming today. In general, these sediments are most abundant in younger Precambrian rocks deposited between about 2,000 and 600 million years ago. Just as today, many of these Precambrian chemical sediments appear to have been deposited by the activities of organisms and, for this reason, we shall consider them further as we discuss Precambrian life.

THE EXPANSION OF LIFE

The principal distinction between Precambrian rocks and those deposited since the beginning of Cambrian time is that the latter contain abundant fossilized remains of animals and plants, whereas such remains are far less conspicuous in Precambrian rocks. Indeed, until quite recently it was generally believed that Precambrian sedimentary rocks contained *no* unequivocal indications of ancient life. Within the past twenty years, however, it has become clear that simple bacteria and seaweedlike algae were abundant in certain favorable environments throughout the long Precambrian interval of earth history (Figure 11.13).

The fossil remains of these simple Precambrian plants are generally quite small and inconspicuous and, for this reason, they long escaped critical study. The contrast between them and the record of Phanerozoic life is still a profound one, however, because near the beginning of Cambrian time many kinds of large, complex, shell-bearing *animals* appear abruptly in the fossil record. Fossil animals have been a common feature in sedimentary rocks ever since (Figure 11.13). In this section we shall first review the Precambrian record of bacteria and algae and shall then briefly consider the principal factors believed to be responsible for changes in the living world, before describing the dramatic expansions of life which brought to a close the long Precambrian interval of earth history.

11.7 Precambrian Life

Precambrian sedimentary rocks contain two fundamentally different kinds of evidence of Precambrian life. The first are mere *chemical* traces of the activities of life; the second are the actual preserved remains of the organisms themselves (Figure 11.13).

Many Precambrian shales contain petroleumlike carbon compounds that probably originated from the

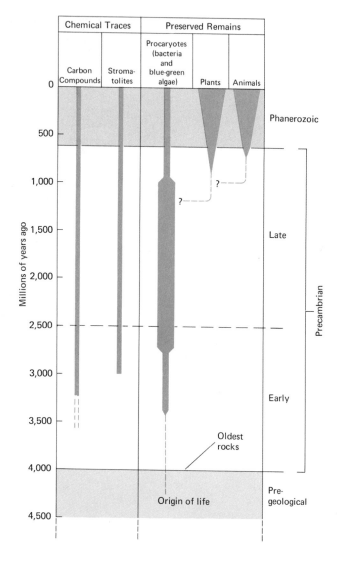

Figure 11.13 *The distribution of life remains in Precambrian rocks. Chemical traces of ancient life as well as fossil procaryotes (bacteria and blue-green algae) are found through much of the Precambrian record, but more advanced plant and animal remains occur only in latest Precambrian deposits.*

decay of organisms, but these reveal very little about the animals or plants that produced them. A more imposing chemical trace of Precambrian life is found in limestones made up of spherical, layered masses known as **stromatolites** (Figure 11.14). Stromatolites are rare in the early Precambrian greenstone belts but are common in younger Precambrian rocks. Similar layered limestone deposits also occur in Phanerozoic sedimentary rocks and are being formed today in shallow, warm oceans by the metabolic activities of simple seaweeds called **blue-green algae**. Because of their resemblance to these younger limestones, Precambrian stromatolites have long been suspected to

(a)

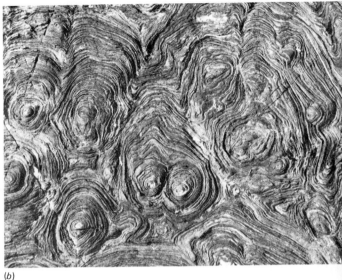

(b)

Figure 11.14 Precambrian stromatolites, layered domelike masses of limestone that are believed to have formed from the metabolic activities of ancient algae. Surface (a) and cross-section (b) views of late Precambrian stromatolites from northern Canada. (Paul Hoffman, Geological Survey of Canada)

be the deposits of Precambrian blue-green algae. Limestones, however, can also originate by direct precipitation of calcium carbonate without the intervention of life and, for this reason, the algal origin of Precambrian stromatolites was long a subject of debate. This uncertainty has been resolved over the past two decades by the discovery of the actual preserved remains of ancient algae and bacteria in Precambrian stromatolites, as well as in the generally older banded iron formations. These discoveries have revolutionized our understanding of the living world through the long Precambrian interval of earth history.

Precambrian fossil algae and bacteria are mostly preserved as tiny, microscopic spheres and threads embedded in the limestones of stromatolites or the cherts of banded iron formations (Figure 11.15). They have now been found in Precambrian sedimentary rocks of almost all ages, and occur in most Precambrian banded iron formations and stromatolites that have so far been investigated. It therefore seems probable that the deposition of both of these Precambrian chemical sediments was closely interrelated with the activities of organisms.

The fact that bacteria and blue-green algae were the only apparent life during much of Precambrian time has great significance because biological studies have shown that present-day representatives of these two groups share unique properties which make them fundamentally different from *all* other kinds of life. So basic are these differences, the most important of which are summarized in

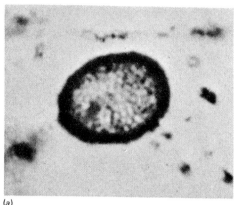

(a)

Figure 11.15 *Precambrian fossil bacterium (a) and blue-green algae (b) from 2,000-million-year-old banded iron formation in southern Ontario. Magnified about 1,500 times. (a, b: Elso S. Barghoorn)*

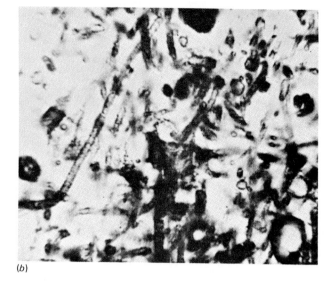

(b)

Figure 11.16 *Procaryotic and eucaryotic cells.*

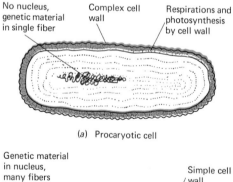

No nucleus, genetic material in single fiber

Complex cell wall

Respirations and photosynthesis by cell wall

(a) Procaryotic cell

Genetic material in nucleus, many fibers

Simple cell wall

Respiration and photosynthesis by small, specialized internal organs

(b) Eucaryotic cell

Figure 11.16, that many biologists now classify bacteria and blue-green algae as a separate kingdom of life, called the **Procaryota,** that is comparable to the familiar plant and animal kingdoms of more complex life. For our purposes, the differences between procaryotes and all other forms of life are most important because of what they imply about Precambrian environments.

Present-day bacteria survive a wider range of environmental extremes than do any other kind of life. Certain forms thrive in polar ice caps and others in the near-boiling water of hot springs. Many are adapted to life without oxygen or sunlight, and some can survive without organic food, using instead the chemical energy of certain sedimentary minerals to drive their life processes. Blue-green algae show almost as wide a range of environmental tolerances with the exception that, like all chlorophyll-bearing green plants, they require sunlight as an energy source for their life processes. In short, bacteria and blue-green algae can exist under a spectrum of environmental conditions that would be intolerable to more advanced organisms. This, in turn, suggests that some sort of change leading to greater overall environmental stability took place near the end of Precambrian time and led to the dramatic expansions of the living world that mark the beginning of Phanerozoic time. We shall consider some speculations about the nature of these specific changes after looking briefly at the more general question of change in the living world.

Man
and the
Early Earth

Many valuable minerals are extracted from rocks formed during the long Precambrian interval of earth history, but by far the most important among these is iron. Iron has long been *the* essential resource of industrialized societies and today accounts for about 95 percent of the total production of industrial metals. Indeed, a significant fraction of the remaining 5 percent is composed of manganese and other metals produced principally to be added to iron in order to increase its strength and corrosion resistance.

Iron is relatively easily produced from any of its oxide minerals, such as magnetite, hematite, or goethite, by intense heating in the presence of carbon (charcoal or coke). During this "smelting," the carbon combines with the oxygen of the oxide mineral to form carbon dioxide gas, leaving behind an accumulation of relatively pure metallic iron. Man has produced iron by this process for at least 4,000 years, and modern steel mills differ from their earlier counterparts only in scale, efficiency, and refinements, rather than in the fundamental process employed.

Because of the large volumes of both iron ore and coal (for coke) consumed in iron production, the traditional centers of its manufacture have been regions, such as the English Midlands, Rhur Valley of Germany, or U. S. Great Lakes, where iron ore and coal are found in close proximity. In recent years, however, the development of cheap, large-scale, ocean transport has reversed this pattern, making it possible to ship the raw materials long distances to be processed. In this way such countries as Japan and the Netherlands have become important producers of iron and steel.

For many years iron was produced principally from ores that contained at least 50 percent iron. These were mostly relatively rare igneous concentrations of magnetite, or goethite-rich lateritic soils (see p. 258). More recently, however, refinements in processing have made it profitable to mine sedimentary hematite occurring in Precambrian *banded iron formations*, most of which contain only 15 to 40 percent iron. This development has enormously increased man's reserves of relatively easily exploited iron, for great volumes of these enigmatic sediments, found only in Precambrian rocks formed between about 3,200 and 1,700 million years ago (see Section 11.6), are found on every continent. Because of these reserves, man is assured of an abundant supply of this most essential metallic resource for many centuries to come.

Close-up (above) and distant view (right) of typical Precambrian banded iron formations, exposed in western Australia. The ore mineral (hematite) is concentrated in the darker layers. (Brian J. Skinner)

Mining iron ore from Precambrian rocks in Minnesota. (Standard Oil Co., N.J.)

11.8   The Nature of Biotic Change

The organisms that have been present on the earth since early Precambrian time can be thought of as a fourth major element in its overall composition, an element comparable to the rocks of the solid earth, waters of the ocean, and gases of the atmosphere. At any one time the total volume of this **biosphere,** as it is sometimes called, is far smaller than the volume of rock, water, or air. It nevertheless profoundly affects these nonliving components of the earth because organisms are small chemical engines that rapidly process many of the elements present in the earth's solids and fluids. This relationship works both ways, for changes in the distribution or composition of the earth's continents, ocean, and atmosphere also cause fundamental changes in its animals and plants. An understanding of these complex interrelationships between life and the earth's physical environments is a central theme of modern research.

Biologists have been convinced for many years that all of the animals and plants present on the earth today have arisen from a very few kinds of ancestral organisms through a process of diversification and change known as **organic evolution.** Among the clearest kinds of evidence for this ancestor-descendant interrelationship are: close biochemical similarities in the fundamental life processes found in *all* organisms; the presence today of relatively simple organisms that appear to be similar to the distant ancestors of more complex animals and plants; and the presence of many transitional fossil organisms that bridge the gaps between dissimilar groups of present-day animals or plants. These and other lines of evidence indicate that the enormous variety of animals and plants present on the earth today has arisen from progressive change in a relatively few, less complex, kinds of ancestors. Modern biological studies have done much to clarify the exact processes which cause such evolutionary change. Some of the books listed at the end of this chapter provide complete introductions to these processes; here we shall be concerned only with those aspects of the subject that are crucial for understanding the earth's history.

The fossil record clearly shows that life has become increasingly complex and diverse through the long course of earth history. This increasing complexity has not, however, come from an accumulation of slow, steady changes taking place at a constant rate. Instead, it is the result of a scattered series of rapid expansions in the living world, which are followed by much longer intervals during which organisms show relatively little change. For example, only bacteria and blue-green algae, relatively simple organisms that differ from all other kinds of life, existed during much of the 3,000 million years of Precambrian time. Rather suddenly, around 600 million years ago, a dramatic expansion took place that gave rise to most of the principal kinds of animals and plants found in the ocean today. Several similar evolutionary expansions, or **radiations** as they are called, have taken place more recently. These have not only profoundly modified life in the sea, but have also led to the enormous diversity of life found today on the surface of the land.

Most of these younger evolutionary radiations were preceded by intervals during which many previous kinds of life rather rapidly became extinct. For example, a rapid expansion of large, land-dwelling hoofed mammals closely followed the extinction of dinosaurs and other large reptiles about 70 million years ago. Between such intervals of extinction and radiation are longer spans of earth history during which the living world shows less dramatic change. In summary, then, the *rate* of evolutionary change in organisms has been far from constant through earth history. Instead, the abundant life that we see around us on the earth today is the product of a series of rather sudden contractions and expansions of the living world. These reorganizations must, in turn, have been closely interrelated with changes in the rock, water, and air of the earth's surface, but the exact nature of these fundamental interrelationships is still unknown.

11.9   Biotic Change and the End of Precambrian Time

The most dramatic and extensive evolutionary radiation in all of earth history took place around 600 million years ago and serves to separate the Precambrian and Phanerozoic intervals of earth chronology. All but a very few of the principal kinds of life found on the earth today originated during this radiation and it will therefore be useful to look more closely at its nature and possible causes before turning, in the next chapter, to the record of the earth's changing life and environments during Phanerozoic time.

All of the major kinds of life present on earth today almost certainly arose from Precambrian procaryote ancestors. The first indication of these more advanced forms occurs in rocks around 1,000 million years old, about 400 million years before the beginning of the Cambrian Period. In a few such rocks some tiny fossils, associated with bacteria and blue-green algae, have been found that may represent *green algae,* a plant group that differs markedly from both bacteria and blue-green algae in showing a cell nucleus and other basic structural features found in all higher plants and animals (Figure 11.16). The first really convincing fossils of more advanced life, however, are not plants but complex invertebrate animals ("invertebrates" are animals that lack the backbone and characteristic internal skeleton found in fishes, amphibians, birds, reptiles, and mammals).

The oldest fossil invertebrates occur as a distinctive grouping of about six kinds of large, soft-bodied, wormlike forms that are found as impressions on sandstones in several parts of the world. These are best preserved in the Ediacara Formation of southern Australia, which contains animals resembling present-day jellyfish, "sea-pens," and segmented worms (Figure 11.17). The first occurrence of this "Ediacara fauna," as it is called, appears to be around 650 to 700 million years old; some workers consider the fauna to be latest Precambrian in age, while others regard it as earliest Cambrian, a distinction which is unimportant here. The marine sedimentary rocks in which the fossils occur show abundant tracks, traces, and burrows made by these animals and, perhaps, by other soft-bodied forms which are not preserved as fossils. Such biogenic structures are abundant in Phanerozoic sedimentary rocks, yet recent studies show that even these *traces* of animal life are absent from Precambrian rocks older than about 700 million years. This indicates that animal life expanded abruptly in latest Precambrian time, rather than having been present, but merely not fossilized, earlier.

Fossil impressions of soft-bodied animals, such as those of the Ediacara fauna, are rare in sedimentary rocks compared to the remains of *shell-bearing* animals which, of course, are much more readily fossilized. About 600 million years ago, relatively soon after the appearance of Ediacara fauna, many kinds of shell-bearing invertebrate animals became abundant in marine sedimentary rocks all over the earth. It is the appearance of these abundant fossil shells that marks the traditional beginning of Cambrian (and Phanerozoic) time.

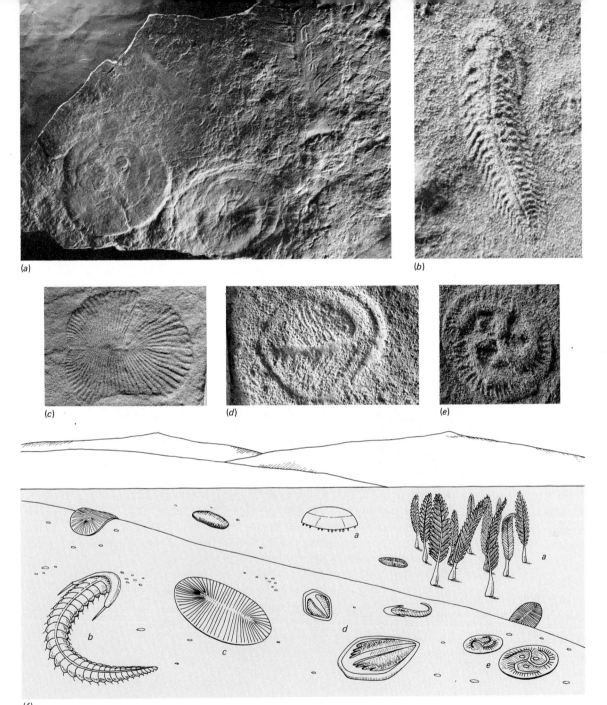

(a)

(b)

(c)

(d)

(e)

(f)

Figure 11.17 *The Ediacara fauna, the oldest known fossil animals, found in latest Precambrian or earliest Cambrian sandstones of southern Australia. (a) Forms similar to jellyfish and sea-pens; (b, c) wormlike forms; (d, e) forms of unknown affinities. (f) Reconstruction of the animals as they might have looked in life. (a–e: Martin F. Glaessner)*

Rather surprisingly, the most abundant animal fossils in Cambrian rocks are not small, simple, one-celled forms as might be expected. Instead, they are many-legged *trilobites* which are distant relatives of modern crabs, and *brachiopods*, two-shelled, superficially clamlike animals (Figure 11.18). The abundance of these relatively advanced animals indicates the initial radiation of animal life rapidly led to complex structures and adaptations. Similar rapid radiations, but on a smaller scale, have occurred several times during subsequent phases of earth history.

The puzzling question about all of this is the *reason* for the sudden appearance of animal life only 600 million years ago, whereas procaryotic plants, from which the animals probably arose, are known to have been in existence through most of the preceding 3,400 million years. Many hypotheses have been put forward to explain this enigma, but most have serious limitations. Recently, several workers have developed a theory relating this event to interactions between the living world and the earth's atmosphere. Although still speculative, these ideas hold promise for clarifying the history of both the biosphere and atmosphere, and we shall review them briefly.

Earlier in this chapter we noted that the earth's early atmosphere was probably made up largely of carbon dioxide, water vapor, and, perhaps, nitrogen released from melting and volcanic outgassing of the solid mantle. The present atmosphere also contains carbon dioxide, water vapor, and nitrogen, but differs profoundly from the early atmosphere in also containing large quantities of gaseous oxygen. Most of this oxygen has apparently been released into the atmosphere through photosynthesis by green plants.

Figure 11.18 Typical trilobites (a, b) and brachiopods (c, d), the most abundant animal fossils of Cambrian rocks.

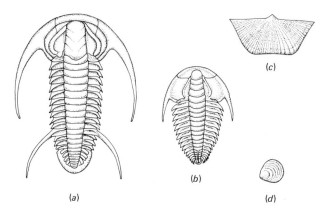

(a) (b) (c) (d)

In the process of photosynthesis, green plants remove carbon dioxide from the atmosphere and use solar energy to convert it, through a complex series of reactions, to *carbohydrates*. These are carbon-hydrogen-oxygen compounds, the most familiar of which are sugars and starches, that provide the energy for all other life processes. The carbon that begins this photosynthetic process as atmospheric carbon dioxide gas ends up as a part of a large carbohydrate molecule that contains, proportionately, *much less oxygen* than did the original carbon dixoide. It is this excess oxygen, released by photosynthesis, that has apparently accumulated throughout earth history to give the atmosphere its present-day abundance of oxygen.

Many geologists now believe that the great expansion of life that ends Precambrian time may reflect the fact that the atmosphere first accumulated abundant free oxygen at about that time. This is suggested by several lines of evidence. Perhaps the most convincing is that many present-day bacteria and blue-green algae, modern descendants of the only kinds of life present through much of Precambrian time, do not require free oxygen, either in the atmosphere or dissolved in the water surrounding them, for their life processes whereas all other organisms *do* require such free oxygen. Photosynthesis by blue-green algae, however, releases oxygen as a waste product (bacteria, even though some are photosynthetic, have different biochemical processes which do not release oxygen).

Initially, any oxygen released by blue-green algae would have quickly combined with dissolved iron and other oxygen-deficient elements and compounds in the surrounding waters, rather than being released to the atmosphere. This process probably accounts for the unique banded iron formations of early Precambrian time. Eventually, however, most of these elements would become oxidized by the algae-produced oxygen. Then free oxygen would begin to slowly accumulate in the oceans, and bubble to the ocean surface to be released into the atmosphere. The development and expansion of oxygen-dependent life late in Precambrian time may mark the time when the atmosphere and oceans first accumulated abundant free oxygen by this process. If so, then during much of Precambrian time only oxygen-independent procaryotic life would have been possible on earth. Nevertheless, the life process of procaryotic blue-green algae, acting over hundreds of millions of years, could have slowly released oxygen to the atmosphere, and thus paved the way for the dramatic expansion of oxygen-dependent life that brings to an end the long Precambrian phase of earth history.

SUMMARY OUTLINE

Pregeological Earth History

11.1 *Differentiation of the solid earth:* the core, mantle, and crust may have separated through early radioactive heating of an initially homogeneous earth, or have accumulated directly from cooling solar gases.

11.2 *Origin of the earth's fluid cover:* the ocean and atmosphere appear to have formed from gases released from the solid earth by local melting and volcanic activity; the atmosphere has probably sharply changed in composition through earth history, while the ocean has remained predominantly water.

11.3 *The beginnings of life:* life is almost as old as the earth itself and probably arose from complex, non-living molecules present in the early ocean and atmosphere.

The Precambrian Earth

11.4 *Distribution of Precambrian rocks:* rocks formed between 4,000 and 600 million years ago are mostly exposed at the earth's surface in the continental shield areas, where they can be interrelated by radioactive dating.

11.5 *Precambrian continental nuclei and mobile belts:* older Precambrian rocks are mostly belts of deep-water sedimentary and volcanic rocks surrounded by large volumes of massive granite; these continental nuclei are, in turn, surrounded by the linear remains of younger Precambrian mountain chains.

11.6 *Precambrian environments:* early Precambrian detrital sediments were mostly deposited in deep ocean basins, whereas younger deposits reflect a spectrum of marine and terrestrial environments; unique Precambrian chemical sediments suggest subsequent changes in atmospheric composition.

The Expansion of Life

11.7 *Precambrian life:* fossilized remains of bacteria and blue-green algae are found in many Precambrian sedimentary rocks along with layered limestone masses and other chemical sediments that are believed to have been formed by the life processes of these primitive plants.

11.8 *The nature of biotic change:* the fossil record suggests that complex life has arisen from simpler organisms in a series of rapid evolutionary expansions and extinctions that are separated by much longer intervals of comparatively little change.

11.9 *Biotic change and the end of Precambrian time:* many kinds of complex, ocean-dwelling animals originated in a rapid evolutionary expansion that begins Phanerozoic time; this expansion may have been caused by a rising concentration of atmospheric oxygen.

ADDITIONAL READINGS

Anhaeusser, C. R., et al.: "A Reappraisal of Some Aspects of Precambrian Shield Geology," *Geological Society of American Bulletin,* vol. 80, pp. 2175–2200, 1969. An advanced review of the distribution and implications of Precambrian rocks.

Anderson, D. L., C. Sammis, and T. Jordan: "Composition and Evolution of the Mantle and Core," *Science,* vol. 171, pp. 1103–12, 1971. An advanced review article.

Barghoorn, E. S.: "The Oldest Fossils," *Scientific American,* vol. 224, no. 5, pp. 30–42, 1971. A popular review of Precambrian bacteria and algae.

Cloud, P. E., Jr.: "Atmospheric and Hydrospheric Evolution on the Primitive Earth," *Science,* vol. 160, pp. 729–36, 1968. An advanced review of the early history of the ocean and atmosphere.

Cloud, P. E., Jr.: "Pre-Metazoan Evolution and the Origins of the Metazoan," in *Evolution and Environment,* Yale University Press, pp. 1–72, 1968. An authoritative, clearly written review of the early history of life.

Dott, R. H., Jr., and R. L. Batten: *Evolution of the Earth,* McGraw-Hill Book Company, New York, 1971. An excellent introduction to all aspects of earth history; Chapters 7 and 8 provide a good summary of the development of the early earth.

*Keosian, J.: *The Origin of Life,* Van Nostrand Reinhold Co., New York, 1968. A brief introduction to the pregeological history of life.

Mason, B.: *Principles of Geochemistry,* John Wiley & Sons, Inc., New York, 1966. Chapters 7 and 8 have good discussions of the origin and development of the ocean and atmosphere.

*Savage, J. M.: *Evolution,* Holt, Rinehart, and Winston, Inc., New York, 1969. An excellent introduction to the mechanisms of biotic change.

* Available in paperback.

12

Evolving Geography, Life, and Environments

The rapid expansion of shell-bearing marine animals in Cambrian time makes it possible to interrelate most Phanerozoic sedimentary rocks by means of the fossil dating technique discussed in Chapter 10. Because of the relative ease and precision with which these rocks can be placed in a chronologic framework, Phanerozoic geography and environments are understood in far greater detail than are the geography and environments of the much longer Precambrian interval of earth history. This chapter briefly summarizes the changing geography, life, and environments of each of the three major subdivisions of Phanerozoic time: the Paleozoic Era, lasting from about 600 to 200 million years ago—a time when geography, life, and environments were very different from today; the transitional Mesozoic Era, from 200 to about 70 million years ago, during which the earth's geography, life, and environments were dramatically reorganized; and the Cenozoic Era, the most recent 70 million years of earth history during which patterns of geography, life, and environment became increasingly like those that we see around us today.

THE PALEOZOIC EARTH

12.1 Paleozoic Geography

During the 400 million years of Paleozoic time (Figure 12.1), the arrangement of continents and ocean basins was unlike the familiar geographic patterns of the present-day earth. As in the preceding Precambrian, relatively little is yet known about the details of early Paleozoic geography. By late Paleozoic time, however, evidence from the distribution of sediments and fossils, combined with paleomagnetic data (Section 3.8), indicate that the continental crust was not broken into several continent-sized

Cambrian trilobites. (Vladimir J. Okulitch)

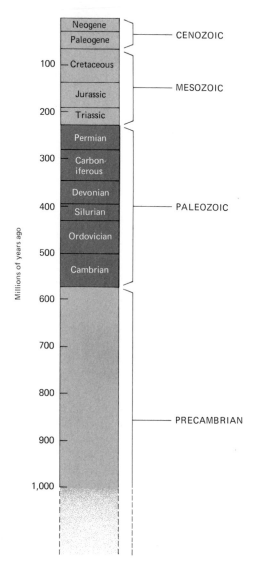

Millions of years ago

Neogene	CENOZOIC
Paleogene	
100 — Cretaceous	
	MESOZOIC
Jurassic	
200 — Triassic	
Permian	
300 — Carbon-iferous	
Devonian	
400 — Silurian	PALEOZOIC
Ordovician	
500 — Cambrian	
600 —	
700 —	
800 —	
	PRECAMBRIAN
900 —	
1,000 —	

Figure 12.1 The Paleozoic interval of earth history (color).

Figure 12.2 Idealized reconstruction of the two late Paleozoic supercontinents, Laurasia and Gondwana; the detailed positions of the continents through Paleozoic time are still uncertain. Eroded remnants of early and late Paleozoic mountain belts are shown in color. (Compiled from various sources)

fragments as it is today. Instead, the present continents were combined into two large "supercontinents" of approximately equal size that were closely adjacent along the margins of present-day North America, Africa, and South America (Figure 12.2). The more northerly of these supercontinents included most of modern North America and Eurasia and is known as **Laurasia.** The more southerly included much of present-day South America, Africa, India, Australia, and Antarctica and is known as **Gondwana.**

Laurasia may have been assembled from several smaller continental fragments through Paleozoic time. Gondwana, on the other hand, has closely similar Paleozoic fossils and sedimentary patterns and may have been a single unit through much of the era. Paleomagnetic studies, however, make it clear that this huge continental mass was not stationary, but sharply shifted its position relative to the rotational poles (Figure 12.3). As in more recent times, plate motions deformed the margins of the Paleozoic continents to produce geosynclines and mountain chains, most of which are still preserved as eroded belts of sharply deformed Paleozoic rocks.

In general, the deformations that produced these belts appear to have been concentrated in two intervals (Figure 12.2). The first group, from about 430 to 375 million years ago, spans the Late Ordovician to Early Devonian Periods of the standard time scale and is known as the *Caledonian* deformation. The second, from about 300 to 250 million years ago, was concentrated in Late Car-

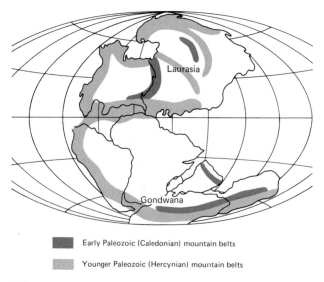

Early Paleozoic (Caledonian) mountain belts

Younger Paleozoic (Hercynian) mountain belts

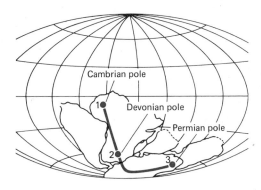

Figure 12.3 Paleozoic motions of the Gondwana supercontinent, as suggested by the magnetic orientations of Paleozoic rocks. (Modified from McElhinny and Luck, Science, *vol. 168, 1970)*

boniferous and Permian time and is known as the *Hercynian* deformation. These intervals almost certainly reflect changes in the amount or direction of lithosphere plate motions, but only a beginning has so far been made on reconstructing these motions by working backward from the deformational history of the continental margins. Such studies, however, promise future clarification of the complex and changing geographic patterns of Paleozoic time.

In addition to differing in the size, shape, and positions of continents and ocean basins, Paleozoic geography was unlike that of the present day in one other important aspect. Throughout much of Paleozoic time the surfaces of the continents stood lower in relation to sea level than they do today, so that shallow seas covered much of the area of the continents. In these shallow seas accumulated thin but widespread layers of fossil-bearing sedimentary rocks that reveal much about the life and environments of Paleozoic time.

In general, there was a progressive decrease throughout Paleozoic time in the proportion of the continents covered by shallow seas (Figure 12.4). For much of Cambrian, Ordovician and Silurian time, only mountain ranges and a few relatively small highland areas appear to have been exposed above sea level. Late in the Devonian Period more of the land surface became exposed and nonmarine sediments accumulated over progressively wider areas until, by Late Permian time, most of the continental surfaces stood above sea level. This relative rise of the continents must, in some fashion, have been related to lateral movements of the continent-bearing lithosphere plates, but the exact nature of this relationship remains unknown.

12.2 The Diversification of Life

The numbers and kinds of living organisms present on the earth expanded enormously during the 400 million years of Paleozoic time. Sea-dwelling algal plants and invertebrate animals dominated the living world at the beginning of the era; then, as through the preceding Precambrian interval of earth history, the land surface was barren and lifeless. Early in the era, however, life expanded onto the land as water-dwelling algae gave rise to simple mosslike and fernlike plants. Some of these, in turn, evolved into the earliest trees and shrubs, which covered much of the land surface by mid-Paleozoic time (Figure 12.5).

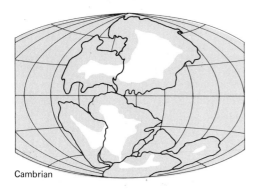

Cambrian

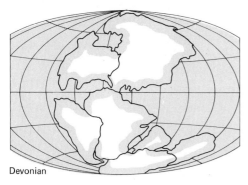

Devonian

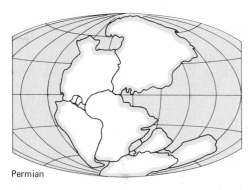

Permian

Figure 12.4 *Schematic representation of changing Paleozoic land-sea patterns. Throughout the era there was a progressive decrease in the proportions of the continental surfaces covered by shallow seas.*

Accompanying the spread of Paleozoic land plants was the rise of the first land animals from water-dwelling ancestors. It is not generally appreciated that only a very few of the principal kinds of animal life have *ever* become adapted for life on land. About 25 major animal groups, called **phyla,** are recognized by biologists; representatives of the most important of these are familiar to everyone and are summarized in Figure 12.6. Each of the animal phyla originated in the ocean in latest Precambrian or early Paleozoic time, and most of them are *still* found only in the ocean. Some representatives of several of the phyla have become secondarily adapted to life in streams, lakes, and moist soil of the land surface, but members of only three phyla have made the much more complex transition to breathing air away from aquatic environments. These three are the Mollusca, Arthropoda, and Chordata (Figure 12.6).

By far the most successful land animals, in terms of number of individuals and species, are the air-breathing *insects* and *arachnids* (spiders and related forms) of the Phylum Arthropoda. Both groups evolved in mid-Paleozoic time along with the first land plants and have been present on land in increasing abundance ever since (Figure 12.5). At about the same time as insects and arachnids were evolving, certain groups of water-dwelling snails (Phylum Mollusca) also developed the ability to breathe air and left the water to feed on land plants where, as all gardeners know, they are still common today.

The final phylum that includes land-dwelling animals, the Phylum Chordata, is the most interesting to us because it contains the many land-dwelling *vertebrate* animals, among them our own species *Homo sapiens*. The earliest fossil vertebrates are primitive fishes found in Ordovician and Silurian deposits (Figure 12.5). During the Devonian Period, at about the same time that plants, insects, arachnids, and molluscs first invaded the land, one group of early fishes developed stubby limbs for moving about on the land surface and simple lungs for breathing air (Figure 12.7). These forms, which also retained the ability to swim, gave rise to the **amphibians,** partially land-dwelling vertebrates that were, and are today, closely tied to water because of their small, delicate eggs which quickly dry out when exposed to air. Well-preserved fossil skeletons of these early amphibians have been discovered and these show that most of them were large, superficially alligatorlike animals quite unlike their specialized surviving relatives, the frogs, toads, and salamanders (Figure 12.7).

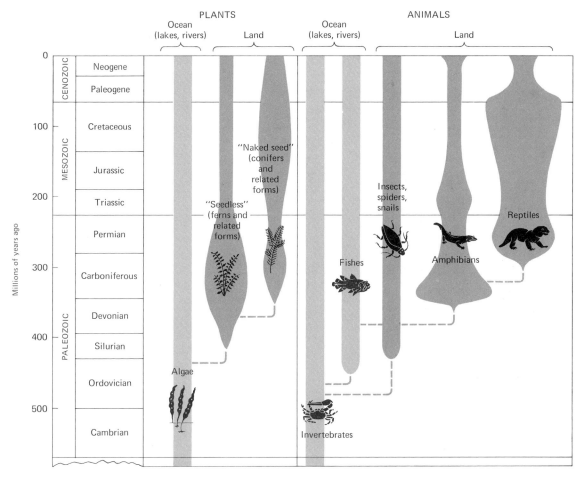

Figure 12.5 The Paleozoic origin and expansion of land plants and animals (color), all of which evolved from water-dwelling ancestors. The width of the vertical columns indicates the approximate abundance of each group. More advanced land dwellers (principally flowering plants, birds, and mammals) arose after the close of Paleozoic time and are not shown.

Amphibians must live near streams, lakes, or very moist soil because of the necessity of protecting their fragile eggs from drying. The final step in vertebrate conquest of the land took place in late Paleozoic time when one group of amphibians developed large eggs with hard coverings to prevent drying, and a supply of yolky nutrients to nourish the developing embryo until it became an air-breathing miniature of the adult. It is this fundamental difference that separates amphibians from **reptiles** and allows the later group to exist over a much wider range of land environments. The first reptiles arose in Carboniferous time and by the close of the succeeding Permian Period a host of large herbivorous and carnivorous reptiles roamed the land (Figures 12.5, 12.7). Among the major vertebrate groups, only birds and mammals, the dominant present-day land vertebrates, had not yet appeared by the end of the Paleozoic Era.

(a)

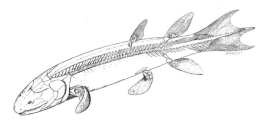

Figure 12.6   The principal subdivisions or "phyla" of animal life. Relatively few groups contain representatives that are fully adapted for life on land; these representatives are shown in color and bold type.

Phyla	Representatives
INVERTEBRATES	
Sarcodina, Mastigophora,* Ciliophora*	Microscopic, single-celled animals
Porifera	Sponges
Cnidaria	Jellyfishes, Corals
Platyhelminthes*	Flatworms
Aschelminthes*	Roundworms
Bryozoa	Moss animals
Brachiopoda	Lamp shells
Annelida*	Segmented worms
Mollusca	Snails, Clams, Cephalopods (squids and related forms)
Arthropoda	Trilobites, Crustaceans, Arachnids, Insects
Echinodermata	Starfishes, Sea urchins, Sea lillies
VERTEBRATES	
Chordata	Fishes, Amphibians, Reptiles, Birds, Mammals

*Predominantly soft-bodied phyla with little or no fossil record.

While life was expanding on land during Paleozoic time, equally fundamental, but somewhat less dramatic, changes were taking place in the ocean. The Cambrian ocean of earliest Paleozoic time was dominated by trilobites, brachiopods, and other invertebrate groups that are now either extinct, or play a minor role in modern seas. In Late Cambrian and subsequent Ordovician time, many of these were replaced by the earliest representatives of groups that dominate the present ocean, animals such as corals, bryozoans, snails, clams, cephalopods, crustaceans, starfish, and sea urchins (Figures 12.8, 12.9). Although there have been many evolutionary changes *within* these groups since they first arose in early Paleozoic time, they have remained continuously important and are still abundant today.

The changes and expansions of Paleozoic life did not take place steadily throughout the era. Instead, the fossil record shows that expansions occurred rather suddenly; furthermore, each expansion usually follows a period of dramatic reduction or extinction of preexisting animals and plants. Several such extinction intervals took place during Paleozic time, the most severe of them near the close of the Cambrian, Ordovician, Devonian, and Permian

(b)

(c)

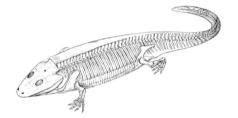

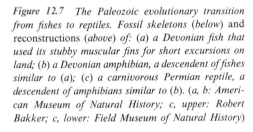

Figure 12.7 The Paleozoic evolutionary transition from fishes to reptiles. Fossil skeletons (below) and reconstructions (above) of: (a) a Devonian fish that used its stubby muscular fins for short excursions on land; (b) a Devonian amphibian, a descendent of fishes similar to (a); (c) a carnivorous Permian reptile, a descendent of amphibians similar to (b). (a, b: American Museum of Natural History; c, upper: Robert Bakker; c, lower: Field Museum of Natural History)

Periods. Each was followed by a relatively rapid evolutionary radiation of new kinds of animals and plants; these relationships are summarized in Figure 12.10. Such intervals of extinction and subsequent expansion of the living world are undoubtedly interrelated with changes in the earth's physical environments but, as yet, there is no clear understanding as to just what these changes were. As in Precambrian time, however, some clues come from studies of the environments preserved in Paleozoic sedimentary rocks.

12.3 Paleozoic Environments

The shallow seas that covered much of the continents during long intervals of Paleozoic time left a widespread record of marine sedimentary rocks. Many of these have escaped subsequent erosion and survive as indicators of Paleozoic environmental patterns. One such pattern, for the unusually well-studied Silurian Period, is shown in Figure 12.11.

Most of the sediments making up Paleozoic sedimentary rocks are similar to sediments being deposited today, yet detailed studies of their distribution in time and space show that some environments were far more widespread and important during certain intervals of Paleozoic

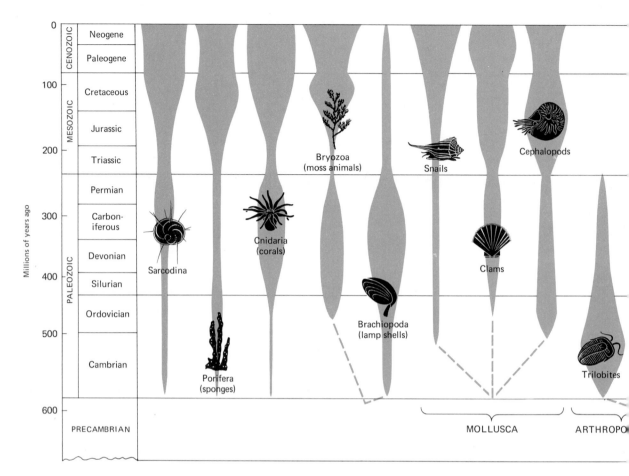

Figure 12.8   The geologic record of the principal ocean-dwelling invertebrate animals. The width of the vertical columns indicates the approximate abundance of each group. Note that most of the dominant groups in the present oceans originated in Late Cambrian and Ordovician time and have persisted ever since. In contrast, trilobites, the dominant Cambrian fossils, became extinct at the close of Paleozoic time.

time than they are at present. In the shallow Paleozoic seas that covered much of the continental surfaces, carbonate shelf and associated evaporite environments were unusually common, especially during early and mid-Paleozoic time when the ocean stood highest relative to the land. As today, most of these carbonate accumulations were made up of fragments of the shells and skeletons of marine life; only a small proportion appear to have been inorganically precipitated from sea water.

There are, however, difficulties in interpreting the exact origin of many Paleozoic carbonate rocks because of chemical and structural changes that occurred as the sediment was buried and converted into rock. Frequently the original calcium carbonate *recrystallized* after burial so that details of the original composition and structure are obscured. A still more puzzling change is the occurrence of large quantities of the mineral *dolomite* among

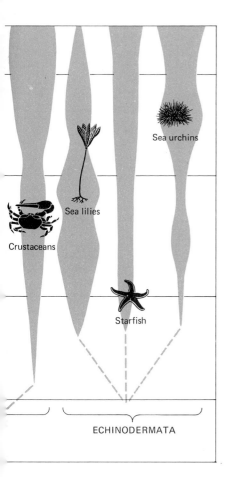

Crustaceans

Sea lilies

Sea urchins

Starfish

ECHINODERMATA

Paleozoic carbonate deposits (Figure 12.12). Dolomite, a carbonate mineral containing about equal quantities of calcium and magnesium in its crystal structure, almost never occurs in present-day carbonate accumulations, and can be produced in the laboratory only at temperatures far higher than those occurring naturally at the earth's surface. Some of the large quantities of Paleozoic dolomite clearly precipitated directly from sea water; the conditions under which this could occur are still mysterious. Probably the bulk of the Paleozoic dolomite resulted, however, from chemical alteration of the original calcium carbonate after

(a)

Figure 12.9 Typical fossil invertebrates from Paleozoic rocks. (a) Carboniferous coral preserved in limestone. (b) Devonian clams preserved in shale. (a: Geological Survey of Great Britain; b: Yale Peabody Museum)

(b)

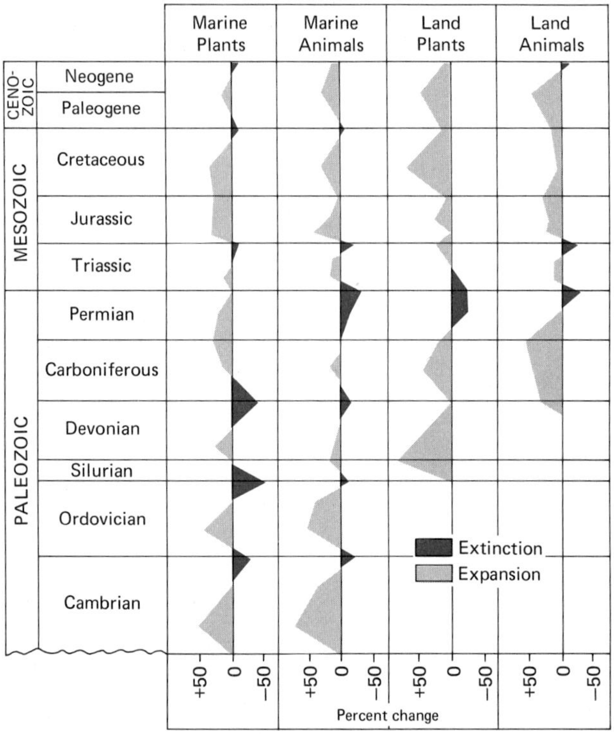

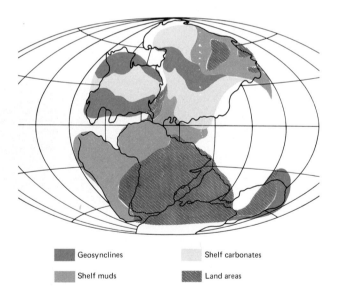

	Marine Plants	Marine Animals	Land Plants	Land Animals

Figure 12.10 *Phanerozoic expansions and extinctions of life. The curves show net expansions (*light color*) or extinctions (*dark color*) of animal and plant families during each Phanerozoic epoch. Paleozoic intervals of general extinction occurred near the end of the Cambrian, Ordovician, Devonian, and Permian Periods; each was followed by a major expansion. (*From Bulletin of Canadian Petroleum Geology, vol. 20, 1973*)*

Figure 12.11 *Sedimentary environments of Silurian time. Shallow seas covered much of the continental surface and thin shelf carbonate and mud sequences accumulated over large areas; thick geosynclinal sequences of mud, sand, and volcanic rocks were deposited along some continental margins. The patterns were reconstructed from scattered and discontinuous present-day exposures of Silurian rocks on each continent. (*Modified from Boucot, Berry, and Johnson, in* History of the Earth's Crust, *1968*)*

burial. Once again, the exact mechanisms responsible for the transformation are poorly understood.

Just as today, evaporites are far less common among Paleozoic sediments than are marine carbonate deposits. During several rather short intervals of Paleozoic

Figure 12.12 Typical Paleozoic dolomite (exposed in northern England). The exact conditions of origin of these widespread magnesium-rich carbonate rocks are unknown (see text). (Geological Survey of Great Britain)

Figure 12.13 Late Paleozoic glacial deposits (color). All are confined to the Gondwana continent.

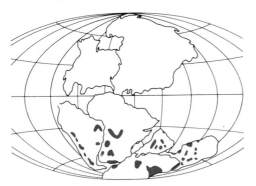

time, however, marine evaporites were particularly widespread. Today, such evaporites form principally in shallow marine bays in the earth's broad zones of arid climate, where evaporation exceeds runoff from land. Presumably, the intervals of widespread Paleozoic evaporites reflect times when atmospheric circulation patterns led to similar conditions over large areas of the continental surface.

In addition to marine carbonates and evaporites, many Paleozoic sediments deposited in other environments are preserved in the continents' sedimentary veneer. Among the most revealing of these are widespread accumulations of swamp-dwelling plant remains that make up extensive coal deposits in Carboniferous and Permian rocks. During this interval of late Paleozoic history, some poorly understood environmental conditions favored the growth and accumulation of swamp plants over large areas of the land surface. Although younger coal-bearing sediments are also common, at no time in earth history did plant remains accumulate on such a scale.

Late Paleozoic coal deposits are known from every present-day continent, indicating that both Laurasia and Gondwana experienced widespread coal-forming conditions. Another environmentally revealing type of non-marine sediment, ice-deposited tillites, occurs in late Paleozoic sequences of South America, India, Australia, Africa, and Antarctica, the regions that were joined as the large Gondwana continent during late Paleozoic time (Figure 12.13). These deposits, combined with paleomagnetic evidence, indicate that late Paleozoic motions moved parts of the continent near the southern pole, causing extensive ice sheets and glaciers. These late Paleozoic ice deposits appear to mark the only interval of widespread glaciation between those of late Precambrian time, and those of the late Cenozoic that are still in progress today.

MESOZOIC REORGANIZATION

The Mesozoic Era of earth history spans about 130 million years, an interval only about one-third as long as the 400 million years of Paleozoic time (Figure 12.14). In spite of its relatively short duration, the era was a time of extraordinary changes in the earth's geography, life, and environments. As it began the earth's surface still reflected the two-continent pattern of the late Paleozoic. When it ended the continental crust was broken into the familiar continents that we know today and birds, mammals, and flowering plants, the most conspicuous kinds of land life

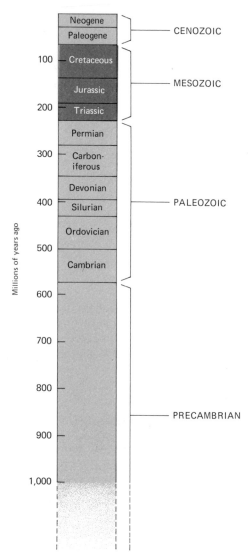

Figure 12.14 The Mesozoic interval of earth history (*color*).

today, had begun their rise to dominance. The Mesozoic Era was thus a time of rapid change and transition between the Paleozoic earth and the familiar patterns of geography, life, and environments of the present day.

12.4   The Separation of the Continents

As the Paleozoic Era drew to a close, the two large continental masses, Laurasia and Gondwana, were probably joined along the margins of present-day western Africa and eastern North America to make a single L-shaped continental mass (Figure 12.15a). This pattern persisted through much of the Triassic Period but, beginning in Late Triassic time, plate movements began to break up this single large continent into the six major continental masses of the present day. Some of these movements can be reconstructed by studies of the age and structure of ocean-floor basalts, the oldest of which were formed by this initial fragmentation. Unfortunately, however, such reconstructions become more and more obscure as they are extended back in time because the older portions of the ocean floor are continuously destroyed in subduction zones as new ocean floor spreads outward from oceanic ridges. Only a relatively few small areas of Mesozoic ocean floor still survive but these, combined with evidence from patterns and times of deformation of the continents, permit the generalized reconstruction of the Mesozoic continental breakup shown in Figure 12.15b and c. The most dramatic events of this wholesale geographical reorganization were the opening of the Atlantic Ocean Basin as North and South America became separated from Africa and Europe, and the fragmentation of eastern Gondwana into four separate blocks which are present-day Africa, India, Australia, and Antarctica.

As with the more poorly understood plate motions of Paleozoic time, Mesozoic continental breakup was accompanied by mountain-building deformations (Figure 12.16). Unlike the preceding Paleozoic deformations, however, which appear to have been much more intense during certain intervals, the Mesozoic deformations lasted over much of the era. These Mesozoic deformations were responsible for, among others, the Rocky Mountains of western North America.

The late Paleozoic trend toward general emergence of the continents above sea level continued in Triassic and much of Jurassic time, but during the Cretaceous Period shallow seas once again covered large areas of the conti-

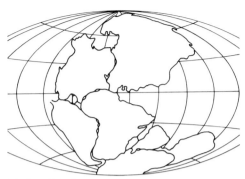

(a) Early Triassic

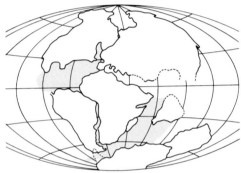

(b) Jurassic

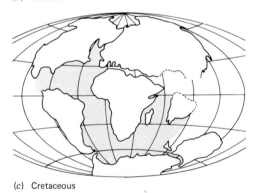

(c) Cretaceous

▨ Mesozoic ocean floor

Figure 12.15 Mesozoic breakup of Laurasia-Gondwana to produce the present-day continents. Areas of new basaltic crust added to the ocean floor are shown in color. (After Dietz and Holden, Scientific American, *vol. 223, 1970)*

nental surface in what was to be the last such widespread flooding of the land (Figure 12.17). As earlier, these world-wide changes in the relative levels of land and sea must have been related, in some as yet unknown fashion, to patterns of growth and movement of the continental masses and ocean floor.

12.5 The Mesozoic Expansion of Life

The close of the Paleozoic Era was marked by extraordinarily severe and widespread extinctions of preexisting life. For reasons that are still obscure, about 30 percent of the families of fossil animals and plants found in Lower Permian rocks became extinct by the close of Permian time and are unknown in younger rocks (Figure 12.10). Marine invertebrate animals were particularly hard-hit. Trilobites became extinct as did many previously abundant groups of corals, bryozoans, brachiopods, and crinoids (Figure 12.8). Life on land was also affected, for many families of land-dwelling plants and reptiles also became extinct. These severe reductions of the living world were followed, in Mesozoic time, by new evolutionary radiations that ultimately led to a far greater diversity of organisms than existed at any time during the Paleozoic Era.

In the ocean, many new invertebrate groups arose in Triassic and Jurassic time to replace those eliminated in the Late Permian extinctions. Most of these, particularly various groups of corals, snails, clams, crustaceans, starfish, and sea urchins, continue as the dominant invertebrates of the oceans today. Paralleling this modernization of marine animal life were changes in ocean-dwelling plant life, particularly in the tiny, floating plants known as **phytoplankton** which far exceed the larger bottom-dwelling seaweeds in total volume and importance as food for marine animals. The fossil record indicates that all three of the dominant kinds of present-day phytoplankton arose in Mesozoic time (Figure 12.18). Fossil phytoplankton from Paleozoic rocks are, in contrast, far less diverse. This suggests that both the ocean's animal and plant life were completely reorganized during Mesozoic time. By the end of the era life in the oceans was, in most respects, closely similar to that existing today.

Perhaps the most dramatic biotic changes of Mesozoic time, at least from our point of view as land dwellers, took place on the continents where new groups of land plants and animals developed. The fossil record shows that the land plants of Paleozoic time were mostly seed-

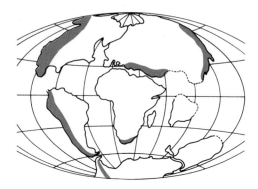

Figure 12.16 Principal Mesozoic mountain belts (color).

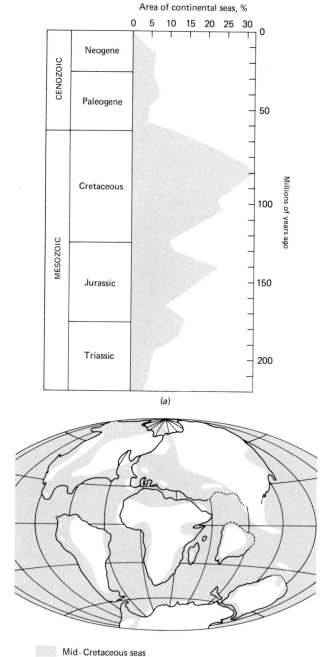

Figure 12.17 Mesozoic and Cenozoic land-sea patterns. (a) Percentage of the continental surfaces covered by shallow seas throughout the two eras. Only in mid-Cretaceous time (b) were there extensive continental seas comparable to those of Paleozoic time. (Compiled from various sources)

Mid-Cretaceous seas

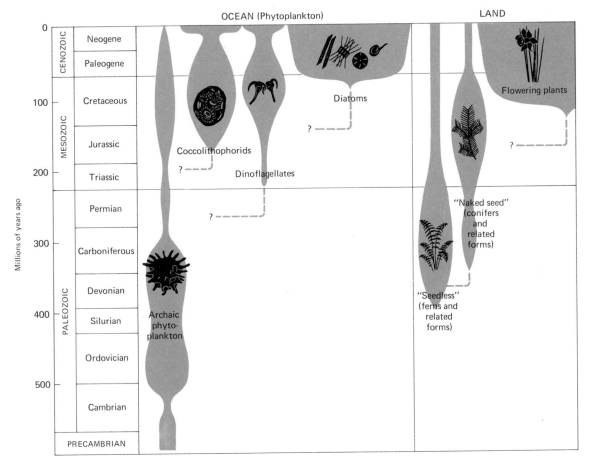

Figure 12.18 *Mesozoic expansions of the plant kingdom. The width of the vertical columns indicates the approximate abundance of each group. In the ocean (left), three groups of small floating plants expanded dramatically to replace the archaic phytoplankton of Paleozoic time. On land (right), flowering plants arose and quickly spread to dominance.*

less, fernlike forms along with primitive seed-bearing relatives of modern conifers (Figure 12.18). These groups continued as the principal land plants until Early Cretaceous time when the more advanced *flowering plants* rapidly expanded to dominance over much of the land surface. This late Mesozoic radiation of flowering plants continued through the Cenozoic Era so that today seedless and primitive seed-bearing plants, such as ferns and conifers, occur only in marginal environments that are cold, wet, shady, or otherwise less desirable than are the vast areas dominated by flowering plants.

Equally dramatic changes took place in land animal life during the Mesozoic Era. As the era dawned in Early Triassic time, two groups of large, superficially alligator-like reptiles were the most advanced land vertebrates (Figure 12.19). These primitive reptiles were to give rise, during Middle and Late Triassic time, to more advanced

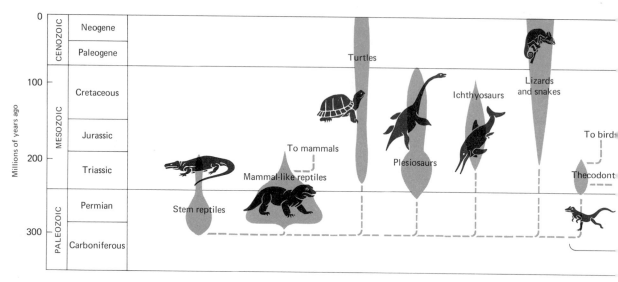

Figure 12.19 *The geologic record of amphibians and reptiles. The width of the vertical columns indicates the approximate abundance of each group. The stem, mammal-like, and thecodont reptiles of Triassic time were replaced, during the Jurassic and Cretaceous Periods, by a host of land-dwelling "dinosaurs," flying pterosaurs, and ocean-dwelling ichthyosaurs and plesiosaurs. Turtles, lizards, snakes, and crocodiles arose at about the same time and are the only major reptile groups to survive into the Cenozoic Era.*

groups that subsequently were to dominate the land. One group, the **archosaur reptiles,** were the principal land animals of Jurassic and Cretaceous time. The other group, the **mammals,** remained small and inconspicuous during Mesozoic time but, as we shall see later, expanded to dominate the land during the Cenozoic Era.

The earliest archosaur reptiles of Late Triassic time were relatively small, superficially lizardlike forms that had the advantage of an upright posture that permitted greater agility and mobility than did the more sprawling posture of their reptilian ancestors. These early archosaurs, known as *thecodonts,* were to give rise to a diverse host of still more advanced reptiles. Two of these groups comprise the familiar "dinosaurs" that were the dominant plant-eating land vertebrates of Jurassic and Cretaceous time. Along with these plant-eating dinosaurs, one group of predatory, carnivorous dinosaurs also arose to prey on the plant-eating forms. Although many dinosaurs remained relatively small animals, some developed into the familiar giants that were the largest land animals ever to evolve (Figure 12.20).

In addition to land-dwelling dinosaurs, several groups of large water-dwelling and flying reptiles were also abundant in Jurassic and Cretaceous time (Figure 12.19). Along with the dinosaurs, these groups became extinct because of some as yet unidentified environmental changes that ended the Mesozoic Era. These extinctions brought to a close the long interval of reptile domination

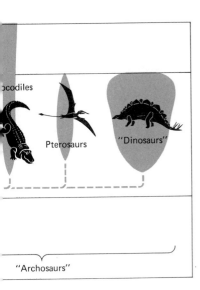

crocodiles

Pterosaurs

"Dinosaurs"

"Archosaurs"

of the land; in the succeeding Cenozoic Era two vertebrate groups that arose from early Mesozoic reptiles, the birds and mammals, expanded to become the dominant larger land animals.

12.6 Mesozoic Environments

Just as in Paleozoic time, the patterns of preserved sedimentary rocks deposited during the Mesozoic Era provide many clues to the earth's changing environments. During Triassic and much of Jurassic time, large parts of the newly separated continental masses stood above sea level with the result that nonmarine sediments covered wide areas. During the Cretaceous Period, in contrast, shallow seas again covered much of the surface of the continents, leading to the deposition of thin but widespread layers of marine sediments. Aside from the more modern aspect of their fossil remains, these marine sediments resemble those deposited during the many intervals of Paleozoic time when seas also covered much of the land surface. The widespread nonmarine sediments of Triassic and Jurassic time present, however, an environmental puzzle.

Unlike most nonmarine sediments forming today, which are usually gray or brown in color, Mesozoic floodplain accumulations of sand and mud are predominantly *red* and, for this reason, are referred to as **redbeds.** Redbeds reached their zenith in Mesozoic time, but were also widespread in late Paleozoic time. In contrast, they are relatively uncommon before the Devonian Period and in Cenozoic deposits.

The red color of redbeds is caused by finely dissipated particles of the iron oxide mineral *hematite,* and the difficulties in understanding the origin of redbeds relate to the chemical conditions under which hematite is formed and preserved. Hematite is the most oxygen-rich of several common iron oxide minerals and forms when oxygen-poor iron-bearing minerals, occurring in either igneous or sedimentary rocks, are weathered by long exposure to atmospheric oxygen. Such conditions exist today in some upland tropical regions where red, hematite-rich soils are forming from intense weathering of the underlying bedrock; similar conditions, covering larger areas of the land, may have provided the original source of the hematite in the Mesozoic redbeds (Figure 12.21).

Although intensive tropical weathering provides a reasonable explanation for the *origin* of the hematite pigment of many redbeds, there are difficulties in under-

(a)

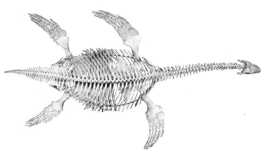

Figure 12.20 Typical Mesozoic reptiles. Fossil skeletons (below) and reconstructions (above) of: (a) plesiosaurs (left) and ichthyosaurs (right); (b) a pterosaur (a condor, the largest present-day flying bird, is shown below for scale); (c) plant-eating (left) and carnivorous (right) dinosaurs. (a, b, lower c: American Museum of Natural History; c, upper: Knight reconstruction, Field Museum of Natural History)

(b)

standing how the hematite was *preserved* during transportation and deposition. Hematite forming in tropical regions today is almost always converted to less oxygen-rich iron minerals (which are brown or yellow in color) as it is transported by streams and deposited in flood plains. This change takes place as the oxygen of the hematite is transferred to oxygen-poor compounds, especially organic materials, in the water and surrounding sediment. Ancient redbeds therefore appear to have required an unusually oxygen-rich environment for their transportation and deposition, an environment unlike any that is common today. The exact conditions that would produce such an environment remain one of the unsolved problems of earth history.

(c)

Figure 12.21 Typical Mesozoic redbeds (exposed in northern Arizona). The exact conditions of origin of these widespread hematite-rich deposits are unknown (see text). (Josef Muench)

Redbeds, like glacial sediments and coal deposited on land, or evaporites and carbonates deposited in the sea, show cyclic changes in abundance through Paleozoic and Mesozoic time. At certain times, one or more of these sediments were particularly widespread, while at other times they were rare or absent. The underlying cause of these cyclic changes must have been variations in some large-scale environmental phenomena operating over much of the earth's surface, and *changing climates* have for many years been evoked as their most probable cause. Carbonate and evaporite deposition, for example, should be favored by widespread warm and dry climates, redbeds and coal deposits by warm and humid climates, and, of course, glacial deposits by cold and moist climates. Recently, however, such strictly climatic interpretations of past environmental cycles have been questioned by some workers who suspect that past changes in the distribution of atmospheric heat and moisture may have been less important in controlling past environments than were changes in the composition of the atmosphere, particularly fluctuations in the amount of atmospheric oxygen. Like the more traditional climatic explanations of environmental cycles, this hypothesis remains highly speculative.

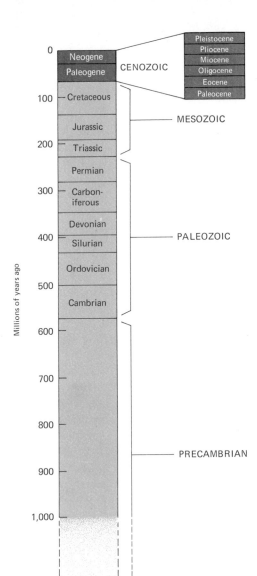

Figure 12.22 *The Cenozoic interval of earth history (color).*

12.7 The Development of Present-Day Geography

When the Cenozoic Era began, about 70 million years ago (Figure 12.22), the Mesozoic breakup of Laurasia and Gondwana had already led to continental masses with about the same shape as today, but their relative positions differed. During this most recent phase of earth history, the changing positions of the continents can be reconstructed with increasing accuracy from the age relations and structures of ocean-floor rocks. Figure 12.23 summarizes our present understanding of these Cenozoic plate movements that have led to the present continental configuration. Note, particularly, the progressive widening of the Atlantic Basin, and the consequent overriding of a former oceanic ridge system by western North America, an event that caused a general elevation of the land surface and intensive marginal deformation (Figure 12.24). At the same time, Africa and India have moved against the southern margin of Eurasia causing the deformations of the Alpine-Himalayan mountain system. The Himalayan Plateau, the highest region of the present continents, appears to have been caused by an unusual doubling of the thickness of the continental crust as the Indian continental fragment was partially overridden by the Asian plate.

Although the direction and general paths of Cenozoic plate motions are reasonably clear, there are still uncertainties about the *rates* at which the movements have taken place. Evidence from dating of ocean-floor rocks suggests that such movements have gone on continuously—at a slow and relatively constant rate. In contrast, mountain-building deformations on the continents, themselves caused by plate movements, appear to have been much more intense at certain intervals, suggesting that the rate of plate motion has varied during the era. Much research is now being directed toward resolving this conflict, for a knowledge of the past rates of plate movements is crucial for understanding many aspects of earth history.

The widespread shallow seas that covered much of the land surface during Cretaceous time represent the last interval of large-scale inundation of the land. By early

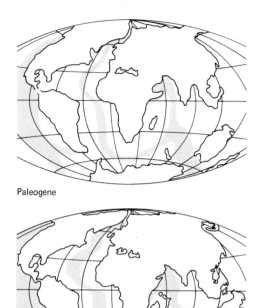

Paleogene

Neogene

Cenozoic ocean floor

Figure 12.23 Cenozoic continental positions. Areas of new basaltic crust added to the ocean floor are shown in color. (After Dietz and Holden, Scientific American, *vol. 223, 1970)*

Figure 12.24 Principal Cenozoic mountain belts (color).

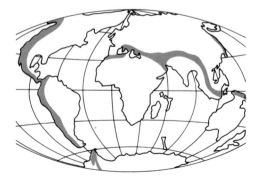

Cenozoic time, the oceans covered only the margins of the continents, but generally extended inland farther than they do today. Throughout the Cenozoic Era there has been a relative rise in the land surface over much of the earth, with shallow seas restricted to progressively narrower bands of the continental margins. As in earlier earth history, the cause of these changes in relative sea level must be interrelated with plate motions, but the exact nature of this relationship remains unknown.

12.8 Changing Life on Land

Life in the oceans was already essentially modern in aspect as the Cenozoic Era began; in contrast, many important changes in land-dwelling life have taken place during the last 70 million years. The rapid expansion of flowering plants, which began in the preceding Cretaceous Period, continued throughout Cenozoic time, leading to the extraordinary diversity of land plants that we see around us on the present-day earth. At the same time, land animal life was also changing dramatically.

The Late Cretaceous extinctions of most of the dominant groups of reptiles set the stage for an explosive evolutionary expansion of mammals early in Cenozoic time. Mammals first evolved from reptilian ancestors during the Triassic Period. These earliest mammals were small, superficially mouselike creatures that had several adaptations not present in most of their reptilian forebears (Figure 12.25). Among the most useful of these were mechanisms, such as insulating fur and a more efficient circulatory system, for keeping the body at a constant temperature regardless of the temperature of the surrounding air. This adaptation makes it possible for mammals to remain active in continuously cold climates or during the winter and on cool nights in temperate climates. Most reptiles, in contrast, have no mechanisms for keeping the body warm as the surrounding air temperatures drop, and can remain active only in relatively warm air.

Because of this adaptive advantage, it is something of a puzzle that mammals did not expand rapidly after they first evolved in early Mesozoic time (Figure 12.26). Instead, they were destined to remain small and relatively uncommon during the Jurassic and Cretaceous Periods when archosaur reptiles dominated the land. Only after the enigmatic reptile extinctions of Late Cretaceous time did mammals diversify and expand to dominate the land.

(a)

(b)

Figure 12.25 The earliest mammals. Fossil jaw bone (a) and reconstruction (b) of a primitive Jurassic mammal about the size of a house cat. Throughout Mesozoic time mammals remained small inconspicuous contemporaries of the dinosaurs. (a: Geological Survey of Great Britain; b: Parker reconstruction, Illustrated London News)

Once the mammalian expansion began it took place rapidly. Early in the Cenozoic Era, several groups of large herbivores and carnivores developed; by the middle of the era, most of the diversely specialized mammalian groups that we know today were established (Figure 12.26). Among these were not only such familiar ground-dwelling herbivores as rodents, horses, camels, pigs, rhinoceroses, and elephants, but also carnivorous cat and bearlike forms and such specialized groups as the tree-dwelling primates, flying bats, and ocean-dwelling seals, porpoises, and whales (Figure 12.26). Indeed, several common mammal groups reached their maximum diversity near the middle of the era and have slowly declined in importance since.

When the reptiles first expanded in late Paleozoic and early Mesozoic time, the present continental fragments were still joined into the two large, interconnected continents of Gondwana and Laurasia. As a result, the early reptiles were able to migrate over much of the land surface and were generally similar everywhere. Likewise, when the *first* mammals arose in early Mesozoic time they were widely distributed over the land surface; in contrast, much of the diversification of mammals took place after the mid-Mesozoic breakup of the continents so that intermingling and migration of mammals between continents was often impeded. As a consequence, mammalian history has followed a somewhat different course on several continents.

North America, Africa, and Eurasia were interconnected through much of the Cenozoic Era and most of the familiar present-day mammal groups arose and were continuously abundant on those continents. In contrast, South America and Australia were isolated from the

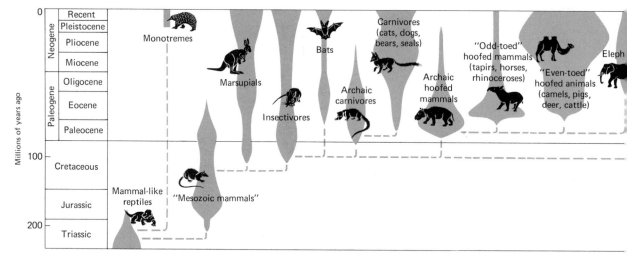

Figure 12.26 *The geologic record of mammals. The width of the vertical columns indicates the approximate abundance of each group. The small, superficially mouselike "Mesozoic mammals" gave rise, in Late Cretaceous and Paleogene time, to the great diversity of present-day mammalian types. Several groups of archaic carnivores and herbivorous hoofed forms that were abundant in Paleogene time are now extinct.*

other continents throughout much of Cenozoic time and each developed its own peculiar mammalian fauna. Mammal history in South America is complicated by a late Cenozoic linkage with North America, which permitted mixing of the unique South American mammals with more advanced forms from the north. Australia, on the other hand, has been continuously isolated and has retained its distinctive mammals. All of these are descendants of primitive *marsupial* ancestors that differed from other mammals principally in their less advanced mode of reproduction. The Australian expansion of these early marsupials led not only to such distinctive mammals as

Figure 12.27 *Two unique primate adaptations, the grasping hand (above) and stereoscopic vision (below), both of which originated to facilitate tree dwelling. Most mammals can grab only by digging in with claws (left), but the opposable primate thumb allows a precise grip. The overlap of right and left eye vision is narrow in most mammals (left), but broad in primates.*

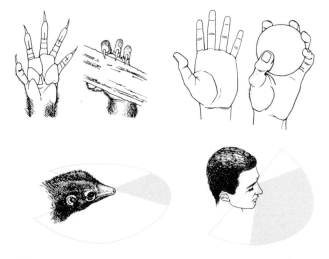

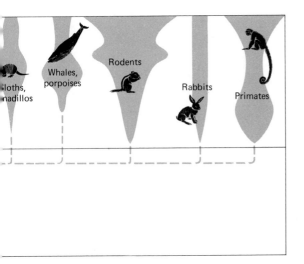

the kangaroo, but also to forms which mimic, in external appearance and habit, such groups as the dogs, cats, and bears that arose on the larger, interconnected continents.

So far we have said very little about the history of the *primates*, the mammalian group that is perhaps of greatest interest for it includes our own species *Homo sapiens*. The first primates of early Cenozoic time diverged from their less-specialized, mouselike ancestors by developing adaptations for living in trees rather than on the ground. Among these adaptations were flexible fingers and an opposable thumb for securely grasping tree limbs, and a shift in the position of the eyes toward the front of the face to afford precise, stereoscopic vision, a prime necessity if a tree dweller is to avoid fatal falls (Figure 12.27). These two key adaptations are characteristics of all primates, including man; indeed, they provided the basis for man's later evolutionary success on the ground, for the grasping hand and precise vision proved ideal for shaping and using tools.

During early Cenozoic time the only primates were several relatively small, primitive types known as **prosimians** ("pre-monkeys") (Figure 12.28). Fortunately we understand a great deal about these earliest primates because some of their relatively unchanged descendants still

Figure 12.28   The geologic record of primates. The width of the vertical columns indicates the approximate abundance of each group. The larger modern apes and men are believed to have arisen from tailless, upright Miocene primates called dryopithecines. The three principal stages in man's evolution are Ramapithecus, Australopithecus, *and* Homo.

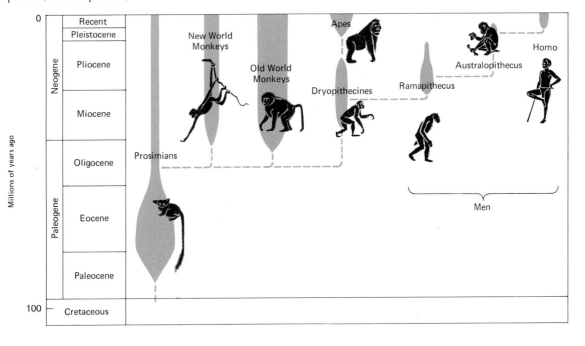

(a)

(b)

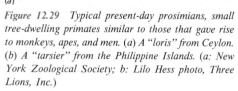

Figure 12.29 Typical present-day prosimians, small tree-dwelling primates similar to those that gave rise to monkeys, apes, and men. (a) A "loris" from Ceylon. (b) A "tarsier" from the Philippine Islands. (a: New York Zoological Society; b: Lilo Hess photo, Three Lions, Inc.)

survive in parts of Africa and southeast Asia (Figure 12.29). In mid-Cenozoic time, three groups of larger and more advanced primates arose from these prosimian ancestors (Figure 12.28). One of these groups, the "new world monkeys," includes the monkeys of South America, all of which developed relatively independently on that isolated continent. A second group of "old world monkeys" developed at the same time in Africa and became widely distributed in both Africa and southern Eurasia. It is, however, the third and final group of advanced primates that arose in mid-Cenozoic time, the **hominoids,** that is of greatest interest to us, for it includes not only our own species but also the four surviving groups of apes (chimpanzees, gorillas, orangutans, and gibbons) and several extinct species of apes and men.

The earliest hominoids of mid-Cenozoic time, called **dryopithecines,** were relatively small animals, about 2 to 3 ft high, which differed from their monkey relatives in lacking a tail and in having a more upright posture (Figure 12.28). Both of these features were probably adaptations for spending more time on the ground than did their mostly tree-dwelling monkey contemporaries. The dryo-

430

pithecines, in turn, gave rise to two main groups of descendants. One group, of less interest to us, led to the larger modern apes; the other led to the earliest close relatives of man (Figure 12.28).

The oldest fossil remains that are clearly manlike are found in sedimentary rocks of Late Miocene and Early Pliocene age and are thus about 10 to 15 million years old. Most of these, unfortunately, are only teeth or fragmentary pieces of jaws and skulls containing teeth (teeth, being the hardest and the most resistant part of the vertebrate skeleton, are the most common type of vertebrate fossil). Even these scattered fragments are sufficient, however, to show that close relatives of modern man have existed for at least 10 million years. Fortunately, Late Pliocene and Pleistocene sediments contain better-preserved bones, skulls, and, occasionally, even complete skeletons of ancient man that give a more complete record of his later evolutionary history. That history is summarized in Figures 12.28 and 12.30.

Modern man differs from present-day apes and the ancestral dryopithecines principally in having a much larger brain and higher intelligence, and the fossil record documents a progressive increase in brain size in late Cenozoic man. Apparently this change was closely interrelated with man's increasing use of carefully fashioned tools, for there is an abundant record of ancient stone tools that show a progressive increase in quality of workmanship (Figure 12.31).

Through most of the 10 million years or so of man's existence he was apparently a migrant gatherer of wild plants and hunter of wild animals which he killed with stone or stone-tipped weapons. In addition, stone tools were used for fashioning shelter, clothing, and tools from animals skins and bones; some present-day groups in Australia and Africa still retain this mode of life. About 7,000 years ago—an extremely short time when measured against the scale of geologic history—some human societies began to domesticate animals and plants so that food could be produced locally without the necessity for migrant hunting, and thus began the first settled agricultural communities. At about the same time, pottery vessels and highly polished stone tools were first used, respectively, for storing and preparing food. In addition, as a result of the efficiency of domestic agriculture, an increased specialization of labor was possible; some men could be spared from the demands of providing food to become

(a)

(b)

Figure 12.30 *Pleistocene men, known from many fossil skulls and skeletal bones. Fossil skulls (below) and reconstructions (above) of:* (a) Australopithecus. *Early Pleistocene men that had the upright posture of modern man but a brain only about half as large;* (b) Homo erectus, *Middle Pleistocene men with a brain intermediate in size between* Australopithecus *and their modern descendents,* Homo sapiens. (*Reconstructions from J. Augusta and Z. Burian,* Prehistoric Man, *Spring Books, London; skulls: American Museum of Natural History*)

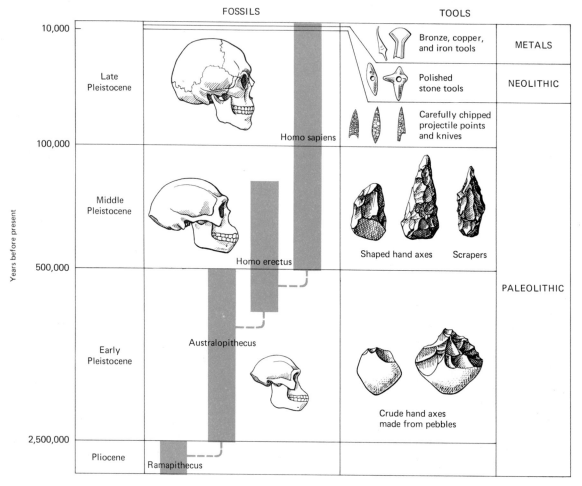

Figure 12.31 *The development of tool making by Pleistocene men. Chipped stone ("Paleolithic") tools of increasing refinement occur throughout the epoch. Polished stone ("Neolithic") tools began only about 10,000 years ago and were rather quickly replaced by metal tools.*

artisans, craftsmen, tradesmen, and priests. It was not long after this that metals were first used for tools and that writing was developed in Mesopotamia and Egypt, an event that led to the earliest written human history bringing an end to the much longer "prehistoric" phase of man's history.

12.9   Ice Shapes the Modern Environment

Unlike the extensive carbonates, evaporites, and coal deposits of the Paleozoic Era, or the widespread redbeds of Mesozoic time, sedimentary rocks deposited during the Cenozoic Era are similar in kind and abundance to those forming today and present few environmental puzzles. We noted earlier that the era is marked by a gradual withdrawal of the seas along the continental margins. For this

Man
and the
Evolving Earth

Just as the search for iron has been a major stimulus to the study of the Precambrian earth, so has much of our understanding of Phanerozoic earth history come from the exploration for mineral fuels which, along with iron, are the most essential resources of all industrialized societies. Both coal and petroleum, the fuels that together supply the bulk of man's present energy needs, are the altered remains of ancient organisms and thus have been formed only since the great expansion of life early in Cambrian time.

Coal deposits are compacted and altered accumulations of ancient land plants and are found in sedimentary rocks of all ages since the first rise of land floras in Devonian time. Coal was the first mineral fuel to be consumed on a large scale and still provides more than half the world's energy. Although enormous quantities have already been extracted, coal-bearing sediments are so extensive that remaining reserves could supply man's energy needs for at least another 250 years. Unfortunately petroleum, which is generally a cleaner and more easily transported fuel, is far less abundant.

Most petroleum is believed to have been derived from small marine animals and plants. When buried and compacted in sedimentary accumulations, their organic remains become altered to less complex liquid or gaseous compounds which may then be trapped in porous surrounding rocks, principally sandstones or limestones, to form petroleum reservoirs. Such reservoirs are most common in Cenozoic rocks and decrease in abundance with increasing age.

Because petroleum accumulations usually are smaller, more deeply buried, and more erratically distributed than are coal deposits, they are much more difficult to find and evaluate. Careful estimates of both known and probable petroleum reserves, however, suggest that only a 50- to 100-year supply remains. For this reason countries, such as the United States, that depend heavily on oil and gas as fuels will be forced relatively soon to seek alternative energy sources. An increasing consumption of more plentiful coal will undoubtedly be one result of declining petroleum production, but still other alternatives exist.

Among the most promising is one based on the heat energy produced by the radioactive decay of uranium and related elements. Because relatively small amounts of uranium fuel can produce large amounts of energy, such "nuclear power" is already replacing coal and petroleum for the production of an increasing fraction of the world's electricity. The main difficulty with nuclear power lies in disposing of its waste products, particularly hot waters which can endanger life in adjacent lakes and rivers, and radioactive wastes which must usually be buried with a subsequent risk of ground-water contamination.

There is, in addition, a far larger and as yet untapped energy source which lacks this disadvantage—the direct radiation of the sun. Calculations show that the solar energy reaching a desert area only about one-tenth the size of Arizona could supply *all* the energy needs of the United States could it somehow be trapped and stored. Many speculative schemes for such direct use of the sun's energy have been proposed, and some of these may become increasingly practical as petroleum reserves decline.

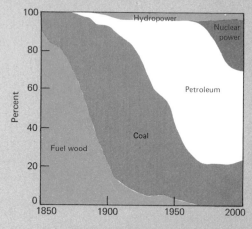

Past and projected energy sources in the United States. (*Modified from Cook,* Scientific American, *vol. 224, 1971*)

A coal mine in Alaska. The coal beds are clearly visible as dark bands along the valley wall. (U.S. Bureau of Mines)

Oil wells in central Kansas (below, left) and liquid petroleum flowing from a "wellhead" (below, right). (Standard Oil Co., N.J.)

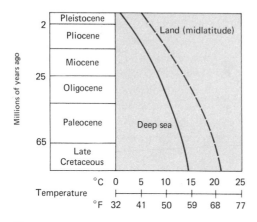

Figure 12.32 Cenozoic decline of average world temperatures, based on animal-plant distributions and oxygen nuclides. (*After Dott and Batten,* Evolution of the Earth, *1971, from various sources*)

Figure 12.33 Alternating glacial and interglacial climates of Pleistocene time, based on fossil and sediment distributions, and oxygen nuclides. (*After Ericson and Wollin,* Science, *vol. 162, 1968*)

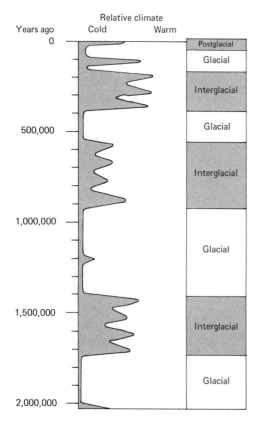

reason, Cenozoic marine sedimentary rocks are mostly confined to the margins of the present continents while the continental interiors have large areas covered by Cenozoic nonmarine deposits. The most environmentally significant of these nonmarine sediments are extensive glacial deposits that cover much of northern North America and Eurasia.

There are many indications that the earth's climates became progressively cooler throughout the Cenozoic Era. In particular, the distribution of fossilized Cenozoic animals and plants, and the oxygen nuclide ratios of Cenozoic fossil shells both suggest that the extensive ice sheets of Pleistocene time were but the most recent evidence of a longer trend toward cooler climates (Figure 12.32). As we noted in Chapter 7, such general climatic change must be caused by changes in the earth's overall budget of solar energy interacting with changes in the positions of the continents and the relative level of land and sea; the exact nature of these interactions remains, however, uncertain.

Whatever their precise cause, the extensive ice sheets of late Cenozoic time have profoundly affected the earth's environments over the past several million years. Careful studies of the distribution and ages of Pleistocene glacial sediments, combined with studies of changes in the oxygen nuclide ratios and fossil remains deposited in the deep ocean floor over the past two million years, show that there were at least four principal intervals of widespread ice sheets during Pleistocene time. These "glacial" intervals were separated by "interglacial" periods during which climates were warmer and ice sheets less extensive (Figure 12.33). The most recent glacial interval ended, in terms of the long span of geologic time, only yesterday. About 20,000 years ago ice sheets covered much of Canada and the northeastern United States, as well as most of northern Europe. A dramatic climatic warming began about 16,000 years ago; by about 9,000 years ago the former ice sheets had shrunk to approximately their present extent; that is, they were confined principally to Greenland in the northern hemisphere and Antarctica in the southern hemisphere.

The expanding and contracting ice sheets of late Cenozoic time led to great changes in the distribution of the earth's animals and plants, which migrated back and forth with the moving ice and changing climates. At the time of maximum ice sheet coverage, for example, much of the central United States beyond the ice margins was barren tundra; today, 20,000 years later, the same regions support lush temperate forests and grasslands. Profound environmental changes also took place along the margins of all the continents as the ice sheets expanded and contracted, for the amount of water removed from the oceans and precipitated as snow to make the ice sheets was sufficient to lower sea level more than 100 m (328 ft), exposing a large part of the presently submerged continental shelves. Similarly, if the present Greenland and Antarctic ice sheets were to melt, as apparently happened during some of the earlier interglacial intervals, sea level would be raised by about 100 m, inundating most low-lying coastal regions.

The rapidly changing environments caused by the contraction and expansion of late Cenozoic ice sheets appear to be relatively unusual events in geologic history for, as we have seen, similar widespread glacial conditions have occurred only once before during Phanerozoic time. It is for this reason that, in many respects, the environmental history of the past few million years is not typical of most of geologic time. These atypical environments had, however, one extraordinarily important result—it was during this interval that most of the evolution of human culture, from early hunters and gatherers using crude stone tools, to the complex cultures of modern man, took place.

SUMMARY OUTLINE

The Paleozoic Earth

12.1 *Paleozoic geography:* through much of Paleozoic time the present continents were assembled into two large supercontinents, Laurasia and Gondwana; shallow seas covered much of the continental surface early in the era but became progressively smaller throughout the era.

12.2 *The diversification of life:* the sea-dwelling algae and invertebrate animals of Cambrian time gave rise to land-dwelling plants, snails, insects, amphibians, and reptiles that dominated the long-barren continental surfaces by the close of the era.

12.3 *Paleozoic environments:* sedimentary carbonate deposits were unusually abundant in shallow Paleozoic seas; many of these are dominated by dolomites, the origin of which is uncertain; widespread coal deposition over much of the continental surface in late Paleozoic time is also an environmental puzzle.

Mesozoic Reorganization

12.4 *The separation of the continents:* plate motions progressively broke Laurasia and Gondwana into the six present-day continental masses during the era; seas were widespread over the continental surfaces only in Cretaceous time.

12.5 *The Mesozoic expansion of life:* severe extinctions near the end of Paleozoic time were followed by an expansion and reorganization of life during the Mesozoic; new groups of planktonic plants came to dominate the ocean and flowering plants the land; a diversity of reptilian types arose.

12.6 *Mesozoic environments:* sediments of the era were similar to those of the present-day except for the wide deposition of red flood-plain deposits whose mode of origin is uncertain.

Cenozoic Modernization

12.7 *The development of present-day geography:* the continents have changed position but not their general form through the era; deformation of their moving margins produced the dominant present-day mountain belts.

12.8 *Changing life on land:* extinctions of most reptile groups at the close of the Mesozoic led to a rapid expansion of land mammals early in the era; among those were simple, tree-dwelling primates that were to give rise to man in latest Cenozoic time.

12.9 *Ice shapes the modern environment:* climates became progressively cooler throughout the era, a trend that culminated in repeated expansion and contraction of continental ice sheets, the most recent of which retreated only 20,000 years ago.

ADDITIONAL READINGS

Beerbower, J. R.: *Search for the Past,* Prentice-Hall, Inc., Englewood Cliffs, N.J., 1968. A well-written intermediate text on the history of life.

*Cloud, P. E., Jr. (ed.): *Adventures in Earth History,* W. H. Freeman and Co., San Francisco, 1970. An excellent anthology of research and review articles on the earth's past.

Dott, R. H., Jr., and R. L. Batten: *Evolution of the Earth,* McGraw-Hill Book Company, New York, 1971. An up-to-date introduction to all aspects of earth history.

Dunbar, C. O., and K. M. Waage: *Historical Geology,* John Wiley & Sons, Inc., New York, 1969. A standard introduction, emphasizing the Phanerozoic history of North America.

Garrels, R. M., and F. T. Mackenzie: *Evolution of Sedimentary Rocks,* W. W. Norton & Company, Inc., New York, 1971. An imaginative intermediate text on the earth's sedimentary history.

Kummel, B.: *History of the Earth,* W. H. Freeman and Co., San Francisco, 1970. A standard introduction, emphasizing worldwide Phanerozoic history.

*McAlester, A. L.: *The History of Life,* Prentice-Hall, Inc., Englewood Cliffs, N.J., 1968. A brief introduction.

*Pilbeam, D.: *The Ascent of Man,* The Macmillan Co., New York, 1972. A lively introduction to human evolution.

Raup, D. M., and S. M. Stanley: *Principles of Paleontology,* W. H. Freeman and Co., San Francisco, 1971. An intermediate-level survey of the principles used in interpreting fossil organisms.

Romer, A. S.: *The Procession of Life,* The World Publishing Company, Cleveland, 1968. An authoritative and readable introduction to life history.

Stokes, W. L.: *Essentials of Earth History,* Prentice-Hall, Inc., Englewood Cliffs, N.J., 1973. A standard introduction, with good discussions of the principles of historical interpretation.

* Available in paperback.

PART 5

The Earth in Space

Parts 1 through 4 have been primarily concerned with the earth as a unit within itself, rather than as one among many accumulations of matter that exist throughout the universe. Part 5 considers the earth in the larger context of these materials that lie beyond it in space, especially those bodies—the sun, moon, and planets—that are closely related to it in origin and history.

Chapter 13 discusses the *sun and its satellites,* stressing the structure and origin of the sun and its relations to the planets, comets, and other materials held by its gravity to form the solar system.

Chapter 14 considers the earth's nearest and best understood neighbor in space—*the moon*.

Chapter 15 surveys *the planets and beyond* with emphasis on similarities and differences between the earth and its planetary neighbors; the chapter closes with a brief introduction to the vast reaches of space that lie beyond our solar system.

The Sun and Its Satellites

The earth is one of nine relatively small and cool accumulations of matter, the *planets,* that orbit in the gravitational attraction of a much larger, burning accumulation, the *sun*. In a still larger perspective, the sun is merely one among billions of similar burning accumulations, called *stars,* that occur throughout the universe. Stars and planets, in turn, are but two of the five principal components of the universe (Figure 13.1). A third component occurs scattered between the stars and planets as relatively small solid particles and thin, widely separated, gas molecules. These are called *interstellar* (''between-star'') *gases and dusts*. A fourth component is *electromagnetic radiation,* principally visible and ultraviolet light, that arises from burning stars to penetrate the entire universe. The final component is the high-energy nuclear fragments known as *cosmic rays* that occur throughout the universe and probably also originate in stars.

From the viewpoint of understanding the earth itself, electromagnetic radiation and cosmic rays are of interest principally as they approach the earth from space and interact with its atmosphere. In this context they were discussed in Sections 5.4 and 10.5. Conversely, the billions of stars, planets, and interstellar dust and gas accumulations that lie beyond the sun's gravitational attraction, although fascinating subjects for study in their own right, tell us relatively little about the earth. It is the earth's own star, the sun, with its satellite planets, dusts, and gases that have determined the history and nature of the earth, and they will be our principal concern in this chapter and those that follow.

THE SUN

The planets, dusts, and gases of the *solar system* orbit around the enormous central sun, which contains 99.9 percent of the mass of the system and provides the gravi-

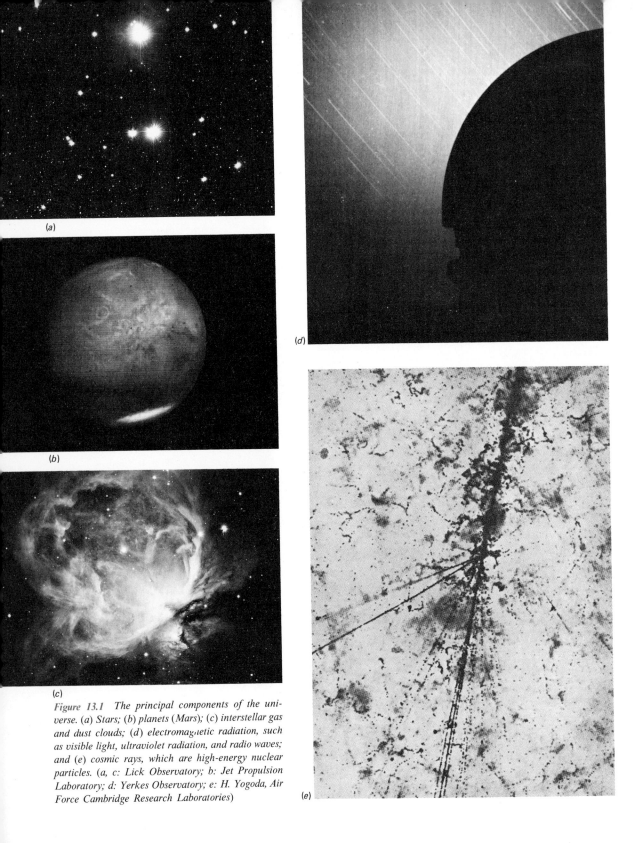

(a)

(b)

(c)

(d)

(e)

Figure 13.1   The principal components of the universe. (a) Stars; (b) planets (Mars); (c) interstellar gas and dust clouds; (d) electromagnetic radiation, such as visible light, ultraviolet radiation, and radio waves; and (e) cosmic rays, which are high-energy nuclear particles. (a, c: Lick Observatory; b: Jet Propulsion Laboratory; d: Yerkes Observatory; e: H. Yogoda, Air Force Cambridge Research Laboratories)

tational attraction that holds it together. In addition to providing this gravitational binding force, the sun's burning interior, fueled by nuclear fusion, creates vast quantities of energy. This energy, when liberated as electromagnetic radiation and rapidly moving nuclear particles, permeates the solar system and profoundly influences all its matter. For these reasons, knowledge of the sun is central to understanding the earth, planets, and all other bodies that surround it.

13.1 Solar Structure

Many properties of the sun have long been known from astronomical calculations. From observations of the orbits of the planets, the sun's gravitational attraction and total mass were determined whereas its diameter, about 1,400,000 km (850,000 mi), was computed from direct angular measurements. From the mass and diameter, its average density was easily determined, which turned out to be only slightly greater than that of water (1.4 gm/cm^3). In spite of its enormous total mass, an average cubic centimeter of the sun therefore weighs far less than does an average cubic centimeter of the solid earth (1.4 versus 5.5 gm).

One of the reasons for the sun's low density is that it is composed predominantly of hydrogen, the lightest element. Theoretical calculations show that the sun's low density is also due to the fact that even its massive interior is made up of matter held in the *gaseous* state by enormously high temperatures. In smaller quantities, gases at such extreme temperatures would rapidly expand and dissipate. In the sun and all other stars, however, this is generally prevented by the opposing gravitational attraction of their enormous mass.

Modern telescopic observations of the sun made from the earth's surface, and also from balloons and artificial satellites, have revealed a great deal about the structure of its heated gases. Specially designed solar telescopes show its visible surface to be a seething, boiling mass of gases with a characteristic "granular" appearance created as hot gases first rise to the surface, and are then pulled by gravity back toward its center (Figure 13.2). This alternate rise and fall takes place in *convection cells* similar to those in a pot of boiling water, and it is the boundaries of such cells that give the surface its granular appearance. The visible, granular surface is called the **photosphere** and is of great importance, for it marks the

(a)

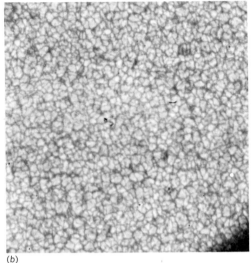

(b)

Figure 13.2 The solar photosphere, the visible surface of the sun. (a) Overall view, showing large-scale granulation; (b) enlarged view showing small-scale granulation caused by convection of solar gases. (a: Observatoire de Paris; b: Project Stratoscope, Princeton University)

Figure 13.3 The solar chromosphere, a narrow zone of seething, transparent gases that overlies the opaque photosphere. (Sacramento Peak Observatory, Air Force Cambridge Research Laboratories)

zone where the densely packed gases of the solar interior expand enough to become transparent to visible light. This expansion creates agitation, turbulence, and energy release, with the result that most of the light and other radiation emitted into space by the sun originates at the photosphere.

Most of the gases that reach the photosphere in convection cells are returned to the interior by the force of solar gravity. A small proportion, however, escapes from the surface to create a transparent atmosphere of hot gaseous particles that surrounds the more dense, opaque gases beneath the photosphere. Telescopic observations

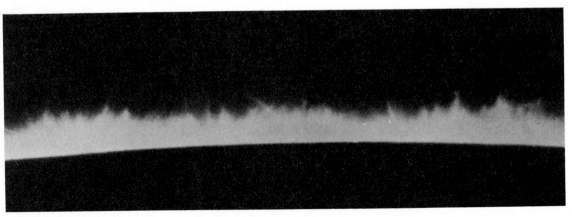

Figure 13.4 *The solar corona, or atmosphere. Only the dense, innermost part of the corona is visible in the photograph; its thin outer gases extend throughout the solar system.* (*Lick Observatory*)

show this solar atmosphere to be made up of two parts. Just above the photosphere is a narrow zone of dense, red-hot gases that seethe and splash upward from its turbulent convective granules. This narrow transitional zone of the solar atmosphere is called the **chromosphere** (Figure 13.3).

Above the chromosphere lies the much larger and less dense portion of the solar atmosphere known as the **corona.** Just as in the earth's atmosphere, the density of the solar corona decreases rapidly away from the surface. Only the innermost part is dense enough to be visible from the earth and even that can only be seen when the much brighter light coming from the photosphere is obscured in special telescopes, or during a solar eclipse (Figure 13.4). Beyond the visible corona, progressively thinner gases of the solar atmosphere extend outward past the earth's orbit into the farthest reaches of the solar system. These widely spaced gas particles make up the *solar wind* which, as we saw in Chapter 5, strongly influences the shape and structure of the earth's outermost atmosphere.

13.2 Solar Disturbances

So far our description has implied that the turbulent solar gases have a rather constant structure and, indeed, for spans of many months or even years this may be the case. Such intervals, known as periods of *quiet sun,* alternate,

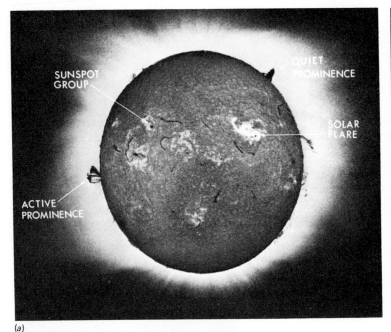

(a)

(b)

(c)

Figure 13.5 *Disturbances of solar structure. (a) General view showing sunspots, flares, and prominences; (b) enlargement of sunspots and adjacent flares; (c) enlargement of a prominence. (a, b: Lockheed Solar Observatory; c: NCAR)*

however, with intervals in which there is considerable disruption in certain aspects of the sun's structure. These periods of *disturbed* or *active sun,* as they are called, reveal much about processes taking place in the solar interior.

The most obvious manifestations of solar disturbances are **sunspots,** dark, cool, roughly circular blemishes that regularly appear on the photosphere (Figure 13.5). Sunspots normally occur in groups, appear first in solar midlatitudes, and then move toward the sun's equator where they dissipate and disappear, usually about 1 to 3 months after they form. When viewed through special telescopes, sunspots can be seen to be dark cavities on the solar surface where, for reasons that are not fully understood, the normal convective motions of the granular surface are inhibited. The number and intensity of observed sunspots reaches a maximum on a regular cycle which averages 11 years in length (Figure 13.6).

Sunspots are the most obvious disturbances of solar structure but are accompanied by other phenomena. Near the sunspots occur **solar flares,** regions of abnormal brightening of the photosphere which last from a few minutes to several hours, and which release unusual amounts of radiant energy and charged nuclear particles (Figure 13.5). During flares the hot, seething gases of the chromosphere frequently extend far out into the less dense corona as

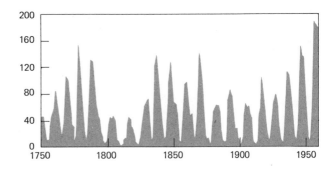

Figure 13.6 *Number of sunspots from 1750 to 1965. Note the regular cycle that averages 11 years in length.*

enormous, spike- or loop-shaped **prominences** (Figure 13.5). At the same time, the corona itself is greatly expanded by the enormous energy released by the flare. Indeed, when some of the charged particles released by flares arrive at the earth in the solar wind, they interact with the charged particles of the earth's upper atmosphere to create *magnetic storms* (see Section 5.9).

Although solar disturbances have been observed for centuries, it is only within the past few decades that theoretical considerations, combined with increasingly refined observations and measurements, have given clues to their cause. It is now clear that they are intimately interrelated with magnetic fields present in the solar gases. When gas atoms and molecules are heated to temperatures of thousands or millions of degrees, the atoms and molecules lose one or more orbital electrons and thus are no longer electrically balanced. Instead, they become charged *ions,* whose motions both create, and are influenced by, magnetic fields (see Section 5.6). Careful measurements of solar radiation show that the sun, like the earth, has an overall magnetic field caused by fluid motions within its dense interior. (In the sun, the field is caused by moving *gas* particles whereas in the earth it is caused by motions of the *liquid* core.) Superimposed on the general solar field are more local magnetic variations that are particularly strong near disturbances. For this reason, sunspots are now believed to be zones where strong magnetic fields *suppress* the motions of charged solar gases, whereas flares and prominences occur when the motions of charged gas particles are *enhanced* by magnetic fields of opposite charge.

If the visible manifestations of solar disturbances are themselves caused by changes in the sun's magnetic structure, then what is the ultimate cause of the magnetic changes themselves? This more basic question is only beginning to be attacked by solar physicists, but one

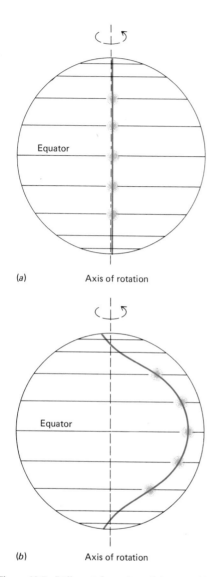

(a)   Axis of rotation

(b)   Axis of rotation

Figure 13.7 Differential rotation of the sun. After a single rotation (a to b), equatorial sunspots can be seen to move more rapidly than do those at higher latitudes.

Equator

Equator

promising idea relates them to the sun's peculiar pattern of rotation. Like the earth, the sun rotates about an axis but, unlike the more rigid earth, solar gases do not all rotate at the same speed. Instead, the gases near the poles require about 35 days to complete one rotation whereas the more rapidly spinning gases near the solar equator reqire only 25 days (Figure 13.7). These differential motions within the dense, charged, solar gases could easily cause complex local magnetic fields that might, in turn, create solar disturbances. The exact nature of such fields is, however, still unknown as is the cause of the puzzlingly regular 11-year cycle of solar disturbances.

13.3   The Source of Solar Energy

So far we have concentrated on the directly observable structure of the sun's surface and atmosphere and have said little about the processes within its hidden interior that release its huge stores of energy. Until the rise of modern nuclear physics there was no known source for the sun's energy, but it is now clear that the solar interior is a nuclear furnace that releases energy in much the same way as do man-made thermonuclear explosions.

Spectrographic measurements of sunlight reaching the earth from the photosphere show that the solar mass is composed predominantly of the two lightest elements—hydrogen, which makes up about 70 percent, and helium, which makes up about 27 percent.[1] Furthermore, theoretical calculations show that at the temperatures and pressures of the solar interior, helium is steadily being produced from lighter hydrogen as four hydrogen nuclei unite to form one nucleus of helium (Figure 13.8). The helium nucleus formed in this fusion process weighs slightly less than the combined weight of the four hydrogen atoms which gave rise to it. This small excess of matter is converted directly to electromagnetic radiation and is the ultimate source of the sun's energy.

THE SOLAR SYSTEM

Most of the mass and energy of the solar system are concentrated in the sun, yet the relatively small amounts of additional material that orbit in its gravitational field

[1] The remaining 3 percent of solar matter is made up of all the other 90-odd elements; the proportions of the elements in this portion resemble the overall composition of meteorites and the earth.

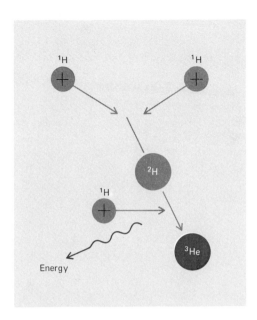

Figure 13.8 Hydrogen "burning" in the sun. Single hydrogen nuclei (protons), made unstable by heat and pressure, first combine to form double hydrogen nuclei; these then unite with a third hydrogen nucleus to form helium-3, with a release of electromagnetic energy.

are far from insignificant, for they include the earth and its planetary neighbors as well as several additional components.

13.4 The Planets

The principal objects held in the gravitational attraction of the sun's enormous mass are the nine planets that orbit around it. These show a regular, geometric spacing away from the sun (Figures 13.9, 13.10, 13.11) and form two distinct groups: the four that are closest to the sun—Mercury, Venus, the earth, and Mars—are small, dense, and largely solid, being surrounded by relatively small amounts of atmospheric gases. These are called the **earthlike** or **terrestrial planets.** In contrast, the **outer** or **major planets**—Jupiter, Saturn, Uranus, and Neptune—are much larger, less dense, and are surrounded by great quantities of atmospheric gases. The outermost planet, Pluto, appears to resemble the terrestrial planets in size and density, but it is so distant from the sun and earth that its properties are uncertain.

All of the planets orbit the sun in the same direction and in approximately the same plane, which is known as the **plane of the ecliptic** (Figure 13.11). In their motions around the sun, however, they move not in precise circles, but in slightly elliptical orbits that place them nearer the sun at some points of their orbits than at others. The time required to complete one orbit, called the *sidereal period,* varies with distance from the sun. Mercury, the innermost

Figure 13.9 Planetary data.

	Planet	Average distance from Sun (earth = 1)	Diameter (earth = 1)	Sidereal Period (earth mo and yr)	Rotational Period (earth hr and days)	Density (water = 1)	Angle of Inclination	Number of Natural Satellites
Terrestrial	Mercury	0.4	0.4	3 mo	59 days	5.2	30° +	0
	Venus	0.7	1.0	7 mo	243 days	5.1	23°	0
	Earth	1.0	1.0	1 yr	24 hr	5.5	23°	1
	Mars	1.5	0.5	2 yr	25 hr	4.0	25°	2
Outer	Jupiter	5.2	11.0	12 yr	10 hr	1.3	3°	12
	Saturn	9.5	9.2	29 yr	10 hr	0.7	27°	10?
	Uranus	19.2	3.7	84 yr	11 hr	1.6	98°	5
	Neptune	30.1	3.5	165 yr	16 hr	2.3	29°	2
	Pluto	39.5	0.5?	248 yr	6 days	?	?	?

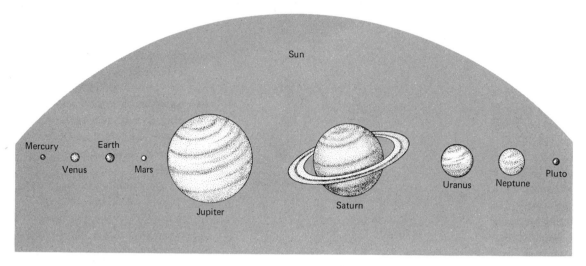

Figure 13.10 Relative sizes of the sun and planets.

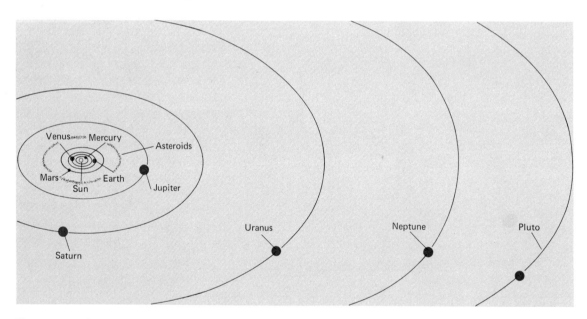

*Figure 13.11 The orbits of the planets and the plane
of the ecliptic; the orbital positions are shown to scale.*

planet, requires only 88 days for a complete orbit whereas
Pluto, the outermost, requires 248 years. The intermediate
earth requires one year of 365 days.

   In addition to orbiting around the sun, each planet
also spins or *rotates* about an axis which defines the *poles*
of the planet. The time required for one rotation, called
the *rotation period,* does not increase regularly away from
the sun as does the sidereal period. Jupiter has the short-

454

est rotation period, 10 hr, and Venus the longest, 243 days. The earth requires one day of 24 hr to complete one rotation. For all of the planets except Venus and Uranus, the direction of rotation is the same as the direction of revolution around the sun (counterclockwise as viewed downward from above the ecliptic plane). For reasons that are uncertain, Venus and Uranus rotate in the opposite direction; such rotation is known as *retrograde motion*.

One of the most important properties of planetary motion is the angle at which a planet's axis of rotation intersects the place of the ecliptic (Figure 13.12a), for this property largely determines the nature and intensity of seasons on the planet as it orbits the sun. When the *angle of inclination* is small, each part of the rotating planet receives about the same amount of solar energy throughout its orbit; when it is large, the distribution of solar energy changes over the planetary surface during different parts of its sidereal orbit, leading to large seasonal effects (Figure 13.12b, c). The angle of inclination (and corre-

Figure 13.12 Planetary angles of inclination (a) and seasonal effects (b, c). The greater the tilt of a planet's axis of rotation with respect to the plane of the ecliptic, the more pronounced its seasonal effects. The seasons shown in b and c apply to the northern (gray) hemisphere and are reversed on the southern (white) hemisphere. On Uranus, with a very steep angle of inclination, the entire winter hemisphere receives almost no solar radiation (c).

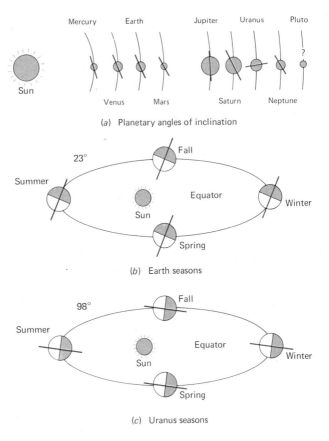

(a) Planetary angles of inclination

(b) Earth seasons

(c) Uranus seasons

sponding seasonal effect) varies from a minimum of 3° on Jupiter, to a maximum of 98° on Uranus; the earth is intermediate with an inclination of 23°.

13.5 Additional Components

The sun and nine planets are the principal objects of the solar system, but they are by no means its only components. Just as the planets are held in orbit around the sun by its gravitational attraction, so too are smaller accumulations of matter, called **planetary satellites,** held in orbit around many of the planets (Figure 13.9). The earth has only one such natural satellite, the moon, whereas Mars and Neptune each have two, Uranus five, Saturn about 10, and Jupiter 12. Mercury, Venus, and Pluto appear to lack satellites.

In addition to the sun, planets, and planetary satellites, there are still other accumulations of matter in the solar system. The regular geometric spacing of the planets shows an anomalous gap between the terrestrial and outer planets. In this zone there occurs not a single, planet-sized accumulation of matter but, instead, a multitude of smaller objects, called **asteroids,** that orbit the sun in a manner similar to the larger planets (Figure 13.11). The visible asteroids range in diameter from less than a kilometer to a maximum of about 800 km (500 mi); some appear to be approximately spherical, but others, including some of the largest, are known from telescopic observations to have highly irregular shapes (Figure 13.13). In all,

Figure 13.13 Names and relative sizes of some of the largest asteroids.

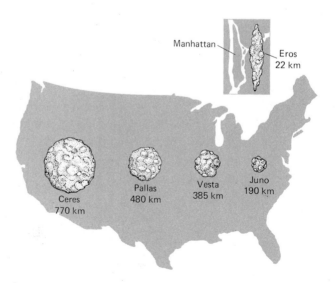

about 30,000 asteroids have been observed from the earth; this total probably includes most of the larger ones, but only a fraction of the smaller bodies in the *asteroid belt,* as the region of the ecliptic plane between Mars and Jupiter is called.

Still other accumulations of matter in the solar system are **comets,** large masses of extremely light, frozen gases made up largely of compounds of hydrogen, carbon, nitrogen, and oxygen. Such frozen masses are believed to be abundant in the outer reaches of the solar system beyond Pluto, the outermost planet. The sun's gravitational influence extends outward at least 1,000 times the distance to Pluto, and thus objects far beyond the visible planets may be held by its gravity. Millions of frozen cometary gas accumulations probably orbit in these cold, distant reaches of the solar system but only rarely do they become evident from earth when, for reasons that are still uncertain, the orbit of one of them becomes disturbed so that it approaches the sun and passes through the inner parts of the system. When this happens they appear as large, rapidly moving, luminous objects in the sky. As they approach the sun the outermost ices are melted and the comet expands enormously in size, leaving behind a long trail of vaporizing gases (Figure 13.14). Some comets are completely vaporized and disappear as they approach the sun. Others swing around the sun and disappear into space in huge elliptical orbits that may bring them back into the inner solar system at long, but predictable, intervals (Figure 13.15).

Besides the larger accumulations of matter seen in planets, satellites, asteroids, and comets, the vast spaces between these objects in the solar system, as throughout the entire universe, also contain smaller gaseous and solid particles, as well as electromagnetic radiation and cosmic rays. Of particular interest are solid particles from interplanetary space that approach the earth, are trapped by its gravitational field, and fall towards its surface. Millions of particles are captured in this way each day, but most are vaporized by the tremendous heat generated as they plunge at high speed into its dense lower atmosphere. A small proportion, however, survive the heat and fall to the surface as *meteorites;* these are of extraordinary interest since they provide direct samples of the interplanetary materials of the solar system. Before looking more closely at the nature and implications of meteorites, however, it will be useful to review briefly some speculations about the ultimate origin of the sun and its satellite materials.

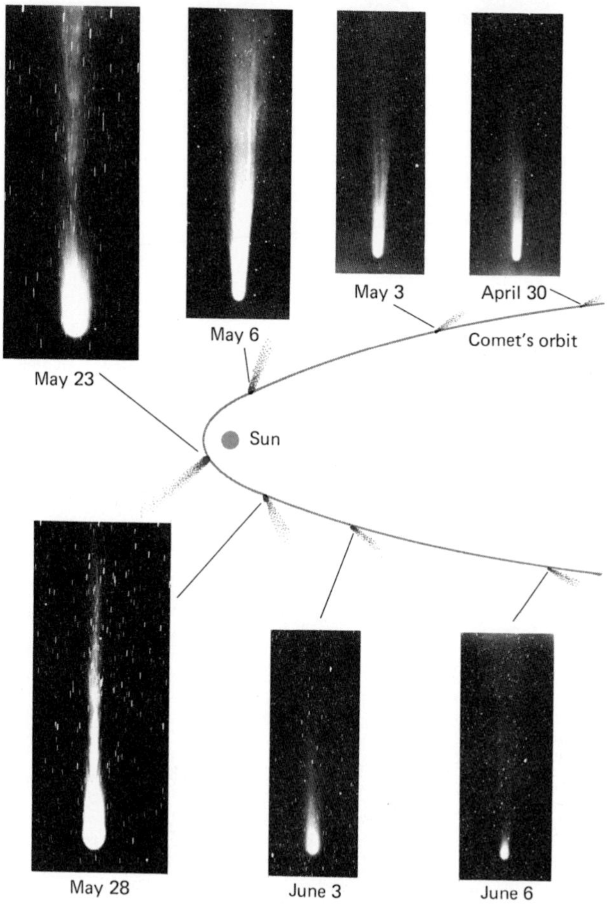

Figure 13.14 *Photographs of a large comet orbiting near the sun. Because of the solar wind, the "tail" of vaporized materials always points away from the sun.*

Figure 13.15 *Probable zone of comet accumulation in the outer solar system. The zone is so distant that its comets cannot be observed, but the orbits of a few become disrupted and irregular (colored path) so that they pass into the inner solar system and are visible from the earth as they are vaporized by the sun.*

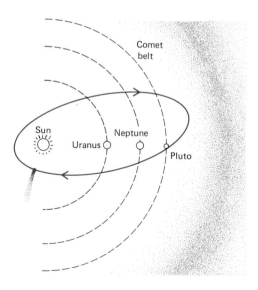

13.6   Origin of the Solar System

Rather surprisingly, it is far easier to decipher the origin and history of the sun itself than to understand the circumstances that gave rise to the planets and other bodies of the solar system. This is true because astronomers can observe, among the billions of visible stars, countless numbers that are similar to the sun but at different stages in their history. By analyzing their color and brightness it has become clear that these *normal stars*, as they are called, originate in enormous clouds made up largely of gaseous hydrogen, but which also contain other gases and heavier dust particles. These massive gas and dust clouds, called *nebulae*, are scattered throughout the universe and some are visible through large telescopes (Figure 13.16).

Astronomers speculate that turbulent eddies sometimes occur within nebulae and tend to concentrate the scattered dust and gases into more dense, swirling

Figure 13.16 A typical nebula, massive clouds of gases and dust that are believed to give rise to stars (see text). (Hale Observatories)

masses. Because of the enormous size of these swirls, the increased concentration of matter within them would lead to strong gravitational forces and cause a still further concentration as gas and dust particles fell towards the center of accumulation. Collision and gravitational compression of these falling particles would create heat energy which, in the early stages of the concentrating cloud, would be dissipated into the otherwise cold and heatless reaches of space. When the condensing mass reached a critical size and density, however, much of this frictional heat could no longer escape. Instead it would become trapped in the center of the spherical cloud and, as a result, the cloud would steadily become hotter.

At first the heating gas cloud, now an embryonic star similar to many that can be observed directly, shines with only a dull red glow (Figure 13.17). Eventually, as the internal temperatures and pressures rise, the hydrogen making up the bulk of the glowing mass begin to fuse into helium atoms and release enormous additional quantities of energy. Once this nuclear furnace is ignited, the star becomes similar to the sun and continues to shine with a hot, yellow-white light for billions of years as its store of hydrogen is slowly converted to heavier helium. Ultimately, as the hydrogen fuel begins to be exhausted and the mass of heavy helium increases, additional nuclear reactions take place that release still more energy, causing the star to rapidly brighten and expand in size. Ultimately, through the series of steps summarized in Figure 13.17, all the potential nuclear energy of the star is consumed, and it fades to a cold, dark ash floating in the vast reaches of space.

The sequence of "stellar evolution" just described is derived from observations of stars examplifying each

Figure 13.17 Size-temperature relations during the life history of a typical star such as the sun. For most of its many-billion-year lifetime the sun will have its present size and temperature but ultimately, as its nuclear fuels approach exhaustion, it will first expand and cool, then contract and become hotter. After the fuels are consumed it will fade to become a cold stellar "ash."

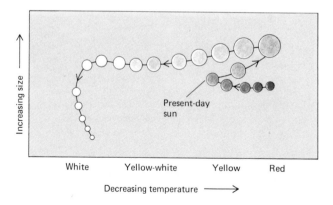

stage of the evolutionary sequence. Theoretical calculations, coupled with observations of composition, brightness, and energy output, show that the sun is a normal, and relatively youthful, star within this evolutionary scheme. It probably formed about 5 billion years ago and rather quickly attained its present brightness. It should continue to burn for several billion years before the supply of hydrogen fuel begins to be exhausted.

The sun's history can be directly inferred from observations of other stars, but no such observations are available to explain the origin of the other bodies of the solar system. Even the closest stars are so enormously distant that any planets they might have are invisible to even the most sensitive and powerful telescopes. For this reason there are many theories about the origin of the planets, but relatively few facts. Some of these theories seek to explain the planets as a result of some extraordinary event, such as a near collision with another star, that drew matter away from the sun to condense in orbit around it. Such theories now seem far less plausible, however, than those which make the origin of the planets, and other matter of the solar system, a normal consequence of the birth and development of stars. We shall describe one such theory that has become among the most widely accepted in recent years.

Observations of very young stars, as well as theoretical considerations, suggest that the original swirling concentrations of dust and gases from which the sun condensed had a flattened, disklike shape (Figure 13.18). Much of the material of this rotating gaseous disk was concentrated in a central bulge which was to become the sun. In the thinner, outer parts of the moving disk, turbulent eddies might have caused small additional concentrations of gas and dust that were ultimately to become the planets. Early in the sun's history, while it was still relatively cool, these "protoplanets" consisted largely of hydrogen and helium as did the sun itself. As the massive central sun heated and hydrogen fusion began, however, its great outpourings of energy drove away the lighter gaseous elements from the protoplanets nearest to it, leaving only relatively small quantities of heavier constituents to orbit as the terrestrial planets (Figure 13.18). Farther from the sun, the large outer planets retained some of their original gaseous matter, which persists today and accounts for their striking contrast with the more dense inner planets. Beyond the outer planets would be enormous numbers of smaller masses of still lighter frozen

Figure 13.18 *Hypothetical origin of the sun and planets from a rotating, disklike mass of dust and gases (see text).*

gases, the comets, that were driven there by the sun's initial heating. In the asteroid zone between the terrestrial and outer planets, smaller and more dense fragments orbit in a region where an initial protoplanet probably never fully consolidated into a single large mass.

This theory of planetary origin accounts not only for the differences between the terrestrial and outer planets, but also for the fact that the planets move in a single plane and direction, which is the plane and direction of movement of the original flattened dust cloud. Furthermore, theoretical considerations suggest that accumulations of matter within such a swirling cloud should have the regular geometric spacing seen today in the orbits of the planets. There is, however, one serious difficulty with the theory. Although the sun contains most of the *mass* of the solar system, most of its *energy of motion* is concentrated in the large outer planets which cover enormous distances as they circle the sun. In the original rotating dust cloud, on the other hand, most of the energy of motion would have been concentrated in the rapidly spinning central sun. Several ingenious schemes for transferring this energy from the early sun to the early protoplanets have been proposed, but all have serious deficiencies. This problem of energy distribution remains an unexplained difficulty in the popular scheme of planetary origin just described.

METEORITES, SAMPLES OF THE SOLAR SYSTEM

The "shooting stars" seen streaking across the sky on any dark night are caused by materials, originally left in space during the formation of the solar system, that have been captured by the earth's gravity. The bright streaks are formed as the materials plunge into the dense atmosphere where they are heated, and usually vaporized, by friction with gas particles at heights of 40 to 100 km (25 to 60 mi). Such objects are called **meteors** if they are completely vaporized and **meteorites** if a part of them survives its fall through the atmosphere and reaches the earth's surface.

13.7   Meteors and Meteorites

Telescopic analysis of the light produced by the fiery paths of meteors shows them to be of two sorts. Most are composed of low-density masses of very light material which leave behind relatively faint trails as they are vaporized in the atmosphere. This most common kind of meteor is believed to be caused by small frozen gas particles left behind by passing comets; this conclusion is strengthened by the fact that such meteors are particularly abundant when the earth's orbit crosses that of a comet (Figure 13.19).

Meteors caused by frozen gases are readily vaporized in the atmosphere and are thus never preserved on the earth's surface as meteorites. The second kind of meteor appears to be caused by much more dense metallic or stony material entering the earth's atmosphere from space. This type of meteor leaves an intense "fire-ball" as it passes through the atmosphere. In contrast to the cometary meteors, a small number of these heavier meteors are not completely vaporized but, instead, reach the

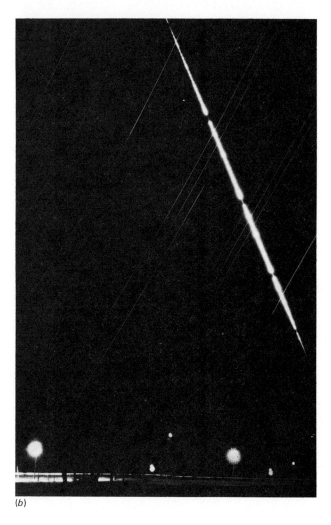

(b)

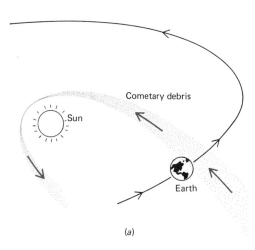

(a)

Figure 13.19 Meteors. (a) Meteors are most common when the earth's orbit crosses that of a comet. Most meteors are believed to be small, frozen masses of cometary debris that fall into, and are vaporized by, the upper atmosphere. (b) Photograph of a typical meteor. (b: Smithsonian Astrophysical Observatory)

earth's surface to become meteorites. We have already made reference to meteorites because of their extraordinary significance in understanding many aspects of the earth. In Sections 2.4 and 2.7 we saw that many inferences about the nature of the earth's interior are based on the overall composition of meteorites, whereas in Section 10.9 we noted that radiometric dating of meteorite minerals gives important clues to the age of the earth. In this section we shall review the characteristics of meteorites in more detail, emphasizing their importance as samples of the solar system beyond the earth.

About 500 meteorites the size of an orange or larger reach the earth's surface each year. Most fall in the ocean; only about 150 descend on land and, of these, an average of only about 4 per year are recovered for study.

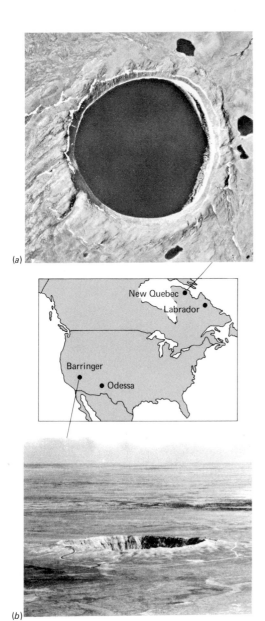

Although most meteorites are less than a meter in diameter, occasionally much larger masses, up to a kilometer or more in diameter, reach the earth's surface where they make huge craterlike holes known as *astroblemes* (Figure 13.20). These are caused by the dissipation of their tremendous energy of motion as they collide with the earth. It has been estimated that such large meteorites fall only about once each 10,000 years.

At least 20 large astroblemes are known on the earth's surface today. Most of these probably formed in the last few million years of earth history, whereas earlier ones have been largely destroyed by erosion. In spite of the large mass of the meteorites that form astroblemes, meteorite fragments are rare in and near the craters. Apparently such meteorites are largely vaporized by the enormous heat created by their impact with the earth. It is, therefore, principally the smaller meteorites which survive their fall to the earth's surface and become available for study.

13.8   Meteorite Composition and Origin

In overall composition, meteorites fall into three groups (Figure 13.21). Some, called **stony meteorites,** are made up entirely of various silicate materials; others, called **iron meteorites,** are composed exclusively of metallic iron and nickel; still others, called **stony-irons,** contain mixtures of both metal and silicate. Most of the 1,600 or so meteorites recovered and preserved in the museums of the world are irons which, after they fall, resist destruction by weathering far more readily than do the silicate minerals of stony meteorites. In addition, masses of metallic iron are far more likely to be recognized and preserved as exceptional materials than are stony meteorites, which are easily mistaken for ordinary rocks. For these reasons, it is impossible to estimate the relative proportion of meteorite types that reach the earth's surface by looking at *all* recovered meteorites. Instead, such estimates are based on studies of the much smaller number of meteorites, about 700 in all, that were *seen to fall* and then collected immediately. About 93 percent of such meteorite *falls,* as they are called, are of the stony type; irons make up most of the remainder with stony-irons accounting for less than 2 percent (Figure 13.21). It thus appears that the material arriving at the earth's surface from elsewhere in the solar system is composed predominantly of silicate minerals just as is the earth itself. The composition and structure

Figure 13.20 *Astroblemes, the impact craters of large meteorites. The map shows sites of four large North American astroblemes that are believed to have been formed within the past 2 million years. Many older craters, now much modified by erosion, are also known. (a) Vertical aerial view of the New Quebec crater, which is about 3 km (1.8 mi) in diameter. (b) Oblique view of the Barringer crater, which is about 1 km (0.6 mi) in diameter. (a: Canadian Department of Energy, Mines, and Resources; b: American Museum of Natural History)*

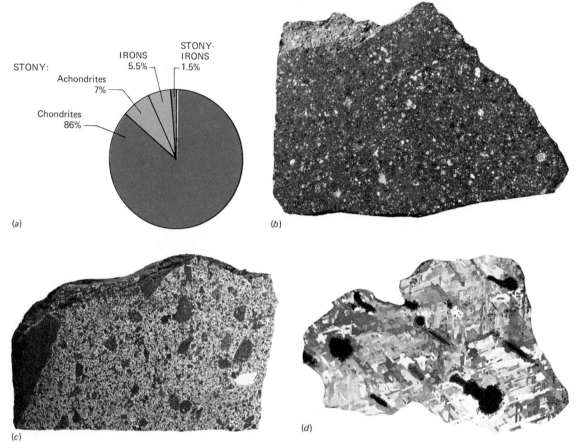

(a)

STONY:

Achondrites
7%

Chondrites
86%

IRONS
5.5%

STONY-
IRONS
1.5%

(b)

(c)

(d)

Figure 13.21 Meteorites. (a) Relative proportions of stony, stony-iron, and iron compositions among meteorites actually seen to fall. Most meteorites are stony "chondrites" containing small, glassy spheres of silicate minerals. (b–d) Cut and polished surfaces of typical meteorites. (b) Stony chondrite. The specimen, found in Mexico, is about 10 cm (4 in.) long. (c) Stony-iron. The dark masses are silicate minerals and the light matrix is iron. The specimen, found in Arizona, is about 5 cm (2 in.) long. (d) Iron. The specimen, found in Arizona, is about 25 cm (10 in.) long. (b–d: Smithsonian Institution)

of these meteoritic silicates differ, however, in many important respects from silicates found on earth.

Most stony meteorites are made up, at least in part, of small, glassy spheres about 1 to 2 mm in diameter (Figure 13.22). These spheres, in turn, are composed of dark magnesium- and iron-rich silicate minerals, particularly olivine and pyroxene (see Section 1.6). These small silicate spheres are called *chondrules,* and the meteorites containing them are referred to as **chondrites.**

The origin of chondrules, which make up a large fraction of the material reaching the earth from space, is uncertain because similar structures are not found among the silicate minerals of the earth. Chondrules look as if they formed from a heating and rapid cooling process that first melted some preexisting material and then quickly "froze" small drops of the melt into spheres. Radiometric dating indicates that most of them solidified about 4,500

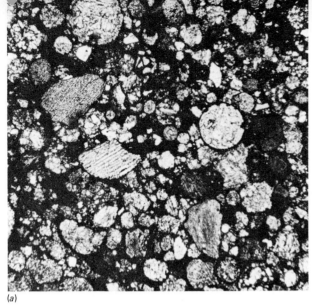

(a)

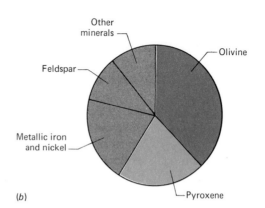

(b)

Figure 13.22 *Chondrites.* (a) *Microscopic view of a typical chondrite showing the characteristic rounded, glassy spheres ("chondrules") made up of silicate minerals. The spheres are 1 to 2 mm in diameter.* (b) *Mineral composition of a typical chondrite. Olivine and pyroxene predominate.* (a: John A. Wood)

Figure 13.23 *The relative compositions of chondritic meteorites and the sun. If the large volumes of solar hydrogen and helium are omitted, the remaining elements show almost identical abundance, a fact which supports the hypothesis that chondrites are fragments of solar matter.* (From Glasstone, Sourcebook on the Space Sciences, *1965, after Chapman and Larson*)

million years ago, at the same time that the earth is believed to have first consolidated. This observation suggests that they may be particles of matter that were melted by the initial solar heating and then "froze" as they moved away from the sun in the early history of the solar system. This suggestion is strengthened by the fact that their composition is remarkably uniform and shows about the same relative proportion of nonvolatile elements as does the sun (Figure 13.23). Indeed, one rare type of chondrule-bearing meteorite, called *carbonaceous chondrites*, even contains some of the more volatile solar elements, such as carbon, hydrogen, and oxygen, chemically combined into silicates and other solid minerals. Both because of their composition and their age, chondritic meteorites,

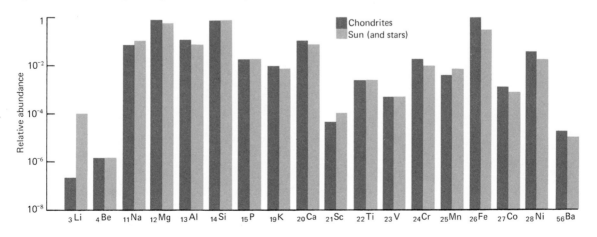

particularly the carbonaceous chondrites, are believed to represent relatively unchanged samples of the primordial material that was initially dispersed from the sun to form the planets and other objects of the solar system.

If chondritic meteorites represent the primoridal building materials of the solar system, then we must still account for the small proportion of meteorites that do not contain chondrules (principally the irons and stony-irons, but also including a few nonchondrule-bearing stony meteorites called *achondrites;* see Figure 13.21). There are two general theories for the origin of these materials. Some workers believe that they condensed directly from solar gases just as did the chondrites. Others feel that they have a secondary origin, from chondrites that at some time were swept together into large, asteroid-sized bodies in which internal melting could take place. Under such circumstances, the heavier iron and nickel content of the silicates might accumulate near the center of the bodies. If such differentiated bodies were later broken up, for example by impact with each other, then the fragments might well resemble the small proportion of iron, stony-iron, and achondritic meteorites that reach the earth.

SUMMARY OUTLINE

The Sun

13.1 *Solar structure:* hot solar gases boil to the surface to produce the granular, turbulent photosphere; beyond are the splashing gases of the chromosphere and cooler gases of the corona which grade outward to the thin solar wind.

13.2 *Solar disturbances:* solar structure is periodically disturbed by cool, dark sunspots and bright flares that release unusual amounts of energy; these disturbances probably result from regular perturbations of the sun's magnetic field.

13.3 *The source of solar energy:* the sun is composed mostly of hydrogen and helium; the helium is produced from a fusion of hydrogen nuclei, with conversion of a part of the hydrogen directly to electromagnetic energy.

The Solar System

13.4 *The planets:* are of two sorts—relatively small and dense terrestrial planets and larger, less dense, outer planets; all orbit the sun in the same plane and direction, but spin on their axes at different rates and angles.

13.5 *Additional components:* planetary satellites, smaller, planetlike asteroids, and large frozen masses of cometary gases are the principal additional components of the solar system.

13.6 *Origin of the solar system:* the sun probably originated from a large, turbulent concentration of matter within a moving cloud of cosmic gases and dusts; the planets may represent smaller accumulations within the spinning, flattened solar

**Meteorites,
Samples of the
Solar System**

cloud; early solar heating may have driven the lighter elements of the cloud away from the inner, terrestrial planets, and concentrated them in frozen, cometary masses in the outer solar system.

13.7 *Meteors and meteorites:* every year millions of frozen cometary fragments and heavier solid particles from space are captured by the earth's gravity, but most are vaporized as they fall through the atmosphere; about 150 meteorites reach the land surface each year, but only about 4 are recovered for study.

13.8 *Meteorite composition and origin:* most meteorites are composed of small glassy spheres of olivine and pyroxene that may represent materials melted by the initial heating of the sun; the smaller quantities of iron and other meteorites may have had a similar origin, or be fragments of larger, differentiated planetary bodies.

ADDITIONAL READINGS

*Berlage, H. P.: *Origin of the Solar System,* Pergamon Press, Oxford, England, 1968. A brief introduction to the many theories of solar and planetary beginnings.

*Brandt, J. C.: *The Sun and Stars,* McGraw-Hill Book Company, New York, 1966. A brief, intermediate-level introduction to solar and stellar physics.

*Ebbinghausen, E. G.: *Astronomy,* Charles E. Merrill Publishing Co., Columbus, Ohio, 1971. A brief introduction to matter beyond the earth, with emphasis on the solar system.

Hartmann, W. K.: *Moons and Planets,* Bogden & Quigley, Inc., Tarrytown, New York, 1972. An up-to-date introduction to the solar system.

Mason, B.: "Meteorites," *American Scientist,* vol. 55, pp. 429–55, 1967. An authoritative and clearly written review article.

*Mehlin, T. G.: *Astronomy and the Origin of the Earth,* William C. Brown Company, Dubuque, Iowa, 1968. A brief introduction, emphasizing the solar system.

Menzel, D. H.: *Astronomy,* Random House, Inc., New York, 1970. A lavishly illustrated popular survey, with much data on the solar system.

*Wood, J. A.: *Meteorites and the Origin of Planets,* McGraw-Hill Book Company, New York, 1968. A readable, intermediate-level survey.

* Available in paperback.

14

The Moon

Mare lavas partially filling highland crater on the lunar farside. (NASA)

For many years meteorites provided the only materials from beyond the earth that were available for direct scientific study. Beginning in 1969, however, samples returned to the earth from the moon have added a new dimension to our understanding of the solar system. Before turning to the nature and implications of these dramatically important lunar materials, it will be useful to review some more general conclusions about the moon that have grown from many years of astronomical calculations and observations of its surface features.

Most of the nine planets have one or more satellites in orbit around them, yet the earth-moon system is unique because of the very large size of the satellite in comparison with its attracting body (Figure 14.1). The moon has a diameter about one-quarter as large as the earth's, whereas most planetary satellites have diameters less than one-twentieth the size of their respective planets. Indeed, the moon is half the size of Mars and two-thirds as large as Mercury; were it in orbit around the sun rather than the earth, it would qualify as a full-scale terrestrial planet.

From astronomical calculations, such as those summarized in Figure 14.2, the moon's diameter, total mass, and distance from the earth have long been established. Knowing the diameter and mass, the average density of lunar materials was easily computed and turned out to be 3.3 gm/cm^3, about the same as the materials of the earth's upper mantle, but far lighter than the earth's *overall* density of 5.5 gm/cm^3. This suggested that the moon lacks the heavy iron core of the earth's interior but is, instead, dominated by silicate minerals throughout. Because it is made up of lighter materials, the total *mass* of the moon is only about one-eightieth that of the earth even though its diameter is one-fourth as great. The earth's far greater mass is the principal reason that a body as large as the

471

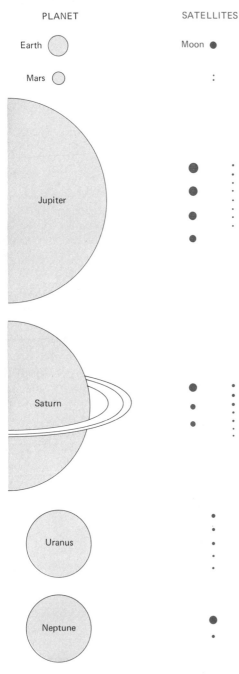

PLANET   SATELLITES

Earth

Mars

Jupiter

Saturn

Uranus

Neptune

Moon

Figure 14.1 The relative sizes of the planets and their satellites. The moon is unique in having a diameter one-fourth as great as its associated planet; most satellites are proportionately far smaller than the planets they orbit. Mercury, Venus, and Pluto appear to lack satellites.

moon can be held in orbit by the earth's gravitational attraction.

The moon completes one revolution around the earth every 27 days and, like the sun and planets, rotates about a central axis (Figure 14.3). Rather surprisingly, however, the period of rotation is the same as the period of revolution, about 27 days, so that the moon always shows the same hemispherical face when viewed from the earth. Because of the tilt of the lunar axis of rotation and, also, because of minor wobbles in its orbital motion, slightly more than half the lunar surface, about 60 percent, can be observed at one time or another from the earth. The remaining 40 percent, called the **farside** to distinguish it from the visible **earthside,** is forever hidden from earth-based observation.

THE LUNAR SURFACE

Before the recent rise of space exploration, information about the lunar surface could only be gathered by earth-based telescopes. Today, however, artificial satellites have provided detailed photographs of the entire surface, including the long-hidden farside.

14.1 Maria and Highlands

From afar, the most prominent features of the lunar earth-side are large dark patches surrounded by lighter areas; both have been directly observed by everyone for they are readily visible to the unaided eye (Figure 14.4). Early tele-scopic observations showed the dark areas to be relatively smooth and flat regions which contrast sharply with the more irregular and rugged terrane of the lighter regions around them. Because they bear a superficial resemblance to the earth's oceans, the dark, smooth areas were long ago named **maria** (Latin for "seas"; the singular is *mare*); the lighter regions of high relief are known as the lunar **highlands.** Maria cover about one-third of the moon's earthside hemisphere and one of the most surprising dis-coveries of recent lunar exploration was that maria are virtually absent from the long-hidden farside (Figure 14.4). Instead, for reasons that are still uncertain, the farside is covered almost entirely by rugged highlands.

On close inspection, the margins of many of the maria show traces of the rugged relief of the surrounding highlands protruding through the flat maria surface and

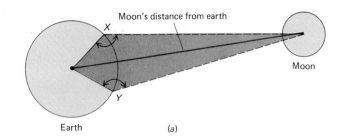

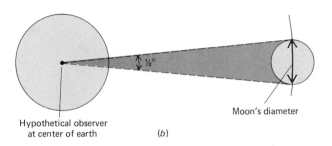

Figure 14.2 Geometrical calculations of the moon's distance and diameter. (a) By simultaneously measuring the moon's angular position at two points on the earth's surface (X and Y), its distance (colored line) can be determined from the shaded triangles (the length of one side of each triangle, the distance from X and Y to the earth's center, is also known). (b) Knowing the moon's distance, its diameter can be calculated from its observed angular width (about ½°).

giving the impression that the maria material rests on top of an older highland surface (Figure 14.5). For this reason, it has long been postulated that the dark maria represent huge lava flows which spread to fill large rugged depressions on the highland surface. This idea has now been confirmed by samples from the maria surfaces, which show their dominant rocks to be dark, basaltic lavas. Except at the maria margins, the lavas completely cover the older highland surface which underlies them. Judging from the exposed highlands, this buried surface probably had mountains at least 6,000 m (20,000 ft) high. The complete burial of such rugged topography indicates that an enormous volume and thickness of lava was generated to form the maria at some stage in the moon's history.

14.2 Craters

The highland regions which cover about two-thirds of the lunar earthside and most of the farside hemispheres are

Figure 14.3 Lunar motions. The plane of the moon's elliptical orbit around the earth is tilted slightly with respect to the plane of the ecliptic. The moon's axis of rotation is inclined only about 7° from its orbital plane.

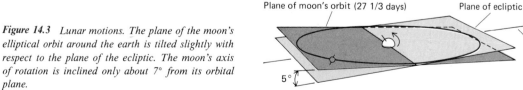

(a)

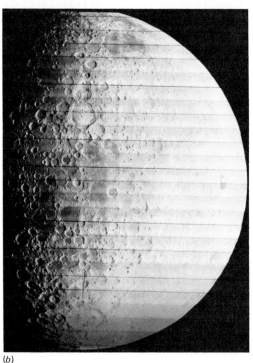

(b)

Figure 14.4 (a) Composite photograph of the lunar earthside, showing the distribution of maria (dark areas) and highlands (light areas). The dots indicate sites of samples returned to the earth by the Apollo 11, 12, 14, 15, 16, 17 and Luna 16, 20 missions. (b) A portion of the lunar farside as photographed from an orbiting satellite. The farside is dominated by highlands and lacks the large maria of the nearside. (a: Lick Observatory; b: NASA)

completely unlike the more familiar mountainous highlands of the earth. Instead of the narrow, linear belts of folded rock that make up the earth's mountainous regions, most lunar mountains appear to be the more-or-less altered remains of circular depressions or **craters** which, next to the larger-scale maria and highlands themselves, are the most prominent features of the lunar surface (Figures 14.4, 14.6).

The millions of craters visible in detailed photographs of the moon's surface come in all sizes and a variety of shapes. The smallest are only centimeters across and the largest have diameters of hundreds of kilometers. Many of the craters are fresh-looking and cup-shaped whereas others are more dishlike with flat bottoms and relatively low, irregular rims that give the appearance of having been modified by erosion. In general, craters with diameters of less than about 8 km (5 mi) tend to be cup-shaped and regular while larger craters show a complete spectrum of shapes, from regular to strongly modified. The overall depths of the craters (distance between the crater floors and the tops of the surrounding rims) tend to vary with crater size; the largest craters, with

(a)

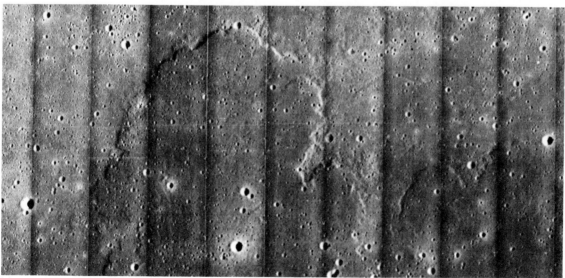

(b)

Figure 14.5 *Mare lavas.* (a) *A mare margin, showing the partial filling and burial of highland craters by flat, dark mare materials. Such evidence first suggested that the maria might be dark lava flows that cover an older highland surface.* (b) *Lobe-shaped pattern, suggesting lava flow, on a mare surface. (a, b: NASA)*

diameters of tens of kilometers, have maximum depths of about 6,000 m (20,000 ft).

Craters occur on the surface of both the maria and highlands, but their size and distribution differ markedly in the two regions (Figure 14.6). Large craters, with diame-

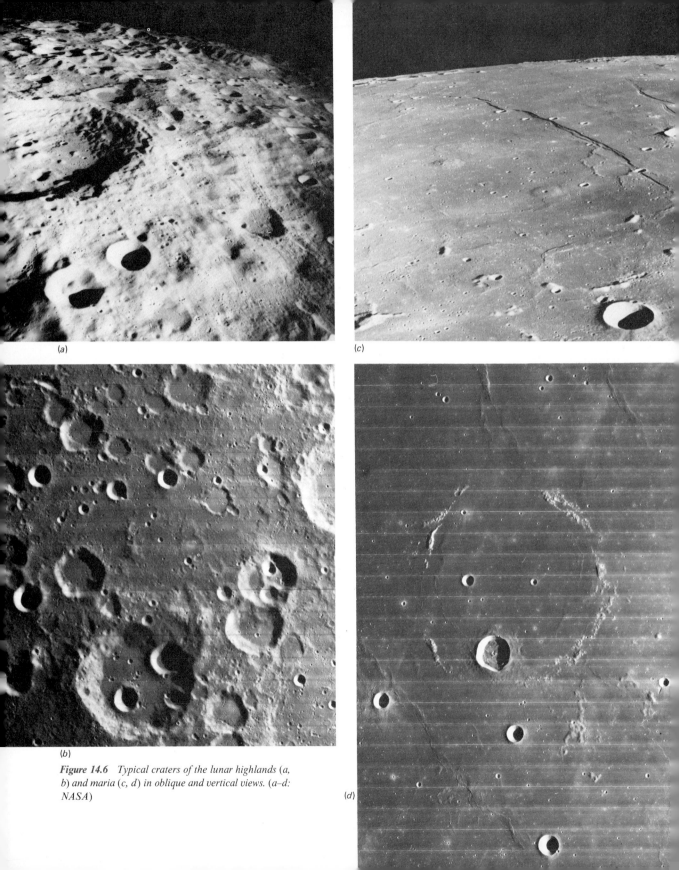

(a)

(c)

(b)

(d)

Figure 14.6 *Typical craters of the lunar highlands (a, b) and maria (c, d) in oblique and vertical views. (a–d: NASA)*

ters of several kilometers or more, are about 10 times more abundant in the highlands than on the maria. It is largely the modified rims of thousands of such large craters that give the highlands their rugged relief.

The origin of lunar craters has long been a subject of debate. On the earth, similar structures are known to form by either of two processes. On the one hand, circular, craterlike volcanoes result from surface outpourings of lava; on the other hand, the impact of large meteorites also leaves craters, called astroblemes, on the earth's surface (see Section 13.8). By analogy with the earth, both internal volcanism and external meteorite impacts have been called on to explain the craters of the moon.

Volcanism has undoubtedly produced some lunar craters, yet recent exploration has made it clear that most were formed by meteorite impacts. On the earth, millions of objects from space are captured each day by its gravitational field and plunge towards its surface. Most of these, however, are vaporized as they enter the earth's dense lower atmosphere and only a few reach the surface. The moon has far less mass and, consequently, less gravitational attracting force for such objects than does the earth, yet lunar gravity is still sufficiently strong to cause a continuous capture of materials. Furthermore, the moon lacks an atmosphere to cause melting and dissipation. Every captured object, both large and small, eventually plunges into its surface at high velocity. Countless millions of such impacts have probably given rise to most of the craters that cover the lunar surface.

14.3 Other Surface Features

In addition to maria, highlands, and craters, there are other, less common, features observable on the lunar surface. Small linear faults can be seen to break the rocks in a few areas, yet the numerous large faults associated with the earth's moving crustal plates are absent. Likewise, the large-scale folded rocks that make up most of the earth's mountain ranges are not seen on the moon, although maria lavas sometimes show small-scale, foldlike structures called **wrinkle ridges** (Figure 14.7). These are believed to have been caused by compression or collapse of the cooling lava.

Two additional features of the lunar surface are of interest because they have few analogues on the earth. The first are **rays,** thin coatings of light-colored dust that radiate for great distances from large, fresh lunar craters

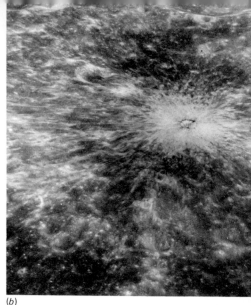

(a)   (b)

Figure 14.7 *Lunar wrinkle ridges (a) and rays (b) (see text). (a, b: NASA)*

(Figure 14.7). Rays, which can be seen from the earth only at times of full moon, apparently form as lunar rocks are pulverized, and the resulting dust scattered, by the force of large meteorite impacts. The second features are **rilles,** long, narrow, troughlike valleys that occur in and around many large craters on the surface of the maria (Figure 14.8). On the earth, somewhat similar valleys, but on a smaller scale, are caused by the shrinkage and collapse of tubular cavities in lava flows. Most lunar scientists believe that rilles had a similar origin.

LUNAR MATERIALS

Telescopes and satellite cameras have provided a wealth of knowledge about features of the lunar surface, yet these instruments supply very little information about the *composition* of the materials that make up the maria, highlands, and other surface features. By analogy with the earth's crust and, also, because the maria resemble huge lava flows, it was long assumed that the lunar surface was dominated by igneous rocks. This has been abundantly confirmed, in the years since 1969, by several hundred kilograms of lunar materials that have been returned to the earth and subjected to exhaustive analyses. Although generally similar to certain igneous rocks of the earth's crust, these lunar samples also have many distinctive characteristics.

(a)

(b)

Figure 14.8 Lunar rilles, troughlike valleys thought to be due to the collapse of tubular cavities in mare lava flows. (a) Aerial view of several rilles near a mare margin. (b) Surface view of a large rille at the Apollo 15 site. (a, b: NASA)

14.4 Lunar Soil

Because the moon lacks an ocean and atmosphere to produce earth-style weathering, its surface was long believed to expose much barren rock. One of the surprising initial findings of lunar exploration was that the moon's solid surface is almost everywhere covered by a deep layer

(a)

(b)

Figure 14.9 Lunar soil. (a) Typical view of the thick layer of rock debris that covers the moon's surface. A sample bag in the foreground and astronaut in the background provide scale. (b) Close-up view of a spacecraft imprint in fine lunar soil material. (a, b: NASA)

of fine-grained "soil" particles, just as are most regions of the earth (Figure 14.9). On the moon, however, this soil debris originates from an entirely different kind of weathering than do earth soils.

In the first place, it is clear that the moon has probably always lacked the significant fluid cover which plays such a fundamental role in weathering and shaping the earth's surface. Just as on the earth, past volcanic outpourings from the interior of the moon undoubtedly released large quantities of water vapor, carbon dioxide, and other gaseous materials. Calculations show, however, that because of the moon's relatively small mass, the normal molecular motions of even the heavier gases would rather quickly cause them to escape its gravitational field and be lost to space. For this reason, it seems improbable that the moon has *ever* had an extensive fluid cover to cause weathering of its surface rocks.

Instead of forming from weathering by an ocean and atmosphere, the debris that covers the lunar surface is the result of **space weathering**—the continuous abrasion of its surface by large and small meteorite impacts. Acting in countless numbers over billions of years, these impacts have pulverized the underlying bedrock and redistributed the resulting debris by "splashing" to produce the moon's soil cover.

The thickness of the layer of lunar soil can be estimated from lunar craters, which tend to show a somewhat different shape as they become deep enough to penetrate the underlying bedrock. Such observations show that the soils are thinner over the maria than over the highlands, which have presumably been subjected to a longer interval of space weathering. On the maria, the soil is usually between 2 and 10 m thick; in the highlands it may reach a maximum thickness of about 20 m.

With one or two possible exceptions, all of the lunar materials so far returned to the earth have been pebble-, sand-, and dust-sized particles from the thick lunar soil, rather than direct samples of the underlying rock. All of this material shows evidence of igneous origin, yet much of it has distinctive secondary textures caused by space weathering within the soil layer (Figure 14.10).

When a meteorite strikes the lunar surface its impact energy melts some of the preexisting soil material at the point of impact and scatters this molten rock over a wide area where it quickly cools to form small spheres of glass. These secondary spheres are a common constituent of lunar soils. Beyond the zone of fusion is a zone in which

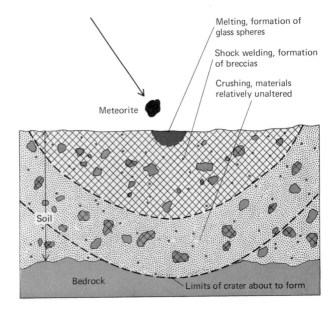

Figure 14.10 *"Space weathering" and the origin of lunar soil materials. Continuous bombardment of the lunar surface by meteorites creates a soil layer composed of small glass spheres (dark color), composite breccia fragments (cross-hatched pattern), and fragments of relatively unaltered bedrock (gray). (After Wood,* Scientific American, *vol. 223, 1970)*

the impact energy does not melt the soil but, instead, compresses and welds together small and large soil fragments to form a heterogeneous solid rock called **breccia.** Like the molten soil, breccia fragments are widely scattered by the energy of impact. Beyond the zone of breccia formation is a third zone in which soil materials and, if the meteorite is a large one, the underlying bedrock, are crushed and scattered with no change in texture other than fragmentation into progressively smaller and smaller particles.

The meteorite impacts of space weathering thus produce three principal textures of materials in the lunar soil: small glass spheres; heterogeneous breccia fragments; and relatively unaltered rock fragments (Figure 14.11). The breccia and unaltered rock fragments may be of various sizes, ranging from dust and sand up to large boulders.

14.5 Lunar Rock

The bulk of the larger rocks returned from the moon are heterogeneous breccias whose included particles show a complex history of fragmentation and welding by meteorite impacts. Analyses of these, supplemented by studies of the less common unaltered rock fragments and sand and dust particles, indicate that the lunar soil is derived from at least three fundamental kinds of underlying igne-

Figure 14.11 Lunar soil constituents. (*a*) *Small glass spheres that average about 0.2 mm in diameter.* (*b*) *Breccia fragments of all sizes. Breccias are composed of heterogeneous rock particles welded together by meteorite impacts. The photograph shows a large breccia boulder, about 1 m (3.3 ft) across, made up of clearly visible light- and dark-colored rock components.* (*c*) *Relatively unaltered bedrock fragments of all sizes. Shown is a large boulder of mare basalt.* (*a: E. C. T. Chao; b, c: NASA*)

(a)

(b)

(c)

ous rock. These are: **mare basalts; premare basalts;** and **anorthosites** (Figure 14.12).

Mare Basalts Most of the lunar samples have come from the soils that cover the mare; they are dominated by dark basalt fragments that apparently come from the underlying lavas. These are made up largely of plagioclase feldspar and pyroxene, as are earth basalts, but differ in containing more iron and far less sodium and certain other minor elements than do most basalts found on earth.

Premare Basalts A second quite different kind of basalt fragment is particularly abundant in soil debris believed to be derived from the highlands that bound and underlie the maria. Compared with the mare materials, these "premare" basalts have a higher proportion of plagioclase feldspar and less pyroxene and other dark minerals. As a result, they are relatively enriched in aluminum and silicon and depleted in iron and magnesium. They also contain much more radioactive potassium and uranium than do the mare basalts, but resemble them in having little sodium. Because of their association with the highlands, these basalts are assumed to antedate those of the mare.

Anorthosites Unlike basalt, the third principal moon rock is of a type that is rare on the earth. This rock, called *anorthosite,* is similar to gabbro, the coarse-grained version of basalt, except that it is light-colored and composed largely of plagioclase feldspar with little additional pyroxene or other iron minerals. Because of the dominance of white feldspar crystals, anorthosites are very light in color and contrast sharply with the dark lunar basalts.

Like the premare basalts, lunar anorthosites appear to be derived from the highlands. Thus the three principal lunar rocks predominate in different regions. The well-sampled maria are composed of iron-rich basalts, while the less intensively sampled highlands apparently contain both aluminum-rich basalts and anorthosites. This general compositional difference, based on a relatively few sample locations, has been confirmed by sensitive instruments orbiting in lunar satellites, which show the highlands to be everywhere more aluminum-rich than the maria. Such instruments also suggest that the radioactive premare basalts may be confined to the northwest quarter of the nearside, with anorthosites dominating other highland regions.

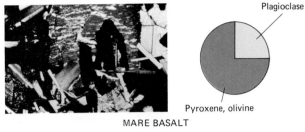

MARE BASALT

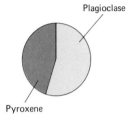

PREMARE BASALT

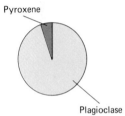

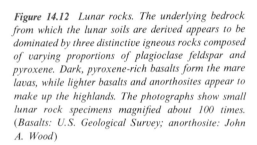

ANORTHOSITE

Figure 14.12 Lunar rocks. The underlying bedrock from which the lunar soils are derived appears to be dominated by three distinctive igneous rocks composed of varying proportions of plagioclase feldspar and pyroxene. Dark, pyroxene-rich basalts form the mare lavas, while lighter basalts and anorthosites appear to make up the highlands. The photographs show small lunar rock specimens magnified about 100 times. (Basalts: U.S. Geological Survey; anorthosite: John A. Wood)

14.6   The Moon's Interior

In addition to providing direct samples of its surface rocks, recent lunar explorations have begun to investigate the materials of its hidden interior. As on the earth, these investigations are based on surface measurements of the moon's internal wave propagation, gravity, and magnetism.

The most revealing glimpses of the lunar interior are provided by "moonquakes" recorded by sensitive seismographs placed on its surface (Figure 14.13). At least three widely spaced seismograph stations are required to accurately locate an earthquake or moonquake and in 1971 the third such lunar station was established. The instruments have since shown that the moon is seismically very quiet compared with the earth. All recorded moonquakes are very small and their total energy release is about a million times less than that of all earthquakes over a comparable period.

(a)

N

(b)   S

Figure 14.13 Moonquakes. (a) *Astronaut setting up
a seismograph at the Apollo 15 site.* (b) *Index map
showing the location of lunar seismic stations established on the Apollo 12, 14, 15, 16, and 17 missions.
Most moonquakes, which are far less energetic than
earthquakes, appear to be concentrated at great depths
beneath the small area shown by the color marker.*
(a: NASA)

The most surprising result of lunar seismic measurements was the discovery that most moonquakes are
sharply concentrated in both time and space. Temporally,
they tend to occur twice each month when the moon is
closest to the earth; thus they must somehow be caused
by tidal effects of the earth's gravity. In addition, about
80 percent of the moonquake energy appears to come
from one very small region at a depth of about 800
km (Figure 14.13). The reason for this peculiar localization
is unknown, but the mere presence of moonquakes at such
a great depth, about halfway to the moon's center, indicates that its deep interior is far more rigid than is that
of the earth. Earthquakes are confined to only the outermost one-tenth of the earth's radius.

Because of their small magnitude and temporal and
spatial concentration, moonquakes reveal relatively little
about the moon's internal structure. Larger shock waves
caused by the impact of abandoned spacecraft, however,
permit the construction of crude seismic velocity-depth
curves (see Section 2.1). These suggest that the lunar
interior is layered as is that of the earth. A rather sharp
discontinuity at about 70 km appears to separate an overlying "crust" from an underlying "mantle." Wave velocities within the crust are about what would be expected
in basalts and anorthosites similar to those exposed on
the lunar surface. Higher mantle velocities suggest more
dense underlying materials, perhaps similar to the ultrabasic materials of the earth's mantle. Little is known of
the deeper interior because the relatively low energy of the
spacecraft impacts, and the location of the seismic stations in relation to them, have so far made it difficult to
measure deep wave velocities.

Satellite measurements of lunar gravity reveal only
one unusual feature—anomalous concentrations of massive materials that floor several of the smaller maria. The
exact nature of these surface concentrations of heavy
materials is unknown, yet they confirm the earthquake
evidence of a relatively strong and rigid lunar interior. On
the more plastic earth, such masses would slowly sink into
isostatic equilibrium rather than remain supported at the
surface.

Unlike the earth, the moon today virtually lacks a
magnetic field. This observation is in harmony with the
moon's overall low density, which strongly suggests that
it lacks the liquid iron core that causes the earth's magnetism. There is, however, one disquieting magnetic puz-

Man and Space

The exploration of the earth's neighbors in space is perhaps the single aspect of geological science that has the least direct relationship to everyday human needs. Even though there has been fanciful speculation about obtaining resources from the moon or planets, it is now apparent that even if these bodies were littered with jewels or gold (as they most certainly are not), they might never be profitably exploited. For example, the total cost of the materials so far returned from the moon is something over $50 million per *kilogram*. This is not to suggest that exploration of space cannot be justified on other grounds, such as intellectual curiosity or national prestige; governments not uncommonly undertake far more expensive endeavors (wars, for example) with far less justification. It merely emphasizes that man's practical concerns are still confined to his home planet.

In this regard, however, there is one aspect of space technology that *is* of direct value—the use of spacecraft not to study other planets but to observe the earth itself. In Chapters 5 and 6 we have already indicated how satellite observations have revolutionized our understanding of both the upper atmosphere and of world weather patterns. In addition, satellite surveys are beginning to find a host of other applications: surveying world food production, detecting the spread of water pollution, monitoring the movements of glaciers and sediments, and charting world patterns of urban growth. In such ways the technology of space exploration is directly aiding man in his efforts to understand and preserve his own planet.

A message from the earth. Contents of an engraved aluminum plate carried aboard a spacecraft, launched in 1972, that will eventually become the first man-made object to leave the solar system. The plate contains a universally decipherable message about the place and time the spacecraft was launched on the remote chance that it may someday be encountered by intelligent life elsewhere in the universe. (a) Position of the sun in the universe shown by its location relative to 14 energetic, starlike pulsars (see Section 15.6); the characteristic radiation patterns of each pulsar are shown schematically. Subsequent changes in these patterns will suggest the time of launch. (b) Schematic of fundamental transition in the hydrogen atom, added primarily to suggest the unit of time employed in (a). (c) Schematic diagram of the solar system with the spacecraft shown leaving the earth. The symbols are a simple code suggesting the distances of the planets from the sun. (d) Generalized spacecraft builders; the outline of the spacecraft provides scale.

Astronaut on "spacewalk" while in orbit high above the earth. (NASA)

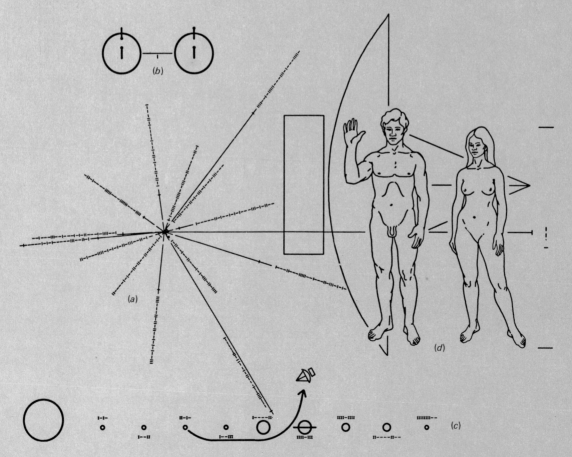

zle. All lunar rocks tend to show strong *remnant magnet-ism*, which indicates that a significant magnetic field was present when they formed. The nature and origin of this earlier magnetic field are but two of the many unsolved problems of the moon's history.

LUNAR HISTORY

As for the earth, a principal reason for studying the present configuration of the moon is to understand its long and complex past. Likewise, our understanding of lunar history, like that of the earth, rests on two principal chronological techniques: *superposition*, or the relative vertical relations of its surface features, and *radiometric dating* of its minerals.

14.7   Surface Chronology

Long before the first lunar samples were returned to the earth in 1969, a large body of information about the moon's history had been compiled by applying the venerable principle of superposition (see Section 10.1) to its surface features. We have already noted the most important conclusion to arise from such study: the edges of the flat maria can be seen to be superimposed on the more rugged topography of the surrounding highlands, thus showing clearly that the maria lavas are younger than the adjacent highlands (see Figure 14.5). Furthermore, careful observations of the maria themselves show some of them to be composed not of single large lava flows, but of several smaller flows that apparently formed at different times. Because the maria lack the abundant large craters of the highlands, it is also clear that many more large meteorites collided with the lunar surface *before* the maria were formed than after.

By the application of similar reasoning it is also possible to determine the relative ages of many large lunar craters, both those of the highlands and the less abundant ones that occur on the maria. Often two craters, or the widely scattered debris thrown from them, can be seen to be superimposed, so that their relative ages are easily determined. More commonly it is necessary to infer the relative ages of craters by their degree of modification by later meteorite impacts (see Figure 14.6). By a careful combination of all such evidence of relative age, lunar scientists have been able to compile, from photographs alone, detailed "geologic" maps of the lunar surface showing the relative ages not only of such major features as the maria and highlands, but of many individual craters and smaller lava flows as well (Figure 14.14).

From such studies a basic fourfold subdivision of major events in lunar history has emerged: first came the solidification and differentiation of the highland rocks; then massive, large-scale cratering of the highlands; next filling of some of the largest highland craters with the mare basalts; and, finally, smaller-scale cratering of both the mare and highland surfaces (Figures 14.14, 14.15).

Figure 14.14 *Lunar surface photograph (a), and a geologic map (b) compiled from it. The region shown, a part of a mare margin and adjacent highlands, contains materials formed during the four principal events of lunar history: highland rocks (oldest); premare craters; mare rocks; and young craters. (a: NASA; b: modified from F. A. Mutch,* Geology of the Moon, *1970, after Titley)*

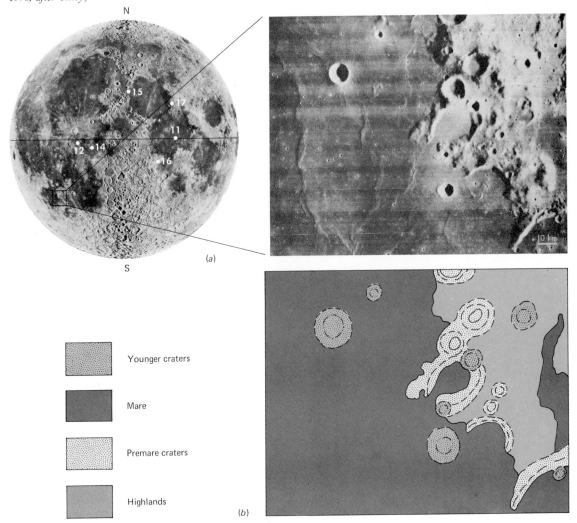

Younger craters

Mare

Premare craters

Highlands

(a)

(b)

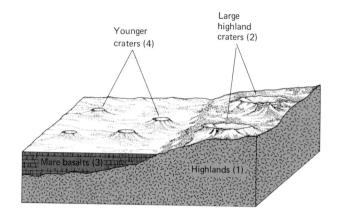

Younger craters (4)

Large highland craters (2)

Large highland craters (2)

Mare basalts (3)

Highlands (1)

Figure 14.15 Schematic diagram of the four principal events of lunar history: (1) formation of highland rocks (oldest); (2) premare highland cratering; (3) partial highland flooding by mare basalts; and (4) younger cratering.

14.8   Radiometric Ages and Thermal History

The major events of lunar history are clearly recorded in its surface features, yet until actual moon rocks were available for radiometric dating, there was no reliable means of determining the absolute ages of these events. Earlier estimates of the age of the maria surfaces had been made from the density of their small craters. By calculating the average number of meteorites reaching the earth today, and making corrections for the moon's smaller mass and other factors, it was possible to arrive at the time required to produce a given density of craters, assuming the rate of meteorite infall to have remained constant. Such estimates first suggested that the maria surfaces were very old in terms of earth history, having formed perhaps 3,000 million years ago. This conclusion has been amply confirmed by radiometric dating (Figure 14.16).

All mare basalts so far dated have ages in the range of 3,000 to 4,000 million years. Thus even these relatively late features in lunar chronology formed near the time of origin of the oldest known earth rocks. Dates from the lunar highlands are only slightly older—around 4,100 million years—indicating that most of the rocks now exposed on the moon's surface originated in a relatively short interval early in its history. Except for continuous cratering, they have remained relatively unchanged for 3,000 million years, or since early Precambrian time on the scale of earth history (Figure 14.16).

In addition to providing absolute ages for the highlands and maria, lunar rocks also raise fundamental questions about the moon's early differentiation and thermal history. Like the rocks of the earth's crust, lunar rocks are quite *unlike* the olivine- and pyroxene-dominated me-

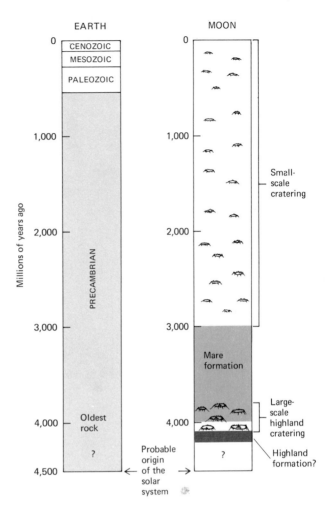

Figure 14.16 *Comparative histories of the earth and moon. The principal events of lunar history, including highland formation, highland cratering, and mare formation, are revealed by radiometric dating to be very old, having taken place during the early Precambrian interval of earth history.*

teorites which are presumably similar to the solar materials from which it consolidated. Lunar surface basalts probably formed from a partial melting of such underlying ultrabasic materials, just as do earth basalts, but there is as yet no satisfactory explanation of the consistent differences in composition between the mare and premare basalts, nor for the fact that both appear to be concentrated on only the earthside of the moon.

A further puzzle is the source of the heat which produced these early internal meltings and then shut down to leave the moon relatively inactive for 3,000 million years. It is unlikely to have been the impact energy of the large meteorites that caused the mare-filled basins, for radiometric dates show that lavas from a single mare formed not all at once, but over several hundred million years of lunar history. Likewise, if the interior were once heated by radioactive decay it should *still* be hot, for there is no way for most of the heat to escape the insulating effect of its outer rock layers. Yet the present moonquakes occurring halfway to the center of the moon show that its interior is now cool and rigid.

Still other puzzles are associated with the light-colored anorthosites that apparently dominate the highlands. Anorthosites are relatively rare rocks on earth, but several regions in which they occur have been intensively studied and their origin is well understood. They form when basaltic magmas are cooled very slowly deep beneath the earth's surface, rather than being extruded as lava to cool relatively rapidly. When cooled slowly, the first crystals to form are heavy, magnesium- and iron-rich silicates that tend to sink to the bottom of the liquid magma, leaving the remaining fluid enriched in lighter calcium and silicon. These elements then crystallize to form an upper layer of feldspar-rich anorthosite. On the earth this process occurs rather infrequently in deep-seated magma chambers, some of which are now revealed at the surface by subsequent erosion. If the lunar highlands consist largely of anorthosite, as now seems probable, then a wholesale melting and differentiation of a thick outer layer of ultra-basic materials must have occurred very early in lunar history. Once again, the source of the enormous heat required for this melting is unknown.

Finally, there is the question of the large-scale meteorite impacts that produced the intensive early cratering of the highland surface. These impacts must have been concentrated in a relatively short interval of lunar history — after the consolidation of the highland rocks more than 4,000 million years ago but before the origin of the mare basalts which fill many of the enormous craters and have maximum ages of almost 4,000 million years. Fragments from the debris of one of the largest mare-filled craters suggest that the material was very hot, even partially melted, for a puzzlingly long interval after it was ejected. Small meteorite fragments in the debris also indicate that the impacting object was made up mostly of iron. Beyond that, little is known of this major phase of lunar history.

14.9   Origin of the Moon

Still other puzzles surround the ultimate origin of the moon and its long-term interactions with the earth. The moon almost certainly consolidated about 4,500 million years ago from smaller masses of cosmic matter, along with the sun, earth, and planets. It may have accreted from fragments initially in orbit around the primitive earth (Figure 14.17a). If so, however, it is difficult to explain the differences of density in the two bodies and the concentration of most of the heavier materials in the earth's core. On the other hand, there are also suggestions that the moon may not have been a satellite of the earth for all of its history. These come from considerations of the motions

Figure 14.17 Hypotheses of lunar origin. (a) Accretion from debris in orbit around the primitive earth. (b) Capture of an independent body by the earth's gravitational field. There are serious objections to both hypotheses (see text).

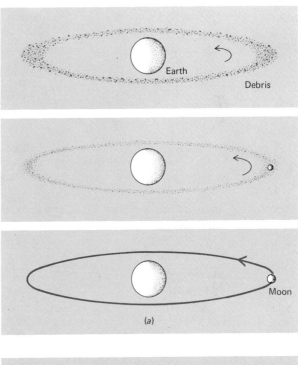

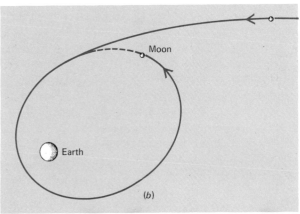

of the two bodies and the interactions of their gravitational attraction.

The gravitational tides caused on the earth by the moon have the effect of gradually slowing down the earth's rate of rotation by transferring some of its rotational energy to the moon's energy of revolution about the earth. This causes the moon to become progressively more distant from the earth as the earth's rate of rotation decreases. Furthermore, the rate at which the moon is now receding from the earth can be calculated from historical records of ancient lunar eclipses. When the rate so obtained is projected into the past, it suggests that the moon was very close to the earth only 2,000 million years ago and thus could not have been a satellite of the earth for all of its history.

From such reasoning, it has been suggested that the moon was originally a separate body, perhaps a fifth terrestrial planet, that was somehow captured by the earth's gravity (Figure 14.17b). There are, however, many difficulties with this idea. Other calculations suggest that if a body as large as the moon approached the earth, it would most probably not be captured but merely have its path deflected. Furthermore, if the moon had been very near the earth as recently as 2,000 million years ago, it would have produced enormous tidal effects and frictional melting of most of the earth's crustal rocks, events for which there is no geological evidence. In short, the question of the origin of the earth-moon system remains one of the most fundamental unsolved problems of planetary science.

SUMMARY OUTLINE

The Lunar Surface

14.1   *Maria and highlands:* the most conspicuous features of the lunar earthside are flat, dark maria and light, rugged highlands; only highlands occur on the long-hidden farside.

14.2   *Craters:* circular depressions of many sizes and shapes cover the maria and highlands; some were probably caused by internal volcanoes, but most are the result of meteorite impacts.

14.3   *Other surface features:* include small faults and foldlike wrinkle ridges, rays of debris from crater-producing impacts, and long, sinuous channels called rilles.

Lunar Materials

14.4   *Lunar soil:* rocks of the lunar surface are almost everywhere covered by a thick layer of pebble-, sand-, and dust-sized debris produced by meteorite impacts.

14.5   *Lunar rocks:* are of three principal types; iron-rich basalts that form the maria, and aluminum-rich basalts and light-colored anorthosites that occur in the highlands.

14.6 *The moon's interior:* shows faint, localized moonquakes at great depths and earthlike outer laying, but presently lacks a magnetism-producing fluid iron core.

Lunar History

14.7 *Surface chronology:* superposition of lunar surface features shows the principle events of lunar history to have been, from oldest to youngest: highland formation; massive cratering; mare formation; less intensive cratering.

14.8 *Radiometric ages and thermal history:* highland differentiation, cratering, and mare filling took place between about 4,100 and 3,000 million years ago, relatively soon after the moon's origin; the source of the heat for highland differentiation and basalt flooding is a major mystery.

14.9 *Origin of the moon:* the moon may have originally consolidated from materials in orbit around the primitive earth, or have been subsequently captured from elsewhere by the earth's gravity.

ADDITIONAL READINGS

Allen, J. P.: "Apollo 15; Scientific Journey to Hadley-Appennine," *American Scientist,* vol. 60, pp. 162–74, 1972. A popular review of the exploration of a mare margin.

Bowker, D. C., and J. K. Hughes: *Lunar Orbiter Photographic Atlas of the Moon,* National Aeronautics and Space Administration, Washington, 1971. A definitive collection of lunar surface photographs.

Dyal, P., and C. W. Parkin: "The Magnetism of the Moon," *Scientific American,* vol. 225, no. 2, pp. 62–73, 1971. A popular review article.

Goldreich, P.: "Tides and the Earth-Moon System," *Scientific American,* vol. 226, no. 4, pp. 43–52, 1972. A popular review of the problem of the moon's origin.

Hinners, N. W.: "The New Moon; A View," *Reviews of Geophysics and Space Physics,* vol. 9, no. 3, pp. 447–522, 1971. A comprehensive review article, advanced but readable.

Mason, B.: "The Lunar Rocks," *Scientific American,* vol. 225, no. 4, pp. 48–58, 1971. A popular review of the first several groups of lunar samples.

Mutch, T. A.: *Geology of the Moon, A Stratigraphic View,* Princeton University Press, Princeton, N. J., 1970. A detailed but readable introduction to lunar surface chronology, with excellent illustrations.

Wood, J. A.: "The Lunar Soil," *Scientific American,* vol. 223, no. 2, pp. 14–23, 1970. A popular review article.

15

The Planets
and Beyond

The moon, now known from many thousands of detailed photographs, rock samples, and surface and satellite measurements, remains a puzzling celestial body. As we go beyond it to even the nearest planets, knowledge decreases sharply and puzzles multiply endlessly.

From our point of view as earth dwellers, it is our closest planetary neighbors—Mercury, Venus, and Mars—that are of the greatest interest because they most nearly resemble the earth in size, composition, and probable history. For this reason, and also because more is known about them, we shall emphasize these earthlike planets before considering, in less detail, the much larger and less dense outer planets and, more briefly still, the vast reaches of space that lie beyond them.

THE EARTHLIKE PLANETS

Until the era of space exploration began in the 1960s, the earth's neighboring planets were known only from astronomical calculations of their sizes, orbital paths, densities, and similar properties, combined with observations of their visible surfaces and radiation through earth-based telescopes (Figure 15.1). For most of the planets we still have only such observations but others, particularly Venus and Mars, the two terrestrial planets that orbit on either side of the earth, have now been reached by instrumented spacecraft. Information sent back to earth from these voyages is beginning to revolutionize our understanding of the inner solar system and thus, indirectly, to reveal much about the earth itself.

15.1 Mercury

Before turning to Mars and Venus we need to say a few words about the third and less well known earthlike planet, Mercury. Among the terrestrial planets, Mercury is the

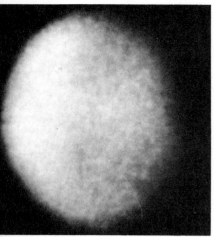

(a) Mercury

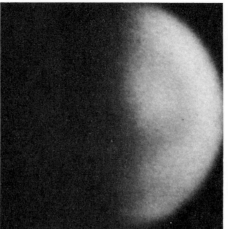

(b) Venus

(c) Earth

Figure 15.1 Photographs of the earthlike planets and the moon, showing the varying degrees of surface detail seen in the best available full-planet views. (a, b: New Mexico State University Observatory; c, e: NASA; d: Lick Observatory)

Figure 15.2 Mercury's position in the solar system, not to scale (see also Figures 13.9–13.11).

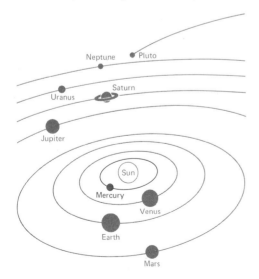

smallest, the closest to the sun, and the most distant from the earth (Figure 15.2). Because of its nearness to the sun, it completes its yearly orbit in only 88 earth days, yet it rotates very slowly on its axis, about once every 59 earth days. Although it has a diameter only about one-third that of the earth, it appears to be composed of relatively heavy materials having the same density as those of the earth—about 5.5 gm/cm³. Because of Mercury's small size and the high temperatures caused by its nearness to the sun, calculations show that any gases released to its surface would be quickly lost into space so that, like the moon, it has probably always lacked an atmosphere. None is evident in telescopic observations which show the planet as a small disk with no clearly distinguishable surface features. Radar observations and analyses of its reflected light vaguely suggest that its surface may resemble that of the moon. Until observations from spacecraft become available, little else is likely to be known about the planet.

15.2 Mars

Mars, which has about half the diameter and one-tenth the mass of the earth, shares with the moon the distinction of being the only bodies beyond the earth with solid, rock-like surfaces visible through earth-based telescopes (Figure 15.3). For this reason Mars, like the moon, has been a perennial subject of fanciful speculation. Powerful telescopes have long revealed three basic features of the

(d) Moon   (e) Mars

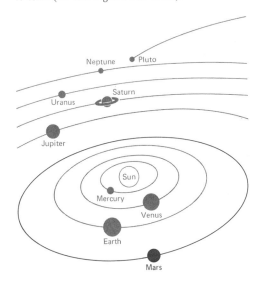

Figure 15.3 Mars's position in the solar system, not to scale (see also Figures 13.9–13.11).

Martian surface: white polar caps that expand and contract with the seasons;[1] large, dark areas that show changes in shape and color; and lighter, reddish-brown areas that show little or no change. In addition, clouds can sometimes be seen above the solid surface indicating that, unlike the moon, Mars has at least some atmospheric gases. Because of Mars's great distance from earth, each of these major features is rather vague and ill-defined under even the best conditions of telescopic viewing, but each has been confirmed by spacecraft observations to be a fundamental feature of the planet (Figure 15.4).

By analogy with the earth and moon, the dark areas of the Martian surface have long been called "maria" and the lighter, reddish-brown regions "deserts." Spacecraft photographs, which now cover most of the planet's surface in varying detail, show the dark "maria" to be covered with craters resembling those not of the lunar maria but of the moon's highlands. As on the moon, some of the craters are fresh and cup-shaped whereas others, perhaps the majority, have indistinct rims and flat bottoms suggesting modification by erosion. This **cratered terrain,** as it is now called in preference to the older term "maria," covers about half the planet and shows craters of all sizes

[1] Mars's axis of rotation has about the same angle of inclination as does the earth's, giving it similar seasonal effects; its period of rotation, about 24.5 hours, gives it days closely comparable to the earth's although it requires about two earth years to complete its longer yearly orbit around the sun.

up to hundreds of kilometers in diameter (Figures 15.4, 15.5). The Martian craters generally appear to be more modified by erosion than are those of the moon, and this is to be expected since Mars, unlike the moon, has an atmosphere to facilitate weathering. As on the moon, Mars's craters are believed to have been formed predominantly by meteorite impacts.

Although the Martian cratered terrain resembles the surface of the moon, the lighter "desert" regions are seen in spacecraft photographs to be quite unlike anything on the lunar surface. Instead, they are large, relatively smooth and flat-surfaced areas that are almost devoid of craters (Figures 15.4, 15.6). Their closest analogue does indeed appear to be the earth's large desert regions but since little is yet known of their origin or composition, it is perhaps best to refer to them merely as **smooth terrain.** Enormous cone-shaped mountains, apparently volcanic, are associated with the smooth terrain in some regions, suggesting that the terrain may represent volcanic ash or lava flows that have covered an underlying cratered terrain as do the lava flows of the lunar maria. The relative lack of

Figure 15.4 *Preliminary map of the surface of Mars, compiled from spacecraft photographs, showing the distribution of the four major kinds of terrain (see text). (R. S. Saunders, NASA)*

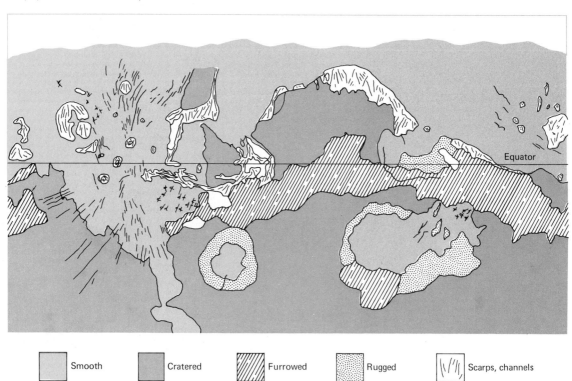

| | Smooth | | Cratered | | Furrowed | | Rugged | | Scarps, channels |

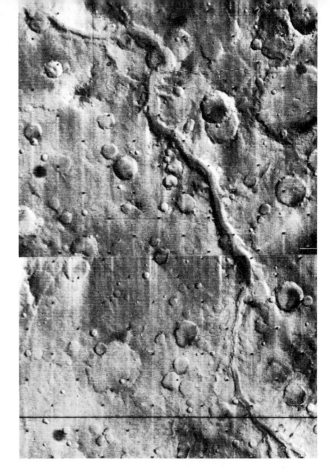

Figure 15.5 Martian cratered terrain. The region shown extends about 700 km (435 mi) in the longest dimension. As on the moon, most of Mars's craters are believed to have been caused by meteorite impacts. The sinuous valley is of uncertain origin, but may be a collapsed lava tube similar to lunar rilles. (Jet Propulsion Laboratory and NASA)

Figure 15.6 Martian smooth terrain. Large, cone-shaped mountains, apparently volcanic, are associated with the smooth terrain in many areas, suggesting that the terrain may be underlain by volcanic ash or lava flows. The huge volcano shown is 500 km (310 mi) across at the base and at least 10 km (6 mi) high. It is twice as large as the entire volcanic mass that forms the Hawaiian Islands on the earth. (Jet Propulsion Laboratory and NASA)

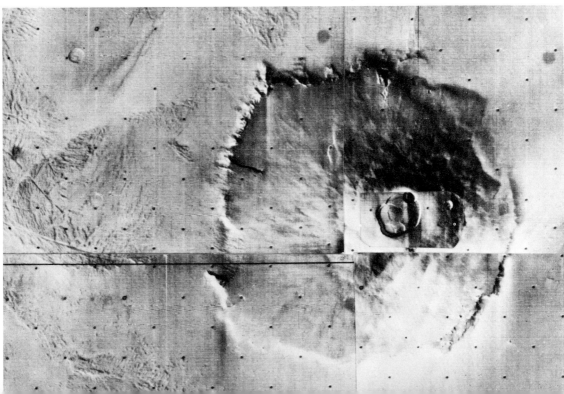

craters on the smooth terrain also indicates, however, that some erosional process is removing younger craters, as on the earth, soon after they are formed. As yet, no process has been suggested that would obliterate craters on the smooth terrain while leaving them intact on the adjacent cratered terrain. In short, the origin of the regions of smooth terrain remains a fundamental puzzle of Martian geology.

In general, smooth terrain predominates in the Martian northern hemisphere and cratered terrain in the southern hemisphere. Between lies a wide equatorial zone of less regular landscape which appears to combine features of both (Figures 15.4, 15.7). Craters are abundant as are long, furrowlike valleys, many of which radiate from volcanolike peaks. Some of these valleys, the largest of which are over 10 km (6 mi) wide and 1,000 km (600 mi) long, resemble lunar rilles and may be modified lava flow channels. Others are dramatically similar to water-eroded valleys on the earth, but their origin remains uncertain. This **furrowed terrain,** as it is called, thus combines craters with volcanic features similar to those of the smooth terrain. There are, as yet, no indications of the reason for the equatorial distribution of this peculiar terrain, nor for the hemispheric dominance of smooth and cratered terrain.

Figure 15.7 Martian furrowed terrain. A broad equatorial zone is dominated by craters and long, furrowlike valleys of unknown origin. The region shown extends about 40 km (25 mi) in the largest dimension. (Jet Propulsion Laboratory and NASA)

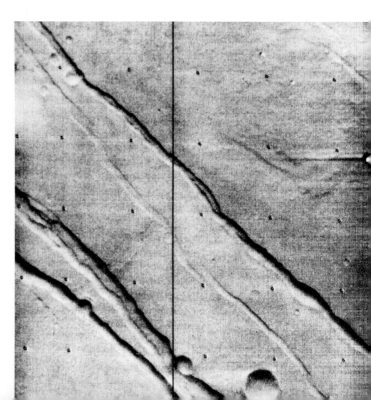

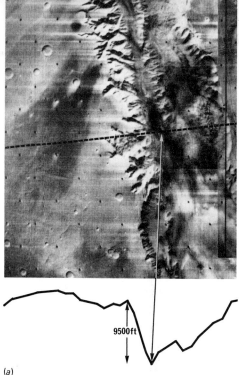

(a)

9500ft

(b)

Figure 15.8   Martian rugged terrain. (a) *Extensive valley system of unknown origin. The cross section below shows a vertical profile of the valleys, the deepest of which (arrow) is twice as deep and six times as wide as the Grand Canyon. The area shown is about 320 km (200 mi) across.* (b) *Pitted topography of unknown origin. The two largest depressions (left) are about 16 km (10 mi) in diameter.* (a, b: *Jet Propulsion Laboratory and NASA*)

In addition to the smooth, cratered, and furrowed terrains, spacecraft photographs show a fourth kind of terrain that is much more localized in distribution. These regions have steep, irregular hills and valleys and are known as **rugged terrain** (Figures 15.4, 15.8). Such regions probably have a variety of origins. Some surround large circular basins and may be modified crater-rim deposits. Others are clearly caused by the partial collapse of dome-like volcanic mountains.

On a somewhat smaller scale, photographs of the Martian surface also show extensive faults and rifts that appear to bound large crustal blocks that may be analogous to the moving plates of the earth's lithosphere. This observation, combined with Mars's complex surface topography, strongly suggests an active planet that resembles the earth more closely than the long-inactive moon.

On the other hand, the few available scraps of evidence concerning Mars's interior suggest it does *not* resemble that of the earth. Astronomical calculations show its density to be about 4 gm/cm^3, far lower than the earth's 5.5 gm/cm^3 and only slightly heavier than the moon's 3.3 gm/cm^3. Furthermore, spacecraft measurements show that Mars, like the moon, lacks the strong magnetic field found on the earth. Both of these observations suggest that Mars's interior lacks the earth's differ-

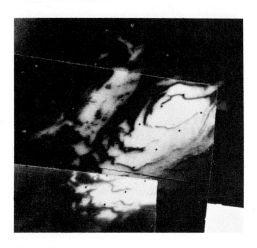

Figure 15.9 *The Martian south polar cap. The cap, seen during the summer, is about 300 km (180 mi) in diameter and is believed to be composed of a thin layer of frozen carbon dioxide, or "dry ice." The concentric patterns apparently reflect the underlying topography. (Jet Propulsion Laboratory and NASA)*

entiated iron core and is dominated by silicate minerals, as is the moon.

A final feature that has long been visible on the Martian surface are white polar ice caps which must be closely interrelated with its atmosphere (Figure 15.9). The atmosphere of Mars, in turn, has been second only to its surface topography as a subject of speculation and investigation. Both spacecraft data and earth-based analysis of light passing through the Martian atmosphere indicate it to be very thin, about one-hundredth the density of the earth's atmosphere. Apparently Mars's low overall mass, about one-tenth that of the earth, allows gases to escape more readily from its gravitational attraction than is the case on the earth.

The Martian atmosphere is not only far less dense than that of the earth but it also has an entirely different composition. Earth-based and spacecraft measurements show it to be made up almost entirely of carbon dioxide with only traces of the nitrogen, oxygen, and water vapor that are dominant constituents of the earth's atmosphere. Because of the relatively low water vapor content, it now appears most probable that the ice caps of Mars are composed not of frozen water, as are those on earth, but of frozen atmospheric carbon dioxide or "dry ice." Spacecraft photographs show the ice caps, in their margins at least, to be only a few centimeters thick for they change configuration very rapidly and cover, but do not obscure, underlying topography. Some of the thin clouds observed occasionally in the Martian atmosphere are also probably made up of frozen carbon dioxide crystals whereas others may be composed of crystals of water ice formed from Mars's small quantities of atmospheric water vapor (Figure 15.10a).

There is a final type of more extensive and persistent cloud that is seen less frequently in the atmosphere of Mars. Telescopic observations have shown that such clouds periodically obscure the surface of the planet for periods up to several weeks long. These planet-wide events were generally assumed to be large-scale atmospheric *dust storms*, a conclusion that was confirmed when an orbiting spacecraft arrived in the midst of one in 1971 (Figure 15.10b). Such storms imply that the surface of the planet is covered with soil debris, as is the earth and moon. They also suggest extremely strong winds in its thin atmosphere, for only at very high velocities could such a thin fluid actively erode and transport even dust-sized surface debris.

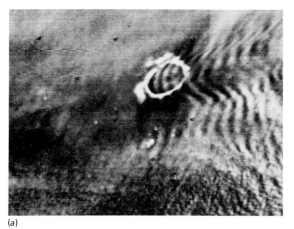

(a)

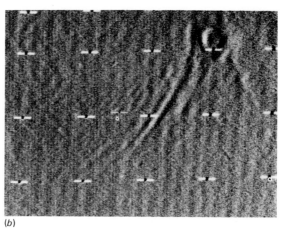

(b)

Figure 15.10 Martian clouds. (a) Clouds in the atmosphere of the winter hemisphere, viewed from above. The only surface feature visible is the frost-rimmed crater, about 90 km (55 mi) in diameter. The clouds, which are probably made up of frozen carbon dioxide, are deflected into a wavelike pattern in the lee of the crater. (b) A portion of a planetwide dust storm, viewed from above. The only surface feature visible is the crater, about 200 km (125 mi) in diameter, at the top. The wavelike pattern may be either atmospheric turbulence or surface sand dunes formed downward of the crater. (a, b: Jet Propulsion Laboratory and NASA)

Figure 15.11 Venus' position in the solar system, not to scale (see also Figures 13.9–13.11).

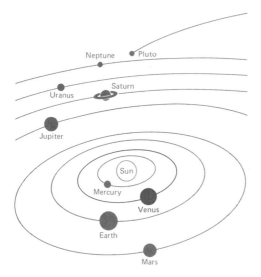

Closely related to the nature of Mars's atmosphere is the question of the presence of living organisms on the planet. All terrestrial life requires liquid water and thus the very small quantities of water observed in the atmosphere of Mars argue against the possibility of life. In addition, Mars's thin atmosphere and greater distance from the sun give it an average surface temperature of $-40°C$ $(-40°F)$, far colder than that tolerated by most life on earth. Although the low temperatures and apparent lack of water now make Martian life appear improbable, we must remember that simple life, particularly bacteria, survive on earth under conditions that are fully as rigorous as those probably found in moist soils of Mars's warmer, equatorial regions. For this reason, we cannot really be sure whether or not life is present until instrumented or manned landings permit an intensive life-search on the surface of the planet.

15.3 Venus

Venus, the final earthlike planet, is the most similar to the earth in many of its characteristics (Figure 15.11). Not only is it the earth's nearest planetary neighbor (it orbits about one-quarter closer to the sun and has a year of 225 earth days), but it also has about the same size and volume. Furthermore, the average density of the materials making up the planet—about 5.1 gm/cm^3—is closely similar to that of the earth (5.5 gm/cm^3). Recent spacecraft exploration has, however, confirmed earth-based observations showing that despite its similar position, size, and density, Venus is decidedly *not* like the earth in many other properties.

When viewed from space, the earth always shows a broken cover of clouds which, at any time, permits at least some of the underlying ocean and continents to be seen. Venus, in contrast, has a dense and continuous cover of thick clouds that extends at least 40 km (25 mi) above the surface of the planet. This cloud cover is thus far deeper than the earth's, for earth clouds seldom form at heights greater than about 15 km (9 mi). Because of this thick cloud layer, almost nothing is known about Venus's solid surface which cannot be seen either with earth-based telescopes or in spacecraft photographs. Indeed, the period of rotation of the solid planet obscured beneath the clouds was uncertain until recent earth-based radar observations showed that Venus rotates very slowly, about once each 243 earth days, in a puzzling, retrograde direction.

Additional radar observations have recently shown the planetary surface to have vague light and dark regions suggestive of those seen on Mars and the moon (Figure 15.12). Spacecraft observations also indicate that in spite of its high average density Venus, like Mars and the moon, does not have a strong magnetic field and thus probably lacks the earth's fluid iron core. Beyond these vague observations, little is known about the structure of the solid surface and dense interior of the planet. Instead, both

Figure 15.12 A radar "map" of the surface of Venus. The light areas may reflect surface topography otherwise obscured by the cloud-filled atmosphere. (Haystack Observatory)

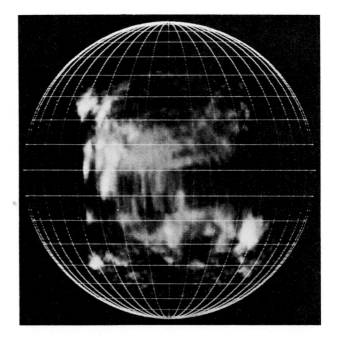

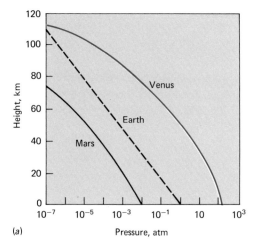

(a)

Pressure, atm

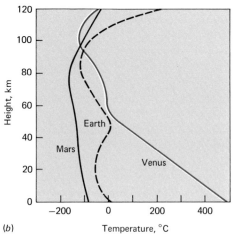

(b)

Temperature, °C

Figure 15.13 Variation of atmospheric pressure (a) and temperature (b) with height in the atmospheres of Venus, Mars, and the earth. The atmospheres of Venus and Mars are composed largely of carbon dioxide in contrast to the nitrogen-oxygen earth atmosphere. (From Eshleman, Scientific American, *vol. 220, 1969)*

earth-based and spacecraft observations of Venus have been primarily concerned with its more readily studied atmosphere and perpetual cover of dense clouds.

Spacecraft observations of the atmosphere of Venus show it to have a composition quite unlike that of the earth but very similar to the atmosphere of Mars. It is composed largely of carbon dioxide; other gases, including water vapor, nitrogen, and oxygen, the dominant gases of the earth's atmosphere, probably make up less than 5 percent of the total bulk. Furthermore, the total mass of Venus's atmosphere is far greater than the earth's—so great, in fact, that the atmospheric pressure on the surface of the planet is about 100 times that of the earth's atmosphere at sea level (Figure 15.13). This pressure is equivalent to that found at depths of about 7 km (4 mi) under the waters of the earth's ocean.

The atmosphere of Venus differs from that of the earth not only in composition and density, but also in temperature. In the upper levels of the atmosphere, near the cloud tops, temperatures are well below freezing as they are in parts of the earth's atmosphere, but they increase rapidly downward through Venus's dense gases to reach a scorching 450°C (842°F) at the planet's surface (Figure 15.13). At such high temperatures no liquid water could exist to form a surface ocean. Most silicate minerals, however, require still higher temperatures before they melt and vaporize and thus the planetary surface, although extremely hot, is probably in large part dry and solid. Both radar observations and abrupt landings by parachuted Soviet spacecraft confirm this inference.

To earth dwellers, Venus appears to be a most inhospitable planet with its very hot surface surrounded by a dense and forever cloudy atmosphere. These striking differences from the earth take on special significance when we remember that Venus is so similar to the earth in position, size, and density. How, then, did the two planets come to have such radically different atmospheres and temperatures? Although the answers to these fundamental questions are still far from certain, some suggestive clues are emerging.

Most probably the atmospheres of Venus and Mars, like that of the earth, have come from volcanic activity which releases gases previously trapped in the planets' solid interiors. We have noted (Section 11.2) that carbon dioxide is a dominant constituent of volcanic gases on the earth just as it apparently is on Venus and Mars. On the earth, however, most of the carbon dioxide released by volcanic activity has not accumulated in the atmosphere but, instead, has dissolved in the liquid ocean where it is later precipitated as solid calcium carbonate and stored in the rocks of the solid crust. The earth's ocean therefore acts as a catalyst for removing carbon dioxide from the atmosphere whereas on oceanless Venus and Mars all the volcanic carbon dioxide remains in the atmosphere. On Mars, because of its small mass, much of the gas escapes into space leaving a thin atmosphere, but Venus, being much more massive, has retained most of the carbon dioxide to create a very dense atmosphere. Interestingly enough, calculations show that if all the calcium carbonate stored in sedimentary rocks on earth were converted back to atmospheric carbon dioxide, the earth's atmosphere would resemble that of Venus. Instead, the sedimentary removal of most of this huge volume of carbon dioxide has left what might otherwise have been a minor atmospheric constituent, nitrogen, as the dominant gas in the earth's much thinner atmosphere.

If the presence of a liquid ocean on the earth is responsible for removing most of the carbon dioxide from its atmosphere, we are still faced with a fundamental problem because, as we saw in Chapter 11, the ocean itself is also believed to have formed from volcanic release of water vapor from the earth's interior. If this is the case, should not similar degassing have led to large quantities of water on Venus and Mars which, as on the earth, would have formed oceans and removed the atmospheric carbon dioxide? Several mechanisms of "water loss" on Venus and Mars have been proposed to get around this difficulty, but none has been generally accepted; the earth's unique ocean remains a puzzling difference between it and its planetary neighbors.

Two additional and interrelated aspects of Venus' atmosphere are also poorly understood—the cause of its high temperatures and the composition of its perpetual cloud cover. The high temperatures have been thought to be due to a sort of a massive "greenhouse effect" similar to that described for the earth's atmosphere in Section 5.5. Carbon dioxide readily transmits short wavelength sunlight but absorbs the long wavelength thermal radiation that would result from solar heating of the planet's surface. Thus a dense carbon dioxide atmosphere might be expected to trap much solar energy leading to high atmospheric and surface temperatures.

A principal difficulty with this idea is Venus's perpetual cloud cover, which might be expected to prevent most sunlight from penetrating to the planetary surface to be reradiated at longer wavelengths. Unfortunately, the composition and general properties of these all-important clouds are still uncertain. They appear to be made up of very small spheres that may be composed of water vapor, ice, frozen carbon dioxide, hydrochloric acid, dust, or all or none of these materials. Whatever their composition, no satisfactory mechanism has yet been proposed for keeping *any* of these materials in continuous and uniform atmospheric suspension so that Venus's clouds, like so many other features of our closest planetary neighbor, remain a fundamental puzzle.

THE OUTER PLANETS

Beyond the four earthlike planets of the inner solar system lies the wide zone which contains no planet but where, instead, millions of smaller asteroids circle the sun. Beyond the asteroid zone, in turn, lie the four giant planets—Jupiter, Saturn, Uranus and Neptune—and beyond these, the small outermost planet Pluto (Figure 15.14). Little is known about tiny Pluto for it is difficult to observe in even the most powerful telescopes; in contrast, the four giant planets and particularly Jupiter, the largest and the closest to the earth, have long been studied by earth-based obser-

Figure 15.14 Photographs of the outer planets at the same scale, showing the varying degrees of surface detail seen in the best available telescopic views. Left to right: Jupiter, Saturn, Uranus (top), and Neptune (bottom). (New Mexico State University Observatory)

vations. In addition, a spacecraft launched in 1972 is scheduled to reach Jupiter late in 1973 to provide the first close exploration of the outer solar system. Further spacecraft explorations planned for the late 1970s should dramatically augment our limited earth-based understanding of the giant outer planets and Pluto.

15.4 Jupiter

Jupiter, the innermost of the four giant planets, is by far the largest body in the solar system other than the sun (Figure 15.15). Indeed its huge volume, about 1,300 times that of the earth, would easily hold *all* the other planets, satellites, and asteroids with room to spare. Because of its great size, and the fact that it is relatively close to the earth, more is known of Jupiter than of the other giant planets which, although smaller, appear to resemble it in composition and structure.

Astronomical calculations show that Jupiter's great bulk is made up of extremely light materials having an average density of only 1.3 gm/cm³, or only slightly greater than the density of water and far lower than the average for the earth (5.5 gm/cm³). Jupiter's materials have about the same density as does the sun suggesting that Jupiter, like the sun, is composed predominantly of the two lightest elements, hydrogen and helium. The only direct evidence for Jupiter's composition comes from spectral analyses of light reflected to earth from the planet. These show hydrogen to be present on Jupiter in at least three forms—as the uncombined element, and in combination with carbon as methane (CH_4) and with nitrogen as ammonia (NH_3). Although spectral studies give only a vague indication of the abundance of these constituents, they clearly suggest that hydrogen is a principal component of Jupiter. By analogy with the sun, it is now assumed that the bulk of the planet is composed of uncombined hydrogen with helium (which cannot be detected by spectral analysis) as the only other dominant constituent. Methane, ammonia, and other materials probably account for only a few percent of the planet's total mass.

Although Jupiter and the sun appear to have similar compositions, their structures are entirely different because the densely packed hydrogen atoms of the more massive sun are "burning" by thermonuclear processes (see Section 13.3). In contrast, the total mass of Jupiter is too small to cause the dense gravitational packing nec-

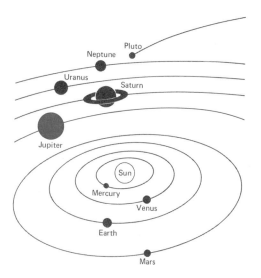

Figure 15.15 Jupiter's position in the solar system, not to scale (see also Figures 13.9–13.11).

essary for thermonuclear burning.[2] Instead, measurements of the heat radiated to the earth from Jupiter's outermost regions show them to be very cold, about −130°C (−202°F). At such temperatures, hydrogen, helium, and methane exist as gases whereas ammonia would be frozen to form tiny ice particles. Optical telescopes show Jupiter, like Venus, to be perpetually covered by an outer layer of dense clouds and these probably represent frozen crystals of ammonia ice floating in an atmosphere of gaseous hydrogen and helium.

Beneath its gaseous, cloud-filled atmosphere, little is known of the structure of Jupiter. The heat radiated by the planet, although insufficient to warm its massive outer atmosphere, can nevertheless be measured and appears to greatly exceed that which the planet receives from the sun. This suggests that the inner regions of the planet are not cold and frozen but are, instead, heated by some unknown source of internal energy. In any event, Jupiter's overall low density indicates that its clouds are unlikely to overlie a solid, rocky planet as do those of the earth, Venus, and Mars. Most probably, they cover a lighter mass of frozen ices or, perhaps, of solid hydrogen which would form a light, metallike material at the enormously high pressures present in the planet's massive interior.

Although we still know very little about the interior of Jupiter, optical telescopes have long provided some intriguing facts about the structure and behavior of its visible, cloud-filled atmosphere (Figure 15.16). Unlike the similarly dense and cloud-filled atmosphere of Venus, which shows almost no visible structure, Jupiter's clouds always appear as irregular, changing, light and dark bands which parallel its equator. Furthermore, observations of irregularities in these bands show that the planet rotates extremely rapidly on its axis—about once every 9 earth hours which, considering the enormous size of the planet, is rapid spinning indeed. Apparently this rapid rotation causes a parallel streaking out of clouds having different brightnesses, thus producing the banded appearance. The cause of these brightness differences, which are not constant but slowly change their shapes and boundaries, is unknown.

[2] Interestingly enough, theoretical calculations suggest that had Jupiter accumulated as only a slightly larger body, its mass would have been sufficient to cause thermonuclear ignition. Had this happened, the solar system would have been very different for it would have been dominated by a twin-star rather than the single sun.

(a)

(b)

Figure 15.16 Two views of Jupiter showing the changing patterns of its parallel cloud bands. The "Great Red Spot" is clearly visible in the upper hemisphere. The shadow of one of Jupiter's larger satellites is visible in the photo on the left. (New Mexico State University Observatory)

The most peculiar phenomenon of Jupiter's atmosphere is an enormous dark region, covering an area larger than the earth and cutting across several parallel cloud bands, which alternately brightens and fades in its southern hemisphere (Figure 15.16). Many of Jupiter's atmospheric bands appear to show faint changing colors—blue, green, yellow, and brown are the most common—but the much larger dark region usually appears to be strongly pink or red. For this reason, it is called Jupiter's "Great Red Spot." The spot generally moves with the rapidly rotating atmosphere but its motions are less regular. The cause of this peculiar feature is unknown.

Finally, one of the most unexpected discoveries concerning Jupiter came in the early 1950s when special telescopes first showed that the planet releases enormous, but erratic, quantities of radio waves similar to those given off by solar disturbances. Jupiter is the only planet showing such radiation which indicates that it, like the sun and earth, has a strong magnetic field caused by fluid movements within its rapidly spinning interior. The exact nature of these motions is obscure even for the earth and sun. For Jupiter, they are completely unknown.

Beyond Jupiter lie three planets—Saturn, Uranus, and Neptune—which resemble it in being large bodies composed of light materials, principally hydrogen and helium. In addition, their rapidly rotating surfaces appear to be dominated by thick clouds similar to those seen on Jupiter. Because these planets are both smaller and far-

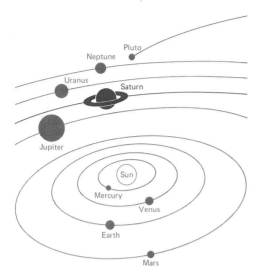

Figure 15.17 *Saturn's position in the solar system, not to scale (see also Figures 13.9–13.11).*

Figure 15.18 *The relative position of Saturn's rings and eight major satellites. The satellites move in the same equatorial plane as the rings. The orbits are to scale, but the satellites are enlarged relative to the planet.*

ther from the earth, they are more difficult to observe and less is known about them. Much of what we have already said about the composition and structure of Jupiter probably applies to them as well, but each shows intriguing differences that we shall briefly summarize.

15.5   Saturn

Saturn, after Jupiter the largest, closest, and best known of the outer planets, has a volume about half that of Jupiter but a mass only one-third as great (Figure 15.17). This is because the materials that make it up have an average density of only 0.7 gm/cm³, versus 1.3 for Jupiter. Saturn's lower density is probably not the result of compositional differences for both planets appear to be composed largely of hydrogen and helium. Instead, Saturn's lesser overall mass probably causes its hydrogen and helium to be merely less densely compacted.

Powerful telescopes show Saturn to be streaked with parallel cloud bands similar to those seen on Jupiter. In addition, however, Saturn shows a unique structural feature seen in no other planet. Extending thousands of kilometers into space from its equator is a thin, disklike zone of ice or dust particles, which appears in telescopes as three flat concentric *rings* around the planet's equator separated from each other by small gaps (Figure 15.18).

Saturn's rings are known to be composed of small separated particles, rather than being a continuous disk of solid or liquid, from several lines of evidence. Theoretical calculations suggest that only small particles could

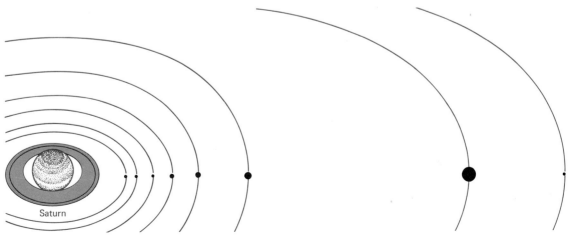

Saturn

remain in a disklike orbit around the planet; more massive material would be pulled to the planet's surface by its strong gravitational attraction. Furthermore, the rings can be seen from the earth to be somewhat transparent and to rotate at different speeds in their inner and outer portions. Both of these observations indicate a composition of small, separated particles. Some observations suggest that the particles in the rings, while covering an enormous overall area of millions of square kilometers, are in the third dimension confined to a very thin zone, perhaps only a few meters thick.

The exact origin of Saturn's unique rings is uncertain but must, in some fashion, be related to the origin of the 10 satellites that circle the planet beyond the rings but in the same equatorial plane (Figure 15.18). So far we have said very little about planetary satellites other than the moon, which is the only conspicuous satellite in the inner solar system. Both Mercury and Venus have none and Mars has only two very small satellites, about 8 and 15 km (5 and 9 mi) in diameter. In contrast, each of the four giant planets has several satellites—Jupiter has 12, Saturn 10, Uranus 5, and Neptune 2. These range in size from only a few kilometers in diameter to somewhat larger than the moon (see Figure 14.1). Some are very light, apparently being composed of materials similar to the parent planets, while others are more dense, suggesting that heavier materials were somehow concentrated to form them. As with the planets themselves as they orbit the sun, the origin of these complex satellite systems is still uncertain. Perhaps the best guess is that they somehow condensed from disklike rings of dust similar to those still present around Saturn.

15.6   Uranus, Neptune, and Pluto

Uranus and Neptune, the next two planets beyond Saturn, are very similar to each other in size and average density (Figure 15.19). Both are considerably smaller than either Jupiter or Saturn but are still about 14 times as massive as the earth. In addition, they have higher average densities than their larger neighbors. Because of their relatively small size and high density it appears that they have a higher proportion of heavier elements than do either Jupiter or Saturn. Both planets appear as only small, vague disks in even the most powerful telescopes. For this reason, little is known of their surface structure.

Uranus has one unique feature which distinguishes it from all the other planets. Its axis of rotation is almost

Figure 15.19   The positions of Uranus, Neptune, and Pluto in the solar system, not to scale (see also Figures 13.9–13.11).

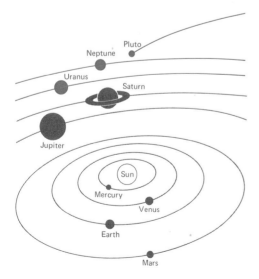

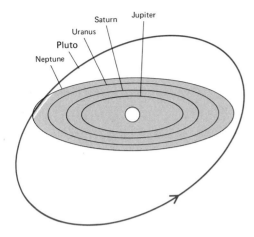

Figure 15.20 *Orbital path of Pluto. Pluto's peculiar orbit is inclined to the ecliptic and intercepts the orbit of Neptune.*

parallel to the ecliptic plane as it travels around the sun, whereas in all the other planets the rotational axis is more-or-less perpendicular to the ecliptic. This unusual inclination causes Uranus to have the strange seasonal effects summarized in Figure 13.12.

Pluto, the tiny outermost planet, is the most enigmatic and difficult to observe of all (Figure 15.19). In size and density it appears to resemble the earthlike planets rather than the giant outer planets. This difference is difficult to explain for, as we saw in Chapter 13, the small size and high density of the terrestrial planets was most likely a result of their nearness to the sun early in the history of the solar system. Perhaps the best explanation for this anomaly is that Pluto is not an original planet but a former satellite of Neptune that somehow escaped the planet's gravitational field to become trapped in solar orbit. This idea is supported by Pluto's peculiar orbital path, which brings it inside the orbit of Neptune during parts of its long journey around the sun (Figure 15.20).

BEYOND THE PLANETS

The sun's influence extends outward to many times the enormous diameter of Pluto's path around the sun and in these farthest reaches of the solar system millions of cometlike masses of frozen gas are believed to be held in orbit by the sun's gravitational attraction (see Section 13.5). Still farther into space lies the nearest star and, beyond it, the billions of other stars and gas and dust accumulations that comprise the visible universe.

Some idea of the immense distances involved as we leave the solar system can be seen in the following comparison. Spacecraft launched from the earth and traveling at tremendous speeds require about 4 days to reach the moon, 4 months to Mars and 1.5 years to Jupiter. But traveling at the same speed, they would require several *million* years to reach even the nearest star; most are infinitely more distant still. Thus as we go from the sun and planets to a consideration of the universe beyond, we enter entirely new dimensions of space and time.

Studies of this extra-solar universe are among the most exciting and challenging aspects of modern science, but are also beyond the outermost limits of the earth-centered geological and geophysical sciences. Here we shall only briefly touch on some of their implications and mysteries as a sort of cosmic conclusion to our survey of the earth and its close relatives in space.

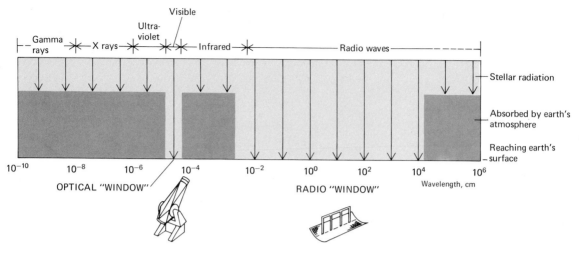

Figure 15.21 *Absorption of stellar radiation by the earth's atmosphere. Only visible light and certain radio wavelengths pass freely through "windows" in the atmosphere.*

15.7 Stars

Because of the enormous distance to even the nearest star, it is extremely unlikely that the universe beyond the solar system will ever be studied through close observation by spacecraft as are the planets. Instead, we shall probably forever know distant stars only by the radiation or particles that they emit and which travel across the vast reaches of space to arrive at the inner solar system where man can observe them.

Most stars, including the sun, radiate large quantities of electromagnetic energy of many different wavelengths, varying from extremely short X rays to very long radio waves. Much of this radiation from distant stars, as from the sun, is absorbed by the earth's atmosphere so that observations of stars from the earth's surface are restricted to those wavelengths that can freely pass through the atmospheric gases. Only two groups of wavelengths have this property—visible light and radio waves (Figure 15.21).

Most of our knowledge of the universe beyond the solar system has come from telescopic observation and analysis of the visible light radiated by distant stars. Within the past 30 years, however, such studies have been augmented by powerful radiotelescopes that receive the second group of wavelengths that pass through the atmosphere to arrive at the earth's surface (Figure 15.22). Finally, balloons, rockets, and satellites are now carrying special telescopes *above* the earth's atmosphere to scan the skies at still other wavelengths, particularly in the X-ray, infrared, and ultraviolet portions of the spectrum. From all of these observations has emerged a wealth of information about the properties of distant stars.

When viewed with the unaided eye, all stars appear as similar points of light except that some are brighter than others. Because of their enormous distance, all stars *still* appear as mere points of light when viewed through even the most powerful telescopes, although their relative brightness is greatly increased and countless additional stars, too dim to be seen otherwise, are visible. By attaching special instruments to telescopes, however, it is possible to analyze the light from an individual star by breaking it into *spectra* of separate wavelengths just as sunlight is broken into separate colored bands by a simple prism of glass (Figure 15.23). Such *spectrographic analyses* show that stars also differ in fundamental properties other than mere brightness.

In some stars the light radiation is concentrated in the *blue* end of the spectrum, which indicates that they are burning at very high temperatures. In others, the concentration is in the cooler, *red* end of the spectrum while still others, the majority in fact, show an intermediate yellow color. It is thus possible to classify stars by *color*, which also implies *temperature,* as well as by their *apparent brightness.*

A third and critically important parameter is a star's distance from the earth. Apparent brightness alone is no indication of distance for a large, hot star might appear bright even when very far away and, conversely, a small, cool star might appear dim even though relatively near. Fortunately, the distance to many nearby stars can be computed geometrically from slight variations in their apparent position as the earth orbits the sun. Knowing the distance, it is then possible to determine the *absolute brightness* and the *size* of the star—an apparently bright but distant star must be large and intrinsically bright, and vice versa.

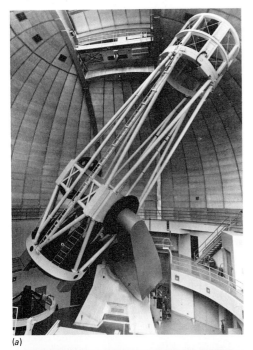
(a)

(b)

Figure 15.22 *Typical optical (a) and radio (b) telescopes used for observing stellar radiation. (a: Lick Observatory; b: U.S. Navy)*

Figure 15.23 *Typical optical spectrum of a star. The dark lines, caused by the absorption of narrow wavelengths by the star's atmospheric gases, can be used to determine the atmospheric composition. The nature of the colored bands themselves indicates the temperature at which the star is burning.*

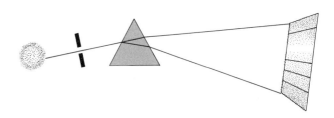

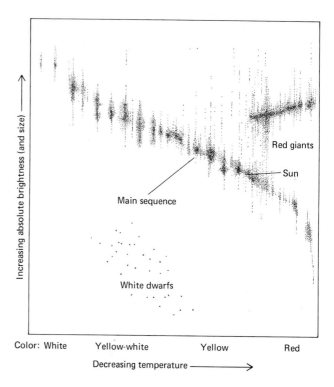

Figure 15.24   *A graph showing the absolute brightness (and size) versus color (and temperature) for about 7,000 stars whose distance can be computed geometrically (see text). Most, including the sun, fall on the "main sequence," which becomes cooler with decreasing size. In addition there are many cool "red giants" and a few very hot "white dwarfs."*

One of the most important summaries of the properties of stars is a simple graph of the absolute brightness (and size) versus color (and temperature) for nearby stars of known distance (Figure 15.24). Such graphs, first compiled in the 1920s, show three general types of stars: a few very small and hot **white dwarfs;** many more very large and cool **red giants;** and the bulk of stars following a pattern called the **main sequence** which becomes steadily cooler with decreasing size. The sun is such a main sequence star of about average size and brightness.

In addition to varying in absolute brightness (size) and color (temperature), many stars have other distinctive characteristics. About half of all stars are not widely separated from the nearest adjacent stars, as is the sun, but are linked in gravitational orbit with one or more neighbors to make *multiple star systems.* Such systems are especially important because they sometimes permit the calculation of a new parameter, a star's total *mass,* from measurements of their separation distance and period of revolution. Such calculations, in turn, show that there is a constant relationship between mass and size. As might be predicted, brighter and larger stars have a greater overall mass than do those that are smaller and less bright.

Figure 15.25 *Supernova remnants and pulsar. The cloud of gaseous debris resulted from a stellar explosion observed by oriental astronomers in 1054 A.D. The arrow points to a faintly visible pulsar, centered in the debris, which is believed to be the energetic core of the exploded star. (Lick Observatory)*

Finally, many stars show intriguing short-term variations in radiation output. In some the brightness changes, either regularly or erratically, over periods of several hours, days, or longer. Analyses of these changes indicate that they are caused by pulsating expansions and contractions, by as much as 10 percent, of the star's volume. More dramatically, once every few hundred years a star can be seen to "explode" with an enormous and sudden increase in brightness, followed by a rapid dimming and an expanding cloud of gas and dust. Several such **supernovae,** as they are called, are recorded in ancient astronomical records, and their expanding remnants are still clearly visible (Figure 15.25).

Perhaps the most interesting variable stars were discovered only in 1967. These are very dim and extremely small stars that do not merely change in brightness, but actually flash *on and off* very rapidly, as frequently as 30 times per second in some. Called **pulsars,** these stars, although very dim at visible wavelengths, emit unusually energetic radio waves which also show the rapid on-and-off pulsation. It was by reception of these radio signals that they were first detected. The pulsed radiation is believed to be due to rapid rotation of the small star, but the cause of the emission of radiation from only a part of its spin-ning surface remains uncertain. Faint pulsars have now been identified in the centers of two expanding supernovae dust clouds, which clearly indicates that some, at least, are the small, energetic cores of exploded stars (Figure 15.25).

Since the early decades of this century, when the great variety of star types was first clearly recognized, astronomers have attempted to interrelate the many kinds into a single sequence of *stellar evolution*, in which the properties of stars show a regular change as they become older. Although there is little direct evidence for such a sequence, because a single star cannot be observed for millions or billions of years, astronomers are in general agreement that meny of the differences seen in stars *are* related to their relative ages. Such inferences are the basis for the postulated history of the sun discussed in Section 13.6, which we shall only briefly review again here.

Stars are believed to form from local concentrations of cosmic gases and dusts, principally hydrogen, which grow by gravitational accretion until their centers are hot and massive enough to cause thermonuclear ignition. When this occurs the gravitational forces drawing the massive star together become balanced by the outward flow of energetic materials from its hot interior and the star enters the main sequence of burning, yellow-hot stars whose exact color-temperatures depend on their initial size of accretion. This stage lasts billions of years as the star consumes its nuclear fuel by the conversion of hydrogen into helium. Most stars, including the sun, are in this extended phase of their evolutionary history.

As the star's nuclear fuel approaches exhaustion, a complex series of changes occurs. With initial cooling the star first contracts, then rapidly expands to become a red giant. As the final fuels are consumed, however, gravita-tional forces take over and the remaining mass contracts, either relatively slowly to form a white dwarf or explosively as a supernova which spreads most of its volume as an expanding gas and dust cloud, leaving behind only a small pulsar where atoms are believed to be so densely packed by gravity that a golfball-sized piece would weigh more than an earthly mountain range!

15.8   Galaxies, Time, and Matter

To the unaided eye, stars appear to be rather randomly scattered about the sky, but telescopes show that there is a concentration of otherwise invisible stars in the disk-

(a)

(b)

Figure 15.26 Galaxies. These disk-shaped accumulations, each containing billions of stars, are the fundamental building blocks of the universe. (a) A typical distant galaxy, viewed edge-on. (b) A relatively nearby galaxy whose component stars can be resolved. (a, b: Hale Observatories)

like plane that makes up the bright "Milky Way" of the night sky. One of the most significant discoveries of large telescopes constructed early in this century was that similar disklike accumulations, each containing billions of stars, were visible in the deep reaches of space, far beyond the relatively nearby stars of the Milky Way (Figure 15.26). Today, billions of these distant star accumulations, called **galaxies,** can be seen and it is clear that they, rather than single stars, are the fundamental building blocks of the universe. Thus the sun is but a single star among billions in the *Milky Way Galaxy* which is, in turn, merely one among billions of similar galaxies that make up the visible universe.

Compared with nearby stars, galaxies are difficult to study because of their enormous distance. In the closer galaxies, individual bright stars can be easily resolved with large telescopes (Figure 15.26*b*), but the more distant galaxies appear only as faint, fuzzy disks or spheres. Unlike individual stars, a galaxy's distance from the earth can be estimated from its apparent overall brightness on the assumption that the billions of stars making it up have about the same *total* brightness as do those in other galaxies. This assumption has been confirmed by direct distance measurements on individual stars within the nearer galaxies.

Such distance calculations show that even the nearest galaxies are enormously distant and the farthest that can be observed are almost inconceivably remote, so much so that the light from them that we observe today has been traveling for *billions* of years. Thus we see them not as they presently exist, but as they were not long after the earth was formed!

As with individual stars, the combined starlight from distant galaxies can be analyzed spectroscopically. Such analyses, first made in the early decades of this century, revealed one of the most fundamental properties of the universe. Light waves, like sound waves from a passing automobile horn, tend to change wavelength slightly depending on whether they are moving towards or away from the observer. The horn appears to change pitch as the car passes and light waves are similarly slightly compressed toward the blue end of the spectrum as they approach and towards the red end as they recede. In analyzing the light from distant galaxies it was observed that *all* showed varying shifts toward the red indicating that they are moving *away* from the earth.

On further consideration it became clear that all galaxies were, in fact, moving away from *each other*, like spots on an inflating balloon. In general, the dimmest and most remote galaxies show the largest spectral shifts indicating that they are receding at the highest velocities. In short, the universe is apparently *expanding* as galaxies move outward into the remote distances of space.

If the present velocities and distances of the expanding galaxies are projected backward in time, they suggest a time, about 20 billion years ago, when all the matter of the now-scattered universe was closely assembled. Most astronomers believe that the present universe originated as this primitive mass of matter was somehow dispersed in an enormous explosion. If so, the universe is about 20 billion years old, or about four times the age of the earth and solar system. Other astronomers reject this "big bang" theory and feel that galaxies, although receding from each other, are somehow being continuously created and destroyed just as are their component stars. Under this "steady state" hypothesis, the universe might have an infinite age.

One of the most dramatic scientific discoveries of recent years bears on these cosmic questions. In the early 1960s astronomers first noticed dim, starlike objects, many of which emit large quantities of radio waves. What makes these objects, called **quasars,** so puzzling is that they show extreme spectral shifts, far larger than any known galaxy, yet they appear to be only single, starlike bodies. The extreme spectral shift shows that they are receding at tremendous velocities, close to the speed of light itself. Furthermore, if their spectra are analogous to those of galaxies, which become more shifted with distance, then quasars are by far the most remote objects of the visible universe and the radiation that we now receive from them began its journey through space as early as 10 billion years ago, early in the history of the universe and long before the origin of the earth and solar system.

A principal puzzle of quasars is the source of their energy, for there is no known mechanism for creating so much energy in such relatively small, starlike bodies if they are really as distant as they appear. At such enormous distances even huge galaxies, composed of *billions* of stars, should be too faint to be visible. Quasars, then, may be revealing some new and strange energy source that was present in the earliest history of the universe.

There is one final question of equally cosmic scope that we shall mention in closing. This is the origin not of stars and galaxies themselves, but of the chemical elements of which they are composed. Today hydrogen, the lightest element, is by far the most abundant in the universe and there is good evidence that it *alone* made up the earliest universe. All the heavier elements are now believed to have been produced as mere "ashes" left over from the nuclear burning of hydrogen in long-extinct stars. As a star declines, some of this heavier material may be

dispersed by exploding supernovae and "recycled" to become a part of a new star as it condenses from cosmic
gases and dusts.

Such a star was our sun and some of the heavier
stellar cinders from which it formed became concentrated
in orbit around it to produce our relatively hydrogen-poor
earth. Thus the familiar elements that dominate the earth—
oxygen, silicon, aluminum, iron, etc.—are themselves the
products of billions of years of processing within the burning interiors of stars. Our brief summary of the universe
beyond the solar system has therefore, as is so often the
case in science, brought us full circle to the very topic with
which we began our survey of the earth in Chapter 1—the
question of the matter and materials that make up our
unique planet.

SUMMARY OUTLINE

**The Earthlike
Planets**

15.1 *Mercury:* is the poorly known, smallest earthlike planet that
orbits nearest to the sun; it has an average density about
the same as the earth's and apparently lacks an atmosphere.

15.2 *Mars:* has distinctive cratered, smooth, and complex surface features, as well as large volcanoes and faults, that
indicate it to be internally active; its thin atmosphere of
carbon dioxide supports polar ice caps and occasional
clouds and planet-wide dust storms.

15.3 *Venus:* is covered by a thick atmosphere of carbon dioxide
that is perpetually filled with dense clouds of unknown
composition; the lower atmosphere and surface are very hot
but little else is known of the solid planet beneath the
obscuring atmosphere.

**The Outer
Planets**

15.4 *Jupiter:* is by far the largest planet but its materials are
very light, being dominated by hydrogen and helium as is
the sun; its rapid rotation causes its atmosphere to be
streaked into parallel, colored bands of unknown composition.

15.5 *Saturn:* resembles Jupiter but has a unique equatorial disk
of dustlike orbiting particles; large satellites orbit in the same
plane but much farther from the planet's surface.

15.6 *Uranus, Neptune, and Pluto:* Uranus and Neptune resemble
Jupiter but are smaller, more distant, and less well known;
Pluto appears to resemble the earthlike planets and may
be a former satellite of Neptune.

**Beyond the
Planets**

15.7 *Stars:* differ principally in brightness, which is proportional
to both size and distance, and in color, which is determined
by temperature; many also show short-term variations
in radiation output; differences in size and temperature
are believed to be related to a star's age, with sharp

changes occuring in old age as the nuclear fuels are exhausted.

15.8 *Galaxies, time, and matter:* stars are localized into billions of spherical or disk-shaped galaxies, each containing billions of stars; all galaxies are moving away from each other and the universe is thus expanding, the expansion apparently having begun about 10 billion years ago; all of the chemical elements of the present universe, except hydrogen, probably originated as "ashes" left over from hydrogen burning in long-extinct stars.

ADDITIONAL READINGS

Abell, G.: *Exploration of the Universe,* Holt, Rinehart & Winston, Inc., New York, 1969. A standard introduction to the solar system, stars, and galaxies.

Dixon, R. T.: *Dynamic Astronomy,* Prentice-Hall, Inc., Englewood Cliffs, N. J., 1971. A brief introduction to the solar system, stars, and galaxies.

Kaula, W. M.: *An Introduction to Planetary Physics; The Terrestrial Planets,* John Wiley & Sons, Inc., New York, 1968. An advanced survey of the earthlike planets.

Lewis, J. S.: "The Atmosphere, Clouds, and Surface of Venus," *American Scientist,* vol. 59, pp. 557–66, 1971. An advanced review article.

Menzel, D. J., F. L. Whipple, and G. de Vaucouleurs: *Survey of the Universe,* Prentice-Hall, Inc., Englewood Cliffs, N. J., 1970. A detailed introduction to the solar system, stars, and galaxies.

Murray, B. C.: "Mars from Mariner 9," *Scientific American,* vol. 228, no. 1, pp. 49–69, 1973. A popular review of recent spacecraft observation.

Ostriker, J. P.: "The Nature of Pulsars," *Scientific American,* vol. 224, no. 1, pp. 48–60, 1971. A popular review of the dense, energetic cores of exploded stars.

Schmidt, M., and F. Bello: "The Evolution of Quasars," *Scientific American,* vol. 224, no. 5, pp. 55–69, 1971. A popular review of the most enigmatic starlike objects.

Index